Biomedical Engineering Handbook

Biomedical Engineering Handbook

Editor: Luke Madison

FA FOSTER
ACADEMICS

www.fosteracademics.com

www.fosteracademics.com

FA
FOSTER
ACADEMICS

Cataloging-in-Publication Data

Biomedical engineering handbook / edited by Luke Madison.
      p. cm.
Includes bibliographical references and index.
ISBN 978-1-63242-874-5
1. Biomedical engineering. 2. Medicine. 3. Bioengineering. 4. Biomedical materials. I. Madison, Luke.
R856 .B56 2020
610.28--dc23

Foster Academics,
118-35 Queens Blvd., Suite 400,
Forest Hills, NY 11375, USA

ISBN 978-1-63242-874-5 (Hardback)

Contents

Preface

Biomedical engineering is the field concerned with the application of engineering design and principles for developing diagnostic, therapeutic and monitoring applications for use in healthcare. It encompasses a number of sub-disciplines with immense medical applications, chief among which are bioinformatics, biomechanics, biomaterials science, neural engineering, tissue engineering, etc. Biomedical engineering has evolved over the years in response to advancements in science and technology. Effective devices for diagnosing and treating diseases, or for rehabilitating, alleviating and compensating for disabilities or injuries are developed and modified owing to advances in biomedical engineering. This book is a valuable compilation of topics, ranging from the basic to the most complex advancements in the field of biomedical engineering. It presents this complex subject in the most comprehensible and easy to understand language. For someone with an interest and eye for detail, this book covers the most significant topics in this field.

This book is a comprehensive compilation of works of different researchers from varied parts of the world. It includes valuable experiences of the researchers with the sole objective of providing the readers (learners) with a proper knowledge of the concerned field. This book will be beneficial in evoking inspiration and enhancing the knowledge of the interested readers.

In the end, I would like to extend my heartiest thanks to the authors who worked with great determination on their chapters. I also appreciate the publisher's support in the course of the book. I would also like to deeply acknowledge my family who stood by me as a source of inspiration during the project.

Editor

Ambulatory blood pressure monitoring in children suffering from orthostatic hypertension

Yang Zhixiang[1], Wang Cheng[2], Xiang Jibing[1], Ge Bisheng[1], Xu Ming[1] and Liu Deyu[1]*

*Correspondence:
lx_ldy@163.com
[1] Department of Pediatrics,
Lixian People's Hospital
in Hunan, Lixian 415500,
China
Full list of author information
is available at the end of the
article

Abstract

Background: It is particularly important to utilize appropriate blood pressure measurement methods to evaluate the changes of orthostatic hypertension (OHT) for children, and this study was designed to analyze the blood pressure type in OHT children with 24 h semiautomatic ambulatory blood pressure monitoring.

Methods: Children who were diagnosed by head-up tilt table test as OHT patients (OHT group) and treated or hospitalized in the syncope specialist outpatient unit of the Second Xiangya Hospital of Central South University mainly for syncope or presyncope with unknown causes during the October, 2009 to September, 2013 were recruited in the study. Healthy children that came to the hospital for physical examination at the same time period according to age and sex were matched as control group. Semiautomatic ambulatory blood pressure monitoring of every child was recorded. The differences of daytime systolic (diastolic) pressure and night systolic (diastolic) pressure were calculated, and the average systolic pressure and diastolic pressure of the entire day, daytime and night were also calculated, respectively.

Results: There were 23 boys and 17 girls in OHT group, aging (11.5 ± 1.9) years. There were 22 boys and 18 girls in the control group, aged (10.6 ± 2.4) years. The difference of daytime systolic pressure and night systolic pressure of the control group was higher than that of OHT group, while the average systolic pressure of the whole day, the average diastolic pressure of the whole day, the daytime average systolic pressure, the daytime average diastolic pressure, the night average systolic pressure and the night average diastolic pressure were higher than that of the control group (P > 0.05). The difference of daytime diastolic pressure and night diastolic pressure of the control group was higher than that of OHT group (P > 0.05). Most children of the OTH group had non-dipper blood pressure type (72.5%), while most children of the control group had a dipper blood pressure type (55.0%). In addition, the time domain SDNN and SDANN in the OHT group were higher than those in the control group (P < 0.01). And, the indicators including TP, ULF, VLF, and LF/HF were higher in the OHT group, when compared with control group (P < 0.01). Besides, in terms of subgroup analysis within the OHT group, the age difference between boys and girls was not statistically significant (P > 0.05). When compared with grils, the time domain SDNN increased (P = 0.003), and the frequency index TP, ULF, and VLF increased in boy group (P < 0.05).

Conclusion: OHT Children's autonomic nervous system showed dysfunction, and differences of systolic blood pressure between day and night were much lower than those of healthy children, and the main blood type was non-dipper blood pressure with circadian rhythm disappearing.

Keywords: Semiautomatic ambulatory blood pressure monitoring, Children, Orthostatic hypertension

Background

To our best knowledge, orthostatic hypertension (OHT) refers to normal blood pressure (BP) in the supine position, and suddenly raised blood pressure when standing or sitting [1]. As known to all, the normal BP is defined as systolic BP (SBP) less than 140 mmHg and diastolic BP (DBP) more than 90 mmHg at three consecutive consultations. And, OHT is defined as a drop of SBP less than 20 mmHg and/or DBP less than 10 mmHg at orthostasis [2]. It is noted that the BP should be assessed after 1 and 2 min of standing from a supine position. With respect to the physiological mechanism of OHT, it is an over-reacting of the sympathetic nervous system, involving a hypersensitivity of vascular baroreceptors in response to orthostasis [3–5]. As far as we are concerned, OHT has potential risk to induce cerebrovascular events and sustained arterial hypertension. According to previous publications, the relevant studies concentrating on OHT mostly focused on the elderly population [6]. Additionally, it is also involved in young and middle-aged people, but rarely reported in children. Concerning the incidence rate of OHT, there is no unified conclusion. In early 1992, Rutan et al. indicated that the incidence rate of OHT in healthy pilots was 4.2% [7]. In 2012, Wang Lili et al. reported that the incidence rate of OHT in elderly patients with hypertension was 9.0% [8]. In 2012, Chinese scholar Du Junbao proposed the diagnostic criteria of OHT in children, and analyzed their clinical features [9]. According to this diagnostic criteria, in 2013, Kang et al. analyzed the clinical information of children (n = 2089, age range 2.0–17.9) with syncope, headache, dizziness, chest tightness, and sigh. And the data illustrated that the total detection rate of OHT was 23.8%, males were higher than females (25.9% versus 21.6%, P < 0.05), 12-year-old group higher than < 12-year-old group (28.1% versus 20.5%, P < 0.01) [10]. Additionally, OHT is closely associated with the occurrence and development of persistent hypertension, cardiovascular events, diabetes, chronic kidney disease, asymptomatic cerebral infarction, and deep white matter ischemic lesions [11]. Therefore, it is particularly important to utilize appropriate blood pressure measurement methods to evaluate the changes of OHT. The 24-h semiautomatic ambulatory blood pressure monitoring can fully reflect the 24-h blood pressure changes and circadian rhythm of subjects, and it is an effective tool for objectively evaluating blood pressure changes. This technique can reduce the measuring error in traditional BP measurement methods, and avoid the influence of "white coat effect" [12]. According to the Consensus of Adolescent Children issued by the American Heart Association in 2014, the 24-h semiautomatic ambulatory blood pressure monitoring is effective in the diagnosis of white coat hypertension, asymptomatic hypertension, prehypertension, isolated diastolic hypertension, and nocturnal hypertension [13]. At present, there are few reports at home and abroad about the application of 24-h semiautomatic ambulatory blood pressure monitoring to evaluate the OHT type in children. In this case–control study, the 24-h semiautomatic ambulatory blood pressure monitoring was utilized to analyze OHT types in the Second Xiangya Hospital of Central South University, which provided reference for the diurnal blood pressure change of OHT in children.

Methods

OHT diagnostic criteria of the identified children

The normal blood pressure measured 10 min after supine in a quiet environment is considered to be basal blood pressure. And the OHT changes were measured 3 min after upright tilt test (HUTT, tilt angle 60). If the systolic blood pressure increased ≥ 20 mmHg, and (or) the diastolic blood pressure increased ≥ 10 mmHg (1 mmHg $= 0.133$ kPa), when compared with basal blood pressure, the patients could be diagnosed with OHT [9].

Inclusion and exclusion criteria of OHT

(1) From October 2009 to September 2013, children at the syncope clinic of our hospital, or hospitalized children with unexplained syncope and premonitory syncope. (2) Patients were diagnosed with OHT via HUTT. (3) Physical examination, blood biochemical examination (fasting blood glucose, myocardial enzyme), conventional electrocardiogram, Holter electrocardiogram, echocardiography, electroencephalogram, cranial CT and MRI were carried out, in order to exclude heart, lung, brain, kidney and thoracic wall diseases.

Study design

In the same period, healthy children in the Children's Health Clinic of Xiangya Second Hospital of Central South University were selected into this research, and were matched with the OHT group (one-to-one based on age and sex). The identified patients were divided into two groups, and received basic HUTT (BHUT) and sublingual nitroglycerin tilt test (SNHUT), respectively. The electric tilting bed (ST-711) manufactured by Beijing Juchi Medical Technology Co., Ltd. and the tilting test monitoring software system (SHUT-100) provided by Beijing STADLEY Technology Inc. were utilized in this study. The specific inspection method refers to the guidelines for children's syncope diagnosis issued by the Chinese Medical Association [14].

Semiautomatic ambulatory blood pressure monitoring

During the semiautomatic ambulatory blood pressure monitoring examination, the subjects were home at 24 h. The ABPM 6100 monitor is provided by Welch Allyn in America, and the detailed information is as follows. (1) Select the same wide cuff and the same arm as semiautomatic ambulatory blood pressure monitoring. (2) Use a mercury sphygmomanometer to measure BP for 2 times, and then record BP value. (3) Initialize the BP monitoring and then enter the subject's personal information. (4) The subjects were required to normatively wear BP monitoring, and manually measure blood pressure for 1 or 2 times. If the difference of systolic blood pressure between mercury sphygmomanometer and electronic sphygmomanometer less than 5 mmHg, the electronic sphygmomanometer could be considered to meet the requirements. Meanwhile, explain to the participants the precautions of wearing BP monitoring, and remind the required work and rest time. (5) After semiautomatic ambulatory blood pressure monitoring for 24 h, the blood pressure information of the subjects was input into the computer [15]. (6) Referring to the revised semiautomatic ambulatory blood pressure monitoring criteria for healthy children in 1997, the ambulatory blood pressure data can be achieved [16]. In this study, 6:00 to 22:00 was set as the daytime awakening time, and 22:00 to 6:00 in the next day was set as the nighttime sleep time. The blood pressure was monitored once every 30 min during the day and night, and the total monitoring time was not less than 23 h throughout the day. The BP judgment standard refers to the authoritative literature [17]. If the number of effective blood pressure less than 80%, it must be repeated every other day. In terms of matters needing attention, choose non-dominant arm to receive BP measuring. Secondly, determine cuff size based on the subject's upper arm circumference. Thirdly, when the semiautomatic ambulatory blood pressure monitoring automatically measures BP, the upper arm of the cuff should be kept as static as possible. Fourthly, during semiautomatic ambulatory blood pressure monitoring, should avoid bathing, agitation, anxiety, strenuous exercise, and ban drinking and other diets that affect autonomic function. Fifthly, during BP monitoring, children's daily routines and activities should be recorded, such as when they wake up, sleep, and take activities. At last, the subjects cannot stop BP monitoring at will [18]. Besides, the parameters of semiautomatic ambulatory blood pressure monitoring are as follows. (1) Average systolic BP (diastolic BP) = total value of systolic or diastolic BP in each time period/number of measuring times during each period. Calculate the mean systolic and diastolic BP at 24 h, in the daytime, and in the nighttime. (2) Diurnal variation of systolic BP = (mean systolic BP in the daytime − mean systolic BP in the nighttime)/mean systolic BP in the daytime × 100%. (3) Diurnal variation of diastolic BP = (mean diastolic BP in the daytime − mean diastolic BP in the nighttime)/mean diastolic BP in the daytime × 100%.

Types of blood pressure

When concerning the types of blood pressure, the thorough explanations are as follows. Under normal circumstances, the human body's BP rises throughout the day and falls during the night, which can be described as "long handle spoon" curve [19]. (1) Dipper BP: Nocturnal BP drop > 10%, when compared with daytime BP, as shown in Fig. 1a. (2) Non-dipper BP: Nocturnal BP drop < 10%, when compared with daytime BP, as shown in Fig. 1b. (3) Over spoon-shaped BP: Nocturnal BP drop > 20%, when compared with daytime BP, as shown in Fig. 1c. (4) Inverse spoon-shaped BP: Nocturnal BP does not drop or increase, as shown in Fig. 1d.

Statistical analysis

The statistical analysis was done by SPSS 21.0 Software. Quantitative variables were illustrated as mean ± SD, the comparisons between groups were carried out via independent sample t test. The qualitative variables as percentages, and the comparisons between groups were conducted with Chi square test. $P < 0.05$ was considered statistically significant.

Results

General characteristics of the identified patients

40 cases in OHT group, 23 boys and 17 girls, age varied from 7 to 14 years old, mean age 11.5 ± 1.9 years old. 40 patients in the control group, 22 boys and 18 girls, age ranged from 4 to 14 years old, mean age 10.6 ± 2.4 years old. There was no significant differences

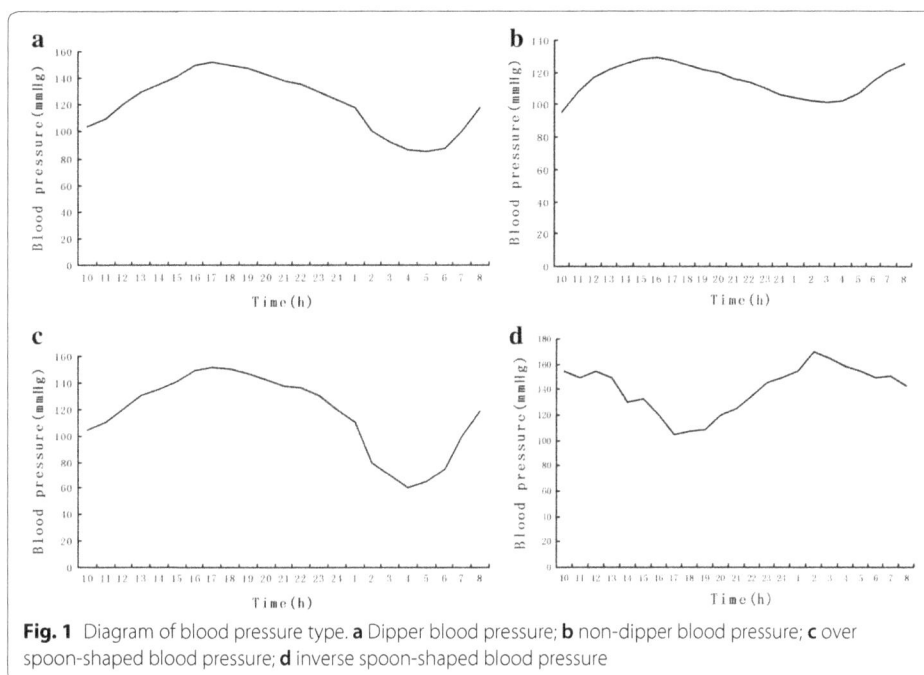

Fig. 1 Diagram of blood pressure type. **a** Dipper blood pressure; **b** non-dipper blood pressure; **c** over spoon-shaped blood pressure; **d** inverse spoon-shaped blood pressure

Table 1 Comparison of ABPM between OHT group and control group ($\bar{\chi} \pm s$, mmHg)

Group	24 h mean SBP	24 h mean DBP	Daytime mean SBP	Daytime mean DBP	Nocturnal mean SBP	Nocturnal mean DBP	SBP circadian difference	DBP circadian difference
OHT (n = 40)	107.2 ± 8.4	58.8 ± 6.1	11.2 ± 8.9	63.2 ± 7.0	101.8 ± 7.8	52.7 ± 5.7	8.4 ± 4.7	16.4 ± 7.1
Control (n = 40)	106.0 ± 8.6	58.2 ± 4.8	111.0 ± 9.2	63.2 ± 6.2	100.1 ± 8.4	52.2 ± 4.4	11.4 ± 3.1	16.9 ± 6.3
t	0.607	0.488	0.124	0.034	0.936	0.396	− 3.391	− 0.363
P	0.595	0.218	0.735	0.702	0.724	0.129	0.033	0.409

BP blood pressure, *SBP* systolic blood pressure, *DBP* diastolic blood pressure

in age and sex between the two groups (P = 0.081 and 0.822, respectively), as shown in Table 1 and Fig. 2. In addition, the primary clinical symptoms of the identified patients in OHT group included cardiovascular system, nervous system, and digestive system, as shown in Table 2. With respect to the common inducing factors that might resulted in OHT, the authors summarized them in this section, including prolonged standing, exercise, posture change, nervousness, prolonged sitting, muggy circumstance, other triggers, as shown in Fig. 3.

Parameter comparison of blood pressure between research group and control group

As shown in Table 3, the average systolic BP and diastolic BP in the OHT group at 24 h, in the daytime, and in the nighttime were slightly higher than those in the control group, but the difference was not statistically significant (P > 0.05). In addition, the difference of systolic BP between day and night in the control group was higher than that of the OHT group (P = 0.033), and the difference of diastolic BP between day and night in the control group was slightly higher than that of the OHT group (P = 0.409).

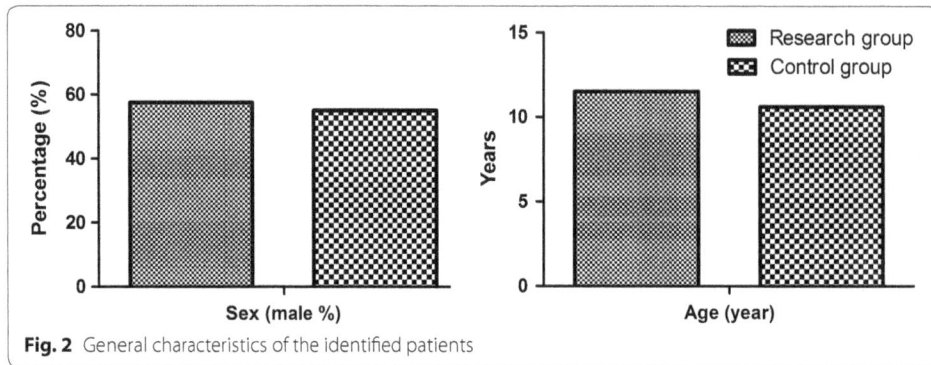
Fig. 2 General characteristics of the identified patients

Table 2 Comparison of heart rate variability between research group and control group

Indicator	Control group	Research group	t	P
SDNN (ms)	121.31 ± 32.54	154.23 ± 9.56	3.795	0.000
SDANN (ms)	122.96 ± 5.40	142.68 ± 3.21	4.127	0.002
rMSSD (ms)	76.08 ± 3.21	77.52 ± 3.62	1.238	0.301
pNN50 (%)	26.30 ± 4.52	27.51 ± 2.36	1.542	0.181
CV	0.12 ± 0.000	0.12 ± 0.000	0.310	0.823
TP	9532.21 ± 452.13	12,311.25 ± 411.23	3.421	0.001
ULF	4231.10 ± 369.21	6987.48 ± 398.41	4.259	0.000
VLF	2896.41 ± 231.41	3961.80 ± 357.12	2.361	0.010

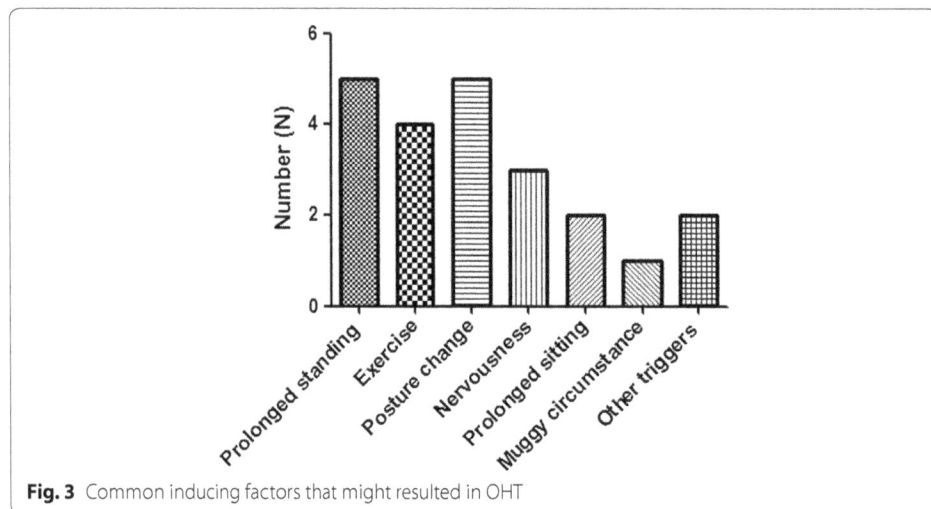
Fig. 3 Common inducing factors that might resulted in OHT

Comparison of blood pressure types between research group and control group

In the OHT group, non-dipper BP accounted for 72.5% (29/40 cases), and dipper BP accounted for 27.5% (11/40 cases). On the contrary, the non-dipper BP accounted for 45.0% (18/40 cases), and dipper BP accounted for 55.0% (22/40 cases). The BP difference between the two groups was statistically significant ($\chi^2 = 6.600$, P = 0.012). No over spoon-shaped BP and inverse spoon-shaped BP could be observed in both OHT and control groups, as shown in Table 4 and Fig. 4.

Table 3 Comparison of heart rate variability between boy and girl in research group

Indicator	Boy group	Girl group	t	P
Age (year)	12.01 ± 0.31	12.89 ± 0.28	1.891	0.089
SDNN (ms)	155.30 ± 3.21	142.58 ± 3.48	3.008	0.003
SDANN (ms)	148.21 ± 3.86	139.11 ± 4.50	0.956	0.451
rMSSD (ms)	79.55 ± 3.85	78.21 ± 3.69	0.781	0.487
pNN50 (%)	31.25 ± 1.26	29.91 ± 1.69	1.112	0.361
CV	0.11 ± 0.00	0.10 ± 0.00	1.460	0.189
TP	11,231.12 ± 445.21	1023.15 ± 397.13	3.921	0.002
ULF	6541.20 ± 451.23	4895.27 ± 401.23	2.361	0.002
VLF	3789.40 ± 230.78	2891.47 ± 309.87	3.561	0.001

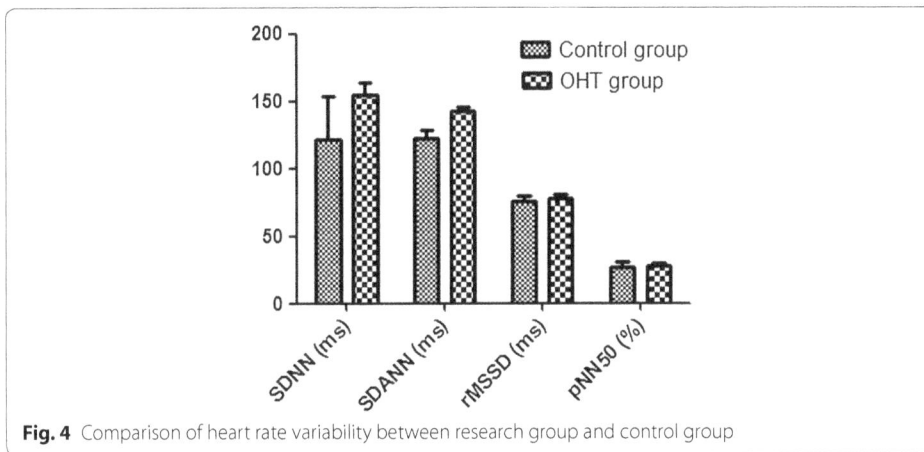

Fig. 4 Comparison of heart rate variability between research group and control group

Comparison of heart rate variability between research group and control group

AS shown in Table 2, the time domain SDNN and SDANN in the OHT group were higher than those in the control group (P < 0.01). In addition, the indicators including TP, ULF, VLF, and LF/HF were higher in the OHT group, when compared with control group (P < 0.01). In addition, in terms of subgroup analysis within the OHT group, the results were illustrated in Table 3. First of all, the age difference between boys and girls was not statistically significant (P > 0.05). When compared with girls,

Table 4 Comparison of heart rate variability between research group and control group

Indicator	Control group	Research group	t	P
SDNN (ms)	121.31 ± 32.54	154.23 ± 9.56	3.795	0.000
SDANN (ms)	122.96 ± 5.40	142.68 ± 3.21	4.127	0.002
rMSSD (ms)	76.08 ± 3.21	77.52 ± 3.62	1.238	0.301
pNN50 (%)	26.30 ± 4.52	27.51 ± 2.36	1.542	0.181
CV	0.12 ± 0.000	0.12 ± 0.000	0.310	0.823
TP	9532.21 ± 452.13	12,311.25 ± 411.23	3.421	0.001
ULF	4231.10 ± 369.21	6987.48 ± 398.41	4.259	0.000
VLF	2896.41 ± 231.41	3961.80 ± 357.12	2.361	0.010

Table 5 Comparison of heart rate variability between boy and girl in research group

Indicator	Boy group	Girl group	t	P
Age (year)	12.01 ± 0.31	12.89 ± 0.28	1.891	0.089
SDNN (ms)	155.30 ± 3.21	142.58 ± 3.48	3.008	0.003
SDANN (ms)	148.21 ± 3.86	139.11 ± 4.50	0.956	0.451
rMSSD (ms)	79.55 ± 3.85	78.21 ± 3.69	0.781	0.487
pNN50 (%)	31.25 ± 1.26	29.91 ± 1.69	1.112	0.361
CV	0.11 ± 0.00	0.10 ± 0.00	1.460	0.189
TP	$11,231.12 \pm 445.21$	1023.15 ± 397.13	3.921	0.002
ULF	6541.20 ± 451.23	4895.27 ± 401.23	2.361	0.002
VLF	3789.40 ± 230.78	2891.47 ± 309.87	3.561	0.001

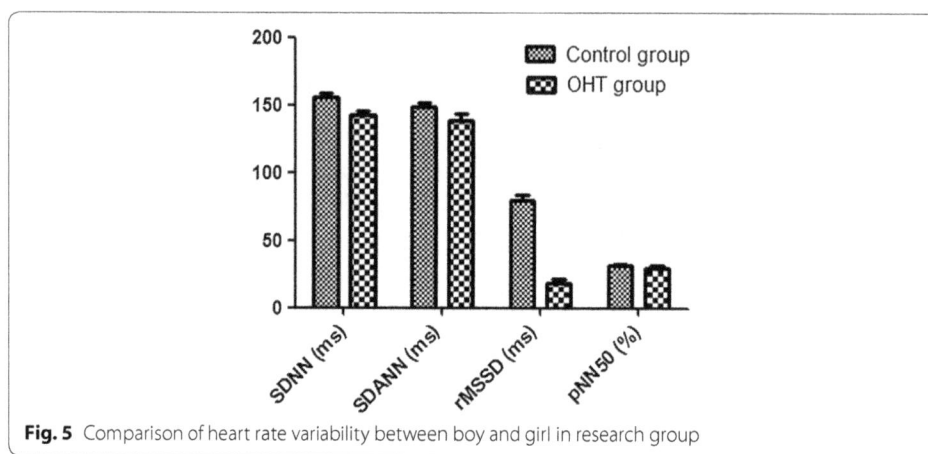

Fig. 5 Comparison of heart rate variability between boy and girl in research group

the time domain SDNN increased ($P = 0.003$), and the frequency index TP, ULF, and VLF increased in boy group ($P < 0.05$), as shown in Table 5 and Fig. 5.

Discussion

Perhaps on account of overcompensation for the gravitational challenge, the BP may rise when standing. In clinical practice, this phenomenon is called OHT, which is drawing increasing attention in recent years. It has been found to be associated with more prominent hypertensive target organ damage, such as silent cerebrovascular lesions. As mentioned above, no consensus has been reached on the definition of OHT, but OHT can be defined as a drop of $SBP \leq 20$ mmHg and/or $DBP \leq 10$ mmHg at orthostasis, according to the American Academy of Neurology. On the basis of our clinical experience, OHT is a common and disturbing problem and is especially prevalent among hospitalized patients. In addition, as a common clinical finding, it is closely associated with an increased risk of cardiovascular mortality, and is also an established marker for cardiovascular risk in patients with hypertension in the general population.

Arterial BP in the human body keeps changing along with blood flow, and there is a significant circadian rhythm in general. In terms of physiological basis, sympathetic nerves predominate during daytime in normal children, which enhances the concentration of catecholamines in plasma. The vagus nerve predominates during sleep at night, and the concentration of catecholamines in blood decreases. In further, when the body is in a supine position during sleep, redistribution of systemic blood flow occurs, and a large amount of blood concentrates in the lower limbs, which subsequently increase blood flow to the center, increased central venous pressure, increase afferent impulses in baroreceptors, and increase inhibitory stimulation of afferent vasomotor centers. Afterwards, the sympathetic excitatory reflex is weakened, the cardiac output decreases when the heart negatively inverts, and then the peripheral vascular resistance decreases and the whole body muscle relaxes, which subsequently results in lower BP levels during the night than during the day.

The circadian rhythm of blood pressure is one of the most important features of the human body. Maintaining a normal circadian rhythm of blood pressure is of great significance for protecting vital organs such as heart, liver and kidney, etc. For normal children, nocturnal BP is lower than daytime BP. Through the BP curve, BP is at its lowest point between 0–3 at early morning, starts to rise at 4–5 in the morning, and peaks between 6 and 8 in the morning. Afterwards, the BP is gradually steady, peaks again at 4–6 in the afternoon, and then slowly decreases. The BP throughout 24 h is presented as "long spoon-shaped curve" with double peaks and one valley. The semi-automatic ambulatory blood pressure monitoring can better reflect the BP fluctuation and circadian rhythm. The "over spoon-shaped" BP is often accompanied by OHT, and the "inverse spoon-shaped" BP is related to continuous sympathetic nerve stimulation, often accompanied by OHT [20]. Our team is dedicated to investigating the ambulatory BP changes in children with OHT and vasovagal syncope, and the corresponding results demonstrated that the BP type in children with OHT is mostly non-dipper BP and is associated with autonomic dysfunction [21]. In this study, the data illustrated that the most frequent BP type was non-dipper BP in children with OHT, which is consistent with the findings of Moriguchi et al. [22]. Wu et al. attempted to investigate the prevalence of hypertension in adults and its related factors (n = 1638), and they put forward the idea that OHT risk was closely related to changes of systolic BP during postural changes [23]. In this study, 80 cases were identified, and they received semiautomatic ambulatory blood pressure monitoring afterwards. The data demonstrated that the average BP in the OHT group was slightly higher than that in the control group, but the difference was not statistically significant. After comparisons between the two groups, it could be claimed that the pathogenetic mechanism of OHT, such as increased sympathetic activity and increased adrenaline boost sensitivity, could result in the increased BP in children with OHT, when compared with healthy children.

Conclusion

In summary, the data in this research revealed that the BP type in children with OHT was mainly non-dipper BP, and the healthy children in the control group were mainly spoon-type BP. The difference between the two groups was statistically significant ($P < 0.05$). Additionally, the difference of systolic BP between day and night in the control group was greater than in the OHT group (P < 0.05). In brief, on account of autonomic dysfunction, the BP type in children with OHT was mostly non-dipper BP, and the circadian rhythm of BP disappeared.

Authors' contributions

Study design: YZ and LD. Literature review: YZ, WC, XJ, GB, and XM. Research performance: YZ, WC, XJ, GB, and XM. Gave advice for setup: LD. Data checking: XJ, GB, and XM. All authors read and approved the final manuscript.

Author details

[1] Department of Pediatrics, Lixian People's Hospital in Hunan, Lixian 415500, China. [2] Department of Pediatric Cardiovasology, Children's Medical Center, The Second Xiangya Hospital of Central South University, Institute of Pediatrics of Central South University, Changsha 410011, China.

Competing interests

The authors declare that they have no competing interests.

References

1. Kario K. Orthostatic hypertension-a new haemodynamic cardiovascular risk factor. Nat Rev Nephrol. 2013;9(12):726–38.
2. Kario K. Orthostatic hypertension: a measure of blood pressure variation for predicting cardiovascular risk. Circ J. 2009;73(6):1002–7.
3. Fessel J, Robertson D. Orthostatic hypertension: when pressor reflexes overcompensate. Nat Clin Pract Nephrol. 2006;2(8):424–31.
4. Bhuachalla BN, McGarrigle CA, O'Leary N, et al. Orthostatic hypertension as a risk factor for age-related macular degeneration: evidence from the Irish longitudinal study on ageing. Exp Gerontol. 2018;106:80–7.
5. Bursztyn M, Jacobs JM, Hammerman-Rozenberg A, et al. Prevalence of orthostatic hypertension in the very elderly and its relationship to all-cause mortality. J Hypertens. 2016;34(10):2053–8.
6. Fan XH, Sun K, Zhou XL, et al. Association of orthostatic hypertension and hypotension with target organ damage in middle and old-aged hypertensive patients. Natl Med J China. 2011;91(4):220–4.
7. Rutan GH, Hermanson B, Bild DE, et al. Orthostatic hypotension in older adults. The Cardiovascular Health Study. CHS Collaborative Research Group. Hypertension. 1992;19(6 Pt 1):508–19.
8. Wang LL, Liu ZD, Zhao YY, et al. Postural change in blood pressure in old-aged hypertensive patients and effects on neurocognitive disorder. J Clin Exp Med. 2012;11(4):241–2.
9. Zhao J, Yang JY, Jin HF, et al. Clinical analysis of orthostatic hypertension in children. Chin J Pediatr. 2012;50(11):839–42.
10. Kang MH, Xu Y, Wang C, et al. Differences of age and gender in children with orthostatic hypertension. Chin J Appl Chin Pediatr. 2013;28(1):24–6.
11. Liu DY, Xiang JB, Wang C. Research progress on orthostatic hypertension. Chin J Pract Pediatr. 2014;29(6):471–6.
12. Norcliffe-kaufmann L, Kaufmann H. Is ambulatory blood pressure monitoring useful in patients with chronic autonomic failure. Clin Auton Res. 2014;24(4):189–92.
13. Flynn JT, Daniels SR, Hayman LL, et al. Update: ambulatory blood pressure monitoring in children and adolescents: a scientific statement from the American Heart Association. Hypertension. 2014;63(5):1116–35.
14. The Subspecialty Group of Cardiology, The Society of Pediatrics, Chinese Medical Association, The Editorial Board, Chinese Journal of Pediatrics. Guidelines for diagnosis of syncope in children. Chin J Pediatr. 2009;47(2):99–101.
15. Xu Y, Lin P, Wang C, et al. Investigation of 24-h blood pressure monitoring for evaluating treatment outcome of nerve-mediated syncope in children. Chin J Contemp Pediatr. 2013;15(5):458–61.
16. Soergel M, Kirschstein M, Busch C, et al. Oscillometric 24-h ambulatory blood pressure values in healthy children and adolescents: a multicenter trial including 1141 subjects. J Pediatr. 1997;130(2):178–84.
17. Zhang KZ, Guo JH, Liu HY, et al. Clinical electrocardiography. Changsha: Hunan Science and Technology Press; 2002. p. 1354.
18. Yang XH, Lu XZ. Adolescent ambulatory blood pressure monitoring. Chin J Hypertens. 2009;17(7):667–9.
19. Tan J, Zhang YY. The mechanism and prevention of postural hypertension. Chin J Pract Intern Med. 2012;32(1):32–5.
20. Xu Y, Kang MH, Wang C. Orthostatic hypertension in children. Chin J Appl Chin Pediatr. 2013;28(13):964–7.
21. Liu DY, Xiang JB, Lin P, et al. 24-h ambulatory blood pressure changes in children with orthostatic hypertension. Chin J Appl Chin Pediatr. 2014;29(22):1731–3.
22. Moriguchi A, Nakagami H, Kotani N, et al. Contribution of cardiovascular hypersensitivity to orthostatic hypertension and the extreme dipper phenomenon. Hypertens Res. 2000;23(2):119–23.
23. Wu JS, Yang YC, Lu FH, et al. Population-based study on the prevalence and correlates of orthostatic hypotension/hypertension and orthostatic dizziness. Hypertens Res. 2008;31(5):897–904.

A novel intensity-based multi-level classification approach for coronary plaque characterization in intravascular ultrasound images

Ga Young Kim[1], Ju Hwan Lee[2], Yoo Na Hwang[1] and Sung Min Kim[1,2]*

From International Conference on Biomedical Engineering Innovation (ICBEI) 2016 Taichung, Taiwan. 28 October–1 November 2016

*Correspondence: smkim@dongguk.edu
[1] Department of Medical Biotechnology, Dongguk University-Bio Medi Campus, 32, Dongguk-ro, Ilsandong-gu, Goyang, Gyeonggi-do 10326, Republic of Korea
Full list of author information is available at the end of the article

Abstract

Background: Intravascular ultrasound (IVUS) is a commonly used diagnostic imaging method for coronary artery disease. Virtual histology (VH) characterizes the plaque components into fibrous tissue (FT), fibro-fatty tissue (FFT), necrotic core (NC), or dense calcium (DC). However, VH can obtain only a single-frame image in one cardiac cycle, and specific software is needed to obtain the radio frequency data. This study proposed a novel intensity-based multi-level classification model for plaque characterization.

Methods: The plaque-containing regions between the intima and the media-adventitia were segmented manually for all IVUS frames. A total of 54 features including first order statistics, grey level co-occurrence matrix, Law's energy measures, extended grey level run length matrix, intensity, and local binary pattern were estimated from the plaque-containing regions. After feature extraction, optimal features were selected using principle component analysis (PCA), and these were utilized as the input for the classification models. Plaque components were classified into FT, FFT, NC, or DC using an intensity-based multi-level classification model consisting of three different nets. Net 1 differentiated low-intensity components into FT/FFT and NC/DC groups. Then, net 2 subsequently divided FT/FFT into FT or FFT, whereas the remainder and high-intensity components were classified into NC or DC via net 3. To improve classification accuracy, each net utilized three different input features obtained by PCA. Classification performance was evaluated in terms of sensitivity, specificity, accuracy, and receiver operating characteristic curve.

Results: Quantitative results indicated that the proposed method showed significantly high classification accuracy for all tissue types. The classifiers had classification accuracies of 85.1%, 71.9%, and 77.2%, respectively, and the areas under the curve were 0.845, 0.704, and 0.783. In particular, the proposed method achieved relatively high sensitivity (82.0%) and specificity (87.1%) for differentiating between the FT/FFT and NC/DC groups.

> **Conclusions:** These results confirmed the clinical applicability of the proposed approach for IVUS-based tissue characterization.

Background

Intravascular ultrasound (IVUS) is a commonly used diagnostic imaging method for coronary artery disease. It takes cross sectional images of the arteries in real time by using catheter with a ultrasound probe and provides diverse information that includes lumen size, plaque rupture, and plaque components. This information is clinically important to determine how to treat a lesion before angioplasty, because different treatment should be apply to each patient according to form of lesion. It also can be used to observe the prognosis after treatment. Additionally, it is useful for early diagnosis of a vulnerable plaque that may cause a stroke or heart attack.

In general, the components of a coronary plaque are manually analysed using visual interpretation method based on grey scale IUVS images. However, each component of plaque shows complicated pattern, which, if not properly recognized, may lead to misdiagnosis of coronary artery disease [1]. Also, diagnostic accuracy mainly depends on the experience of the individual reviewing the images. Therefore, many studies have been automatically classify the plaque components to improve diagnostic results. Among the various methods, virtual histogram (VH), which uses radio frequency (RF) signal data, is regarded as the gold standard in diagnosing coronary artery disease.

VH characterizes the plaque components into fibrous tissue (FT), fibro-fatty tissue (FFT), necrotic core (NC), or dense calcium (DC) based on a combination of information that includes envelope amplitude and underlying frequency content of the RF signal [2]. This information is provided as a colour map, and it can be applied to the diagnosis of various coronary artery diseases that involve a thin-cap fibroatheroma. VH showed high clinical effectiveness for the classification of plaque components, and it has been verified in many previous studies [3, 4]. However, VH showed two main limitations. First, VH can acquire only a single-frame image in one cardiac cycle, because it uses electrocardiogram-gated acquisition [5]. This limits the number of images that can be taken as well as the longitudinal resolution. Also, specific software is needed to obtain the RF signal data [1]. Therefore, it is difficult to apply the VH into the existing equipment.

This study proposed a novel intensity-based multi-level classification model to classify the components of coronary plaque. The proposed method extracts six texture feature sets from the plaque region of an IVUS image. These features were optimized using principle component analysis (PCA). Then, the proposed classification model characterizes plaque components into FT, FFT, NC, or DC.

Main contribution of this study is to improve the classification ability of plaque components in IVUS image by using the new concept of a model, intensity-based multi-level classifier. In addition, extensive features were analysed and optimal feature set was selected among these features. Proposed approach achieved significant classification results to differentiate plaque components. The rest of this paper is laid out as follows. "Methods" section describes the methods, which consist of image acquisition, feature extraction, feature selection, classification, and performance evaluation. Then, "Results" section conveys the experimental results of the proposed method, and "Discussion" section provides the discussion. Finally, "Conclusions" section presents the conclusion of this study.

Methods
Image acquisition
This study acquired sequential IVUS images from 11 coronary artery disease patients. It was performed with a 20 MHz 2.9 F phased-array transducer catheter (Eagle Eye, Volcano Corp., Rancho Cordova, California), and grey scale IVUS wasacquired at a constant speed of 0.5 mm/s using a pullback device. The IVUS images consist of 252 frames of 400×400 images. The regions of plaque between the intima and media-adventitia were manually segmented by an expert to obtain information regarding the plaque. Segmented images include 1,230,159 pixels that consist of FT, FFT, NC, and DC tissue. This study was approved by Institutional Review Board of Ulsan University Hospital. Written informed consent was obtained from all patients.

Feature extraction
Feature extraction is the important process of analysing the texture data in images and obtaining meaningful information for the diagnosis of diseases. To obtain sufficient information of pixel, each feature was extracted from a 5×5 window region. However, a 3×3 window was used to establish a local binary pattern (LBP) in order to analyse the feature information based on neighbour pixels adjacent to the central pixel. Table 1 shows first order statistics (FOS), grey level co-occurrence matrix (GLCM), Law's energy measures (LEM), extended grey level run length matrix (GLRLM), intensity, and LBP features that were extracted from each mask region.

First order statistics
FOS [6] analyzes an original image based on the gray-scale value of each pixel. Unlike second order statistics, it extracts features without considering the relationship between pixels. This study extracted 5 features, mean, variance, standard deviation, kurtosis, and skewness, from 5×5 window region. Each feature was calculated using Eqs. (1–5).

$$Mean = \frac{1}{I \times J} \sum_{i=1}^{I-1} \sum_{j=1}^{J-1} G(i,j) \tag{1}$$

$$Variance = \frac{1}{I \times J} \sum_{i=1}^{I-1} \sum_{j=1}^{J-1} \left(G(i,j) - Mean\right)^2 \tag{2}$$

$$Standard\ deviation = \sqrt{\frac{1}{I \times J} \sum_{i=1}^{I-1} \sum_{j=1}^{J-1} \left(G(i,j) - Mean\right)^2} \tag{3}$$

Table 1 Total feature set obtained in the process of feature extraction

Feature set	Feature	Feature set	Feature
FOS	Mean	LEM	MSS S5S5
	Variance		MSS R5S5/S5R5
	Standard deviation		MSS R5R5
	Kurtosis		MAS E5L5/L5E5
	Skewness		MAS S5L5/L5S5
GLCM	Autocorrelation		MAS R5L5/L5R5
	Contrast		MAS E5E5
	Cluster prominence		MAS S5E5/E5S5
	Cluster shade		MAS R5E5/E5R5
	Dissimilarity		MAS S5S5
	Energy		MAS R5S5/S5R5
	Entropy		MAS R5R5
	Homogeneity	Intensity	Intensity
	Maximum probability	GLRLM	Short run emphasis
	Variance		Long run emphasis
	Sum average		Grey level nonuni-formity
	Sum variance		Run length nonuni-formity
	Sum entropy		Run percentage
	Difference variance		Low grey level run emphasis
	Difference entropy		High grey level run emphasis
	Information measure of correlation		Short run low grey level run emphasis
	Normalized inverse difference moment		Short run high grey level run emphasis
LEM	MSS E5L5/L5E5		Long run low grey level run emphasis
	MSS S5L5/L5S5		Long run high grey level run emphasis
	MSS R5L5/L5R5		
	MSS E5E5		
	MSS S5E5/E5S5		
	MSS R5E5/E5R5		
		LBP	Basic LBP
			Uniform LBP

$$Kurtosis = \left[\frac{1}{I \times J} \sum_{i=1}^{I-1} \sum_{j=1}^{J-1} \left(\frac{G(i,j) - Mean}{\sqrt{Variance}} \right)^4 \right] - 3 \tag{4}$$

$$Skewness = \frac{1}{I \times J} \sum_{i=1}^{I-1} \sum_{j=1}^{J-1} \left(\frac{G(i,j) - Mean}{\sqrt{Variance}} \right)^3 \tag{5}$$

where $G(i, j)$ is the grey scale value of each pixel in the IVUS image. I and J are the dimensions of the matrix.

Grey level co-occurrence matrix

Haralick [7] suggested the GLCM, which extracts features based on information on the spatial relationship between pixels. GLCM generates a co-occurrence matrix $P(i, j|d, \theta)$ from the original image by calculating the frequency of a pair of pixels with distance d

and angle θ [8]. In this procedure, i and j represent the gray-scale values of two pixels. In this study, distance 1 was adapted to minimize computational time. The angle was set as 135°. By this procedure, 17 features were extracted from each window, as shown in Table 1.

Law's energy measures

LEM [9, 10] extracts texture information from the original image based on the texture energy transform. To analyze the image, LEM uses the five Laws' vectors, which are $L5$, $E5$, $S5$, $R5$, and $W5$. These vectors represents the level, edge, spot, ripple, and wave, respectively.

The Laws' vectors are multiplied with one another, and 5×5 image masks are thusly generated. In this procedure, the mean values of image masks were used for the transpose matrixes, such as $E5L5$ and $L5E5$. Image masks were convoluted with the window of the IVUS image, and then nine texture images were acquired as defined in Eq. (6).

$$
\begin{aligned}
&E5L5/L5E5 \ R5E5/E5R5 \\
&S5L5/L5S5 \ S5S5 \\
&R5L5/L5R5 \ R5S5/S5R5 \\
&S5E5/E5S5 \ R5R5 \\
&E5R5
\end{aligned}
\tag{6}
$$

Finally, the mean value of the square sum (MSS) and the mean value of the absolute sum (MAS) were calculated from texture images using Eqs. (7) and (8).

$$
MSS = \frac{1}{I \times J} \sum_{i=1}^{I-1} \sum_{j=1}^{J-1} G(i,j)^2
\tag{7}
$$

$$
MAS = \frac{1}{I \times J} \sum_{i=1}^{I-1} \sum_{j=1}^{J-1} |G(i,j)|
\tag{8}
$$

where $G(x, y)$ is the grey scale value of each pixel in the IVUS image. In this study, 18 LEM features were extracted from each window of the IVUS image.

Intensity

Intensity is a simple texture feature that signifies the grey scale value of a pixel. Each plaque component commonly shows a different intensity distribution (Fig. 1). NC and DC are associated with higher intensity than are FT and FFT. In particular, DC involves the highest intensity components, because it is echogenic on ultrasound. In this study, intensity was extracted for each plaque component in order to improve classification accuracy.

Extended grey level run length matrix

GLRLM [11–15] extracts higher order statistical texture information. It reconstitutes the original image into a two-dimensional matrix based on the grey scale values of the

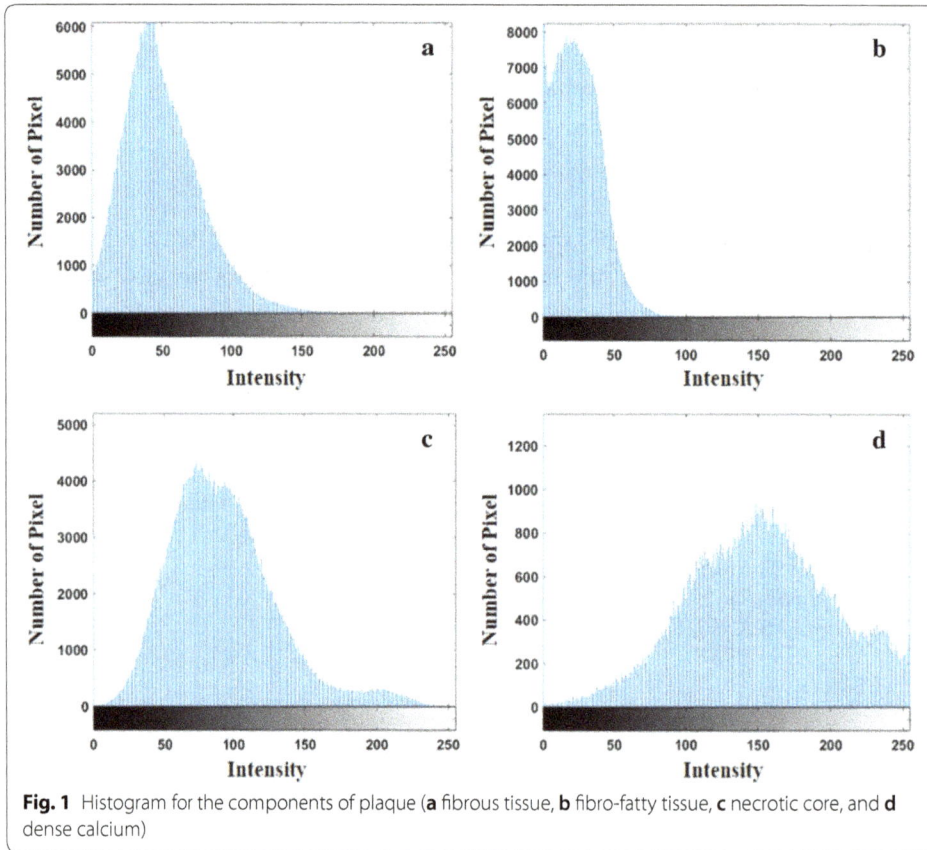

Fig. 1 Histogram for the components of plaque (**a** fibrous tissue, **b** fibro-fatty tissue, **c** necrotic core, and **d** dense calcium)

pixels. Run length matrix $P(i, j|\theta)$ was calculated by counting the repeated number of grey scale value i with run length j in condition of angle θ. In many previous studies, five conventional GLRM features, short run emphasis, long run emphasis, grey level non-uniformity, run length non-uniformity, and run percentage, were generally used for analysis of plaque components. However, previous experimental results were not satisfactory, because they only considered the length of runs when analysing the texture information of IVUS images [1]. Therefore, in this study, the extended GLRLM features that include six additional features compared with conventional GLRLM were extracted to obtain diverse texture information from IVUS images (Table 1).

Local binary pattern

LBP [1, 12] analyses the binary pattern based on local structural information of the original image and detects uniform texture features. To accomplish this, LBP appoints the circular symmetric neighbourhood pixels that are at distance of radius R from the central pixel. Then, a binary digit value is obtained by subtracting the central pixel from each neighbour pixel as shown in Eq. (9). This value was passed to Eq. (10), and the basic LBP was calculated.

$$s(x) = \begin{cases} 1, & if\ x \geq 0 \\ 0, & if\ x < 0 \end{cases} \tag{9}$$

$$LBP = \sum_{n=0}^{N-1} s(G_n - G_c)\, 2^n \qquad (10)$$

where G_n and G_c represent the grey scale values of the neighbour and central pixel, respectively. N is the number of neighbour pixels and $s(x)$ is the thresholding function.

The rotation invariant uniform LBP (LBPriu2) that was defined by Ojala [13] was also calculated based on Eq. (11) [14]. The function U counted the number of transitions between 0 and 1 in the binary digit, as shown in Eq. (12).

$$LBP^{riu2} = \begin{cases} \sum_{n=0}^{N-1} s(G_n - G_c), & \text{if } U(LBP) \le 2 \\ N+1, & \text{if } U(LBP) > 3 \end{cases} \qquad (11)$$

$$U(LBP) = |s(G_{N-1} - G_c) - s(G_0 - G_c)| + \sum_{n=0}^{N-1} \left| s(G_n - G_c) - s(G_{n-1} - G_c) \right| \qquad (12)$$

In our study, two LBP features were extracted form a 3×3 window of an IVUS image through the process discussed above.

Feature selection

Feature selection is the process that chooses the optimal feature set to improve the classification accuracy for the lesion. If the whole of features that is acquired in the process of feature extraction is used as input for the classification, computational time will be unnecessarily long. Furthermore, a large amount of the data may occur the curse of dimensionality that decreases the classification accuracy. Therefore, this study applied PCA to select optimal features.

PCA [15–17] decreases the dimensionality of the feature space and optimizes the feature set. For this purpose, PCA was used to identify the direction vector that has the greatest variance from the feature data. Then, the principle component (PC) was acquired by projecting the original data along the direction vector. PC has reduced feature value dimensionality compared to the original data, because overlapping data was removed. Through the above process, PCA reduced the mean square error variance and provided better information on each plaque component. In this study, the optimal feature set was selected from the original feature set by using PCA for tissue characterization.

Classification

In the process of the classification, data is assigned to a pre-defined class based on the knowledge obtained during training [18]. The proposed method classified plaque components into FT, FFT, NC, or DC through the below four steps (Fig. 2).

First, the plaque components were divided into two groups based on intensity. Each plaque component shows a different intensity distribution as shown in Fig. 1. NC and DC shows higher intensity compared with FT and FFT. A key point of the proposed method is that the plaque components with higher intensity values than FT are involved

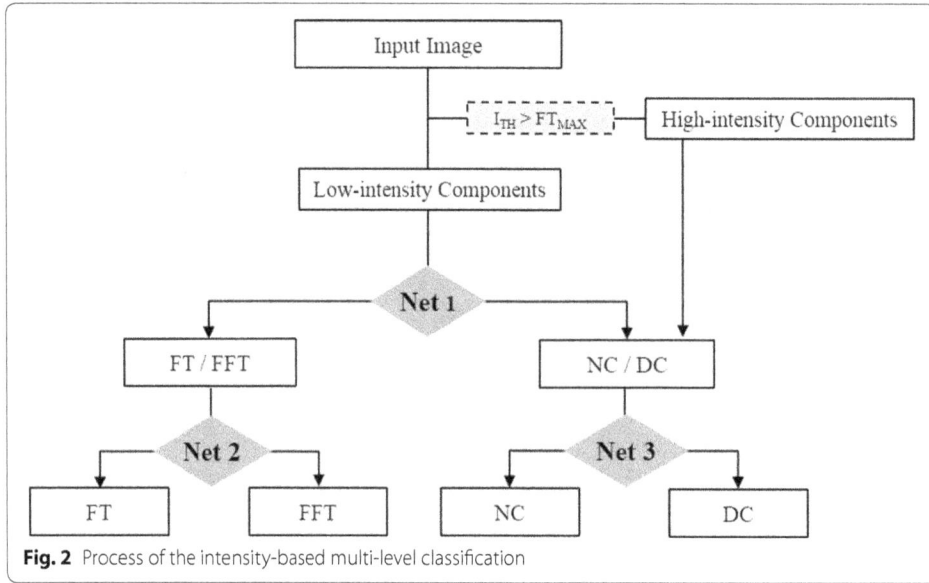

Fig. 2 Process of the intensity-based multi-level classification

in NC or DC. Therefore, the maximum value of FT was appointed as a threshold, and the plaque components were divided into low- or high-intensity components based on this threshold. High-intensity components consist of NC and DC, and the remainder consist of components with lower intensity value than the designated threshold. Next, low-intensity components were divided into the FT/FFT or NC/DC groups using net 1. Then, the net 2 classified the FT/FFT group into FT or FFT. Finally, the remaining and high-intensity components were differentiated into NC or DC via net 3. Each net was trained based on the feature set that was selected by PCA. A random forest operated by the construction of multiple decision trees was used as a classifier. To prevent over-fitting of the classifier and decrease computational time, depth and the number of trees were set at 10 and 100, respectively.

Performance evaluation

To evaluate the performance of the proposed method, sensitivity, specificity, accuracy, and receiver operating characteristic (ROC) curve were applied. Sensitivity, specificity, and accuracy assess the ability of a classification model based on the error rate that includes true positive (TP), true negative (TN), false positive (FP), and false negative (FN). Each evaluation index was calculated using Eqs. (13) to (15).

$$Sensitivity = \frac{TP}{TP + FN} \tag{13}$$

$$Specificity = \frac{TN}{FP + TN} \tag{14}$$

$$Accuracy = \frac{TP + TN}{TP + TN + FN + FP} \tag{15}$$

Table 2 Selected feature set for net 1

Feature set	Feature	Feature set	Feature
FOS	Mean	LEM	MSS S5E5/E5S5
	Variance		MSS R5E5/E5R5
	Standard deviation		MSS S5S5
GLCM	Autocorrelation		MSS R5S5/S5R5
	Variance		MAS E5L5/L5E5
	Sum average		MAS S5L5/L5S5
LEM	MSS E5L5/L5E5		MAS R5L5/L5R5
	MSS S5L5/L5S5		MAS E5E5
	MSS R5L5/L5R5		MAS S5E5/E5S5
	MSS E5E5		MAS R5E5/E5R5
			MAS S5S5

Table 3 Selected feature set for net 2

Feature set	Feature	Feature set	Feature
FOS	Mean	LEM	MSS R5L5/L5R5
GLCM	Autocorrelation		MSS R5R5
	Variance		MAS E5L5/L5E5
	Sum average		MAS S5L5/L5S5
	Sum variance		MAS R5L5/L5R5
			MAS R5S5/S5R5
LEM	MSS E5L5/L5E5		MAS R5R5
	MSS S5L5/L5S5	Intensity	Intensity

Sensitivity indicates the percentage of data that were correctly classified as positive, while specificity indicates the percentage of data that were correctly classified as negative. Accuracy measures the ability of the proposed method to identify total data [19].

ROC curve is a graphical plot that represents 1-specificity and sensitivity on the x and y axes [19]. In this study, area under the curve (AUC) was additionally used as an index to evaluate the performance of the proposed method. AUC represents the value between zero and one, and a higher AUC value means a higher classification performance.

Results

Selected feature sets for each net by PCA

Different feature sets were optimized for the three nets by using PCA method. For net 1, 21 features were selected as the optimal feature set; they consist of the FOS, GLCM, and LEM (Table 2). Intensity was selected for nets 2 and 3 in addition to those three kinds of features. As shown in Tables 3 and 4, the feature set of net 2 includes 15 components, while 18 components were selected for net 3. GLRLM and LBP was not chosen for all nets. On the other hand, 10 features were common to the three nets, and they consist of mean, autocorrelation, variance, sum average, MSS E5L5/L5E5, MSS S5L5/L5S5, MSS R5L5/L5R5, MAS E5L5/L5E5, MAS S5L5/L5S5, and MAS R5L5/L5R5. For all three nets, most LEM features were selected as the optimal value. In particular, 15 features of 18 LEM features were chosen for net 1.

Table 4 Selected feature set for net 3

Feature set	Feature	Feature set	Feature
FOS	Mean	LEM	MSS R5S5/S5R5
GLCM	Autocorrelation		MSS R5R5
	Variance		MAS E5L5/L5E5
	Sum average		MAS S5L5/L5S5
	Sum variance		MAS R5L5/L5R5
LEM	MSS E5L5/L5E5		MAS S5S5
	MSS S5L5/L5S5		MAS R5S5/S5R5
	MSS R5L5/L5R5		MAS R5R5
	MSS S5E5/E5S5	Intensity	Intensity

Table 5 Classification results of the proposed method

Net	Sensitivity (%)	Specificity (%)	Accuracy (%)	AUC
Net 1	82.0	87.1	85.1	0.845
Net 2	81.2	59.6	71.9	0.704
Net 3	80.6	75.9	77.2	0.783

Table 6 Classification results of net 1 according to different feature selection methods

Selection method	Sensitivity (%)	Specificity (%)	Accuracy (%)	AUC
w/o selection	82.0	86.9	85.0	0.845
GA	81.9	86.9	84.9	0.844
PCA	82.0	87.1	85.1	0.845

Table 7 Classification results of net 2 according to different feature selection methods

Selection method	Sensitivity (%)	Specificity (%)	Accuracy (%)	AUC
w/o selection	80.8	60.1	71.9	0.705
GA	80.8	59.7	71.7	0.703
PCA	81.2	59.6	71.9	0.704

Classification results of proposed method for each net

Table 5 shows the results of the tissue classification for each net produced by using proposed method. Classification accuracy was in the order of net 1 > net 3 > net 2. Net 2 showed relatively low classification results, especially for specificity, which was identified as 59.6%. On the other hand, in net 1, sensitivity, specificity, and accuracy were higher than 80.0%. The proposed method also presented a high AUC of 0.845 for net 1. Furthermore, net 3 showed a relatively high classification accuracy, 77.2%.

Comparison of the classification results according to different feature selection methods

To evaluate the significance of the feature set that was selected using PCA, the classification results were compared with a genetic algorithm (GA). GA selected 25, 23, and 25 feature components for nets 1, 2, and 3, respectively. PCA showed higher classification ability than GA for all nets as shown in Tables 6, 7, and 8. According to applying the

Table 8 Classification results of net 3 according to different feature selection methods

Selection method	Sensitivity (%)	Specificity (%)	Accuracy (%)	AUC
w/o selection	80.4	75.2	76.7	0.778
GA	80.2	75.8	76.6	0.780
PCA	80.6	75.9	77.2	0.783

Table 9 Classification results of net 1 according to different classifiers

Classification method	Sensitivity (%)	Specificity (%)	Accuracy (%)	AUC
DNN	79.4	86.5	84.0	0.829
FFNN	78.0	87.5	84.2	0.827
Proposed method	82.0	87.1	85.1	0.845

Table 10 Classification results of net 2 according to different classifiers

Classification method	Sensitivity (%)	Specificity (%)	Accuracy (%)	AUC
DNN	2.1	99.9	44.1	0.510
FFNN	81.3	57.9	71.2	0.696
Proposed method	81.2	59.6	71.9	0.704

Table 11 Classification results of net 3 according to different classifiers

Classification method	Sensitivity (%)	Specificity (%)	Accuracy (%)	AUC
DNN	79.2	79.9	77.5	0.780
FFNN	78.5	76.0	76.7	0.772
Proposed method	80.6	75.9	77.2	0.783

GA, the accuracy and AUC slightly decreased in net 1 and 2 than was the case without feature selection. On the other hand, PCA improved the classification results for all nets. Especially, classification accuracy of net 3 increased from 76.7 to 77.2%.

Comparison of the classification results according to different classifiers

Tables 9, 10, and 11 show the classification results of the coronary plaque components according to different classifiers. The proposed method showed slightly lower accuracy in net 3 compared with a dropout neural network (DNN). However, the proposed method achieved much higher classification performance in net 2 than the DNN. In particular, the proposed method showed a significantly high sensitivity of 81.2%, while a very low sensitivity of 2.1% was observed when using the DNN. It also represented higher accuracy, of about 27.8%, than did the DNN for net 2. Furthermore, the proposed method showed slightly higher accuracy and AUC for all nets in comparison with a feed forward neural network (FFNN).

Discussion

In order to differentiate the coronary plaque components in an IVUS image, VH, which analyzes an image based on the RF signal, has been widely applied. However, it has limitations in longitudinal resolution, and it requires specific software to acquire the RF signal data. In this study, FOS, GLRLM, LEM, intensity, extended GLRLM, and LBP features were extracted, and they were optimized using PCA. Then, an intensity-based multi-level classification model was used to classify the coronary plaque components into FT, FFT, NC, and DC based on the texture information in the IVUS image.

In the present study, three different feature sets were individually selected by applying PCA to each net. The results showed that a large number of LEM features were selected for all nets. In particular, LEM features accounted for more than 70% of 21 selected features in net 1. LEM provides diverse texture information that includes level, edge, spot, ripple, and wave for each window. Therefore, many LEM features were considered as the most significant features for classification of coronary plaque components. If an MSS feature was selected as part of the optimal feature set, the MAS feature of that image mask was also chosen. This is because the MSS and MAS have a high correlation to the same image mask. Also, PCA commonly selected 10 features in all nets, which are the mean, autocorrelation, variance, sum average, MSS E5L5/L5E5, MSS S5L5/L5S5, MSS R5L5/L5R5, MAS E5L5/L5E5, MAS S5L5/L5S5, and MAS R5L5/L5R5. Thus, 10 features were identified as essential indicators for classification of plaque components. However, GLRLM and LBP were not selected for all nets. These two types of features were affected by noise within the IVUS image. Noise may occur in the process of receiving, coding, and transmission of IVUS image [20]. Binary data generated in the process of extracting LBP features is sensitive to noise [21]. GLRLM is also extremely susceptible to noise in IVUS images [22]. For this reason, GLRLM and LBP did not show high significance in the analysis of IVUS images.

The proposed method classified the coronary plaque components based on the selected optimal feature sets, and the results showed a high accuracy (85.1%) and AUC (0.845) for net 1. This is because of the obvious difference between the FT/FFT and NC/DC groups, as shown in Fig. 1. On the other hand, the grey scale based classification method has some technical limitations in differentiating between FT and FFT, because both are medium echo reflective and show similar characteristic in IVUS images [23, 24]. Therefore, the classification results presented relatively low accuracy for net 2 (71.9%). Nevertheless, it showed a comparatively high sensitivity of 81.2%, which indicates that the proposed method has a high ability to classify FFT. Therefore, it is expected to improve overall classification accuracy by increasing ability to accurately identify for the FT.

By comparing classification accuracy according to different feature selection methods, PCA showed the highest performance in terms of computational load. Using all features, which consisted of 54 components, resulted in longest test time for the proposed classification model, 673 s. It was slightly decreased to 624 s by applying GA. By comparison, PCA greatly improved test time to 189 s, which shows its high effectiveness in terms of computational time.

The proposed method showed higher classification performance than did other classifiers including FFNN and DNN. In particular, the proposed method showed a relatively high accuracy of 71.9% in net 2, while the lowest accuracy, 44.1%, was identified for DNN. These results occurred because of the inherit characteristics of the random forest. The random forest presents high classification accuracy for data that includes noise, because it consists of multiple decision trees that have different properties. For this reason, the proposed method showed high classification ability for FT and FFT, which are difficult to differentiate because of their similar grey scale, compared with other classifiers.

This study has two main limitations. First, it did not acquire significant classification results compared with VH. It is considered that the results were affected by the quality of IVUS images used in the study. The proposed method is greatly influenced by image resolution, because it classifies plaque components based on a grey scale of image. Therefore, classification accuracy may be improved by using IVUS images with higher frequency than 20 MHz. Furthermore, the proposed method showed relatively low classification performance for high-intensity components. This seems to be affected by the amount of NC data that is included in the high-intensity components. High-intensity components are mostly composed of DC, and NC is included in relatively low proportions. The classifier may be not properly trained to identify NC, which shows high-intensity values owing to the low amount of NC data. Therefore, classification accuracy may be improved by supplementing the data on NC.

Conclusions

This study proposed a novel intensity-based multi-level classification model to classify coronary plaque components in IVUS images. The proposed method selected 10 features that included FOS, GLCM, and LEM as key indicators for the characterization of plaque components. Quantitative results indicated that the proposed method showed significantly high classification accuracy for all tissue types. Net 1, 2, and 3 classifiers revealed classification accuracies of 85.1%, 71.9%, and 77.2%, respectively, and the areas under the curve were 0.845, 0.704, and 0.783. In particular, the proposed method achieved relatively high sensitivity (82.0%) and specificity (87.1%) for differentiating between FT/FFT and NC/DC groups. These results confirmed the clinical applicability of the proposed approach for IVUS-based tissue characterization. To improve classification accuracy, future studies should include additional experiments with a greater amount of NC data. Tissue characterization of 45 MHz IVUS images needs to be validated with various textural feature sets. Moreover, another experiment on controlling the level of noises in the various patterns will be performed.

Declarations
Authors' contributions
GYK conducted the study include feature set and classification model, and drafted the manuscript. JHL designed the study and provided comments on manuscript. YNH designed the study and analysed the data. SMK participated in overall design and revision of the work, and was the research manager. All authors read and approved the final manuscript.

Author details
[1] Department of Medical Biotechnology, Dongguk University-Bio Medi Campus, 32, Dongguk-ro, Ilsandong-gu, Goyang, Gyeonggi-do 10326, Republic of Korea. [2] Department of Medical Devices Industry, 26, Pil-dong 3-ga, Jung-gu, Seoul 04620, Republic of Korea.

Competing interests
The authors declare that they have no competing interests.

Consent for publication
All authors consent for publication.

Funding
This work and the publication were supported by the Dongguk University research fund of 2016 (S-2016-G0001-00015).

References

1. Taki A, Hetterich H, Roodaki A, Setarehdan SK, Unal G, Pieber J, Navab N, Konig A. A new approach for improving coronary component analysis based in intravascular ultrasound images. Ultrasound Med Biol. 2010;36:1245–58.
2. Garcia-Garcia HM, Gogas BD, Serruys PW, Bruining N. IVUS-based imaging modalities for tissue characterization: similarities and differences. Int J Cardiovasc Imaging. 2011;27:215–24.
3. Nasu K, Tsuchikane E, Katoh O, Vince DG, Virmani R, Surmely JF, Murata A, Takeda Y, Ito T, Ehara M, Matsubara T, Terashima M, Suzuki T. Accuracy of in vivo coronary plaque morphology assessment—a validation study of in vivo virtual histology compared with in vitro histopathology. J Am Coll Cardiol. 2006;47:2405–12.
4. Nair A, Kuban BD, Tuzcu EM, Schoenhagen P, Nissen SE, Vince DG. Coronary plaque classification with intravascular ultrasound radiofrequency data analysis. Circulation. 2002;106:2200–6.
5. Giannoglou VG, Stavrakoudis DG, Theocharis JB: IVUS-based characterization of atherosclerotic plaques using feature selection and SVM classification. In: Proceedings of the IEEE 12th international conference on bioinformatics & bioengineering (BIBE), Larnaca; 2012. p.715–20.
6. Srinivasan GN, Shobha G. Statistical texture analysis. Proc World Acad Sci Eng Technol. 2008;36:1264–9.
7. Haralick RM, Shanmugam K, Dinstein I. Textural features for image classification. IEEE Trans Syst Man Cybem. 1973;3:610–21.
8. Zulpe N, Pawar V. GLCM texture features for brain tumor classification. Int J Comput Sci Issues. 2012;9:354–9.
9. Vince DG, Dixon KJ, Cothren RM, Cornhill JF. Comparison of texture analysis methods for the characterization of coronary plaques in intravascular ultrasound images. Comput Med Imaging Graph. 2000;24:221–9.
10. Valavanis IK, Mougiakakou SG, Nikita A, Nikita KS. Evaluation of texture features in hepatic tissue characterization from non-enhanced CT images. In: Proceedings of the 29th annual international conference of the IEEE engineering in medicine and biology society. Lyon; 2007. p. 3714–44.
11. Garcia G, Maiora J, Blas MD. Evaluation of texture for classification of abdominal aortic aneurysm after endovascular repair. J Digit Imaging. 2012;25:369–76.
12. Escalera S, Pujol O, Mauri J, Radeva P. Intravascular ultrasound tissue characterization with sub-class error-correcting output codes. J Sign Process Sys. 2009;55:35–47.
13. Guo Z, Zhang L, Shang D. A completed modeling of local binary pattern operator for texture classification. IEEE Trans Image Process. 2010;19:1657–63.
14. Dan Z, Chen Y, Yang Z, Wu G. An improved local binary pattern for texture classification. Optik-Int J Light Electron Opt. 2014;125:6320–4.
15. Balasubramanian D, Srinivasan P, Gurupatham R. Automatic classification of focal lesions in ultrasound liver images using principle component analysis and neural network. In: proceedings of the 29th annual international conference of the IEEE engineering in medicine and biology society. Lyon; 2007. p. 2134–7.
16. Amer HM, Abou-Chadi FEZ, Obayya MI. A computer-aided system for classifying computed tomographic (CT) lung images using artificial neural network and data fusion. Int J Comput Sci Netw Secur. 2011;11:70–5.
17. Sachdeva J, Kumar V, Gupta I, Khandelwal N, Ahuja CK. Segmentation, feature extraction, and multiclass brain tumor classification. J Digit Imaging. 2013;26:1141–50.
18. Padma A, Sukanesh R. Automatic classification and segmentation of brain tumor in CT images using optimal dominant gray level run length texture features. Int J Adv Comput Sci Appl. 2011;2:53–9.
19. Padma A, Sukanesh R. A wavelet based automatic segmentation of brain tumor in CT images using optimal statistical texture features. Int J Image Process. 2011;5:552–63.
20. Lazrag H, Naceur MS. Despeckling of intravascular ultrasound images using curvelet transform. In: Proceedings of the 6th international conference on sciences of electronics, technologies of information and telecommunications (SETIT). Sousse; 2012. p. 365–9.
21. Fu X, Wei W: Centralized binary patterns embedded with image Euclidean distance for facial expression recognition. In: Proceedings of the 4th international conference on natural computation. Jinan; 2008. p. 115–9.
22. Whelan PF, Molloy D. Machine vision algorithms in java: techniques and implementation. London: Springer; 2001.
23. Athanasiou LS, Karvelis PS, Tsakanikas VD, Naka KK, Michalis LK, Bourantas CV, Fotiadis DI. A novel semiautomated atherosclerotic plaque characterization method using grayscale intravascular ultrasound images: comparison with virtual histology. IEEE Trans Inf Technol Biomed. 2012;16:391–400.
24. Hiro T, Leung CY, De Guzman S, Caiozzo VJ, Farvid AR, Karimi H, Helfant RH, Tobis JM. Are soft echoes really soft? Intravascular ultrasound assessment of mechanical properties in human atherosclerotic tissue. Am Heart J. 1997;133:1–7.

3

Pulse wave response characteristics for thickness and hardness of the cover layer in pulse sensors to measure radial artery pulse

Min-Ho Jun, Young Ju Jeon, Jung-Hee Cho and Young-Min Kim[*]

*Correspondence:
irobo77@kiom.re.kr
Future Medicine Division,
Korea Institute of Oriental
Medicine (KIOM), 1672
Yuseongdaero, Yuseong-gu,
Deajeon 34054, Republic
of Korea

Abstract

Background: Piezo-resistive pressure sensors are widely used for measuring pulse waves of the radial artery. Pulse sensors are generally fabricated with a cover layer because pressure sensors without a cover layer are fragile when they come into direct contact with the skin near the radial artery. However, no study has evaluated the dynamic pulse wave response of pulse sensors depending on the thickness and hardness of the cover layer. This study analyzed the dynamic pulse wave response according to the thickness and hardness of the cover layer and suggests an appropriate thickness and hardness for the design of pulse sensors with semiconductor device-based pressure sensors.

Methods: Pulse sensors with 6 different cover layers with various thicknesses (0.8 mm, 1 mm, 2 mm) and hardnesses (Shore type A; 30, 43, 49, 71) were fabricated. Experiments for evaluating the dynamic pulse responses of the fabricated sensors were performed using a pulse simulator to transmit the same pulse wave to each of the sensors. To evaluate the dynamic responses of the fabricated pulse sensors, experiments with the pulse sensors were conducted using a simulator that artificially generated a constant pulse wave. The pulse wave simulator consisted of a motorized cam device that generated the artificial radial pulse waveform by adjusting the stroke of the cylindrical air pump and an air tube that conveyed the pulse to the artificial wrist.

Results: The amplitude of the measured pulse pressure decreased with increasing thickness and hardness of the cover layer. Normalized waveform analysis showed that the thickness rather than the hardness of the cover layer contributed more to waveform distortion. Analysis of the channel distribution of the pulse sensor with respect to the applied constant dynamic pressure showed that the material of the cover layer had a large effect.

Conclusions: In this study, in-line array pulse sensors with various cover layers were fabricated, the dynamic pulse wave responses according to the thickness and the hardness of the cover layer were analyzed, and an appropriate thickness and hardness for the cover layer were suggested. The dynamic pulse wave responses of pulse sensors revealed in this study will contribute to the fabrication of improved pulse sensors and pulse wave analyses.

Keywords: Piezo-resistive pulse sensor, Thickness and hardness of cover layer, Artificial pulse wave simulator, Radial artery pulse, Pulse wave response

Background

Recently, carotid artery measurement studies for diagnosis of cardiovascular disease have been investigated [1–3], and computational approaches and modeling studies of blood vessels for vascular stenosis and blood flow analysis have been increased [4–7] in western medicine. On the other hand, many studies of eastern medicine have aimed to objectify, quantify, and automate wrist pulse diagnosis by employing modern sensors for data acquisition, data processing, and pattern classification [8]. Precise and accurate pulse wave measurements should be prioritized to diagnose diseases or to determine pulse patterns through pulse wave analysis [9]. Piezo-resistive pressure sensors are commonly used to measure pulse waves of the human radial artery, which represents the most prominent method for precisely measuring radial artery pulse [10]. In-line array pressure sensors are a better alternative to tonometry sensors, which have only one pressure sensor, for conveniently obtaining the spatial information. Jeon et al. [11] describe a seven-channel pressure sensor array for measuring pulse width. Similarly, Chung et al. [12], Choi et al. [13], and Kim et al. [14] use a two-dimensional pressure sensor array to extract spatial pulse features. Hu et al. [15] present a sensor probe consisting of a capacitive array sensor with 12 sensing points to determine the optimal pulse-taking position. Peng and Lu [16] introduce a flexible 5×5 capacitive pressure sensor array based on flexible printed circuit boards and integrated CMOS switched capacitor readout circuits for determining pulse patterns. In Chang's study [17], a 9-channel sensing probe based on piezoelectric PVDF sensors is described for collecting pulse patterns. Xu et al. [18] introduce a sensor system with a strain cantilever beam transducer as the main sensor and an array of 7 additional sensors for detecting pulse width.

To protect the pulse sensor damage, a cover layer on the pulse sensor is required because the pressure sensors of the pulse sensor directly contact the skin near the radial artery with considerable force. However, this cover layer coated on the pulse sensor has an adverse effect on sensor performance. A thick cover layer on the pulse sensor has been reported to affect other channels in the pulse sensor array due to the force distribution, and signal distortion may also occur because of the temperature difference between the skin surface and the pulse sensor [19, 20]. Although some studies on pulse sensor cover layers have reported static characteristics according to thickness or temperature [21], no study has evaluated the dynamic pulse wave response according to the thickness and hardness of the cover layer.

Dynamic pulse waves in the radial artery have been studied and used in various applications, such as arterial stiffness assessments [22, 23], cardiovascular disease diagnoses [24], central BP monitoring [25–27], and pulse wave analysis [28]. It is necessary to accurately reflect the dynamic response of the pulse waves to apply pulse sensors with a cover layer for these various uses. The dynamic response of the pulse sensor for monitoring radial pulse waves can be altered by the thickness and hardness of the cover layer. Therefore, the changes in pulse wave dynamics must be studied according to the thickness and hardness of the cover layer of the pulse sensor.

The heart rate of normal adults is usually normalized to 75 bpm (normal range of heart rate: 60–90 bpm), and the difference between diastolic blood pressure and systolic blood pressure of normal adults in the radial artery is known to be ~40–50 mmHg [29–31]. However, measurement of pulse wave in the radial artery of normal adults is too volatile to analyze the sensor performance. Figure 1 shows the signal differences in the pulse wave measured on the radial artery of a human body according to the various thicknesses and hardnesses of the cover layer in a pulse sensor. T is thickness (0.8, 1, and 2 mm), and H is hardness (Shore A: 30, 43, 49, and 71) in Fig. 1. The difference of the pulse wave signals is distinctly shown in Fig. 1 when the thickness of the cover layers ranges from 0.8 to 2 mm and the hardness of Shore type A ranges from 30 to 71. Therefore, it is necessary to study the variation of the pulse waveform according to the thickness and hardness of the cover layer in the pulse sensor. However, it is difficult to analyze the pulse sensor characteristics by measuring the pulsation of the radial artery in the human body because the pulse wave signals are highly volatile according to the subject's condition. Therefore, a simulator that can produce a constant heart rate and blood pressure difference was fabricated and used in the experiment to analyze the sensor performance.

In this study, pulse sensors with a cover layer coated with materials of different thicknesses and hardnesses were fabricated. To evaluate the influence of the thickness and hardness of the cover layer on pulse sensor performance, the pulse wave dynamic responses of the fabricated pulse sensors were analyzed using simulated pulse waves generated by a simulator. Analysis of the results revealed the appropriate thickness and hardness of the cover layer for use in pulse sensors. These findings provide guidance on the thickness and hardness of the cover layer that should be used when fabricating pulse sensors.

Methods

Fabrication of the pulse sensor

The fabricated pulse sensor consisted of an arrangement of 6 in-line sensor cells, Piezo-resistive pressures sensors, and 1 temperature sensor, i.e., a thermistor. A total of 7 sensor cells were mounted on the PCB, and the PCB electrodes and sensor cells were connected by wire bonding. The wire bonding area was coated with a polymer to prevent breakage of the connected wire. Finally, the pulse sensors and the aligned sensor cells were coated with 3 types of silicone with different hardnesses to prevent

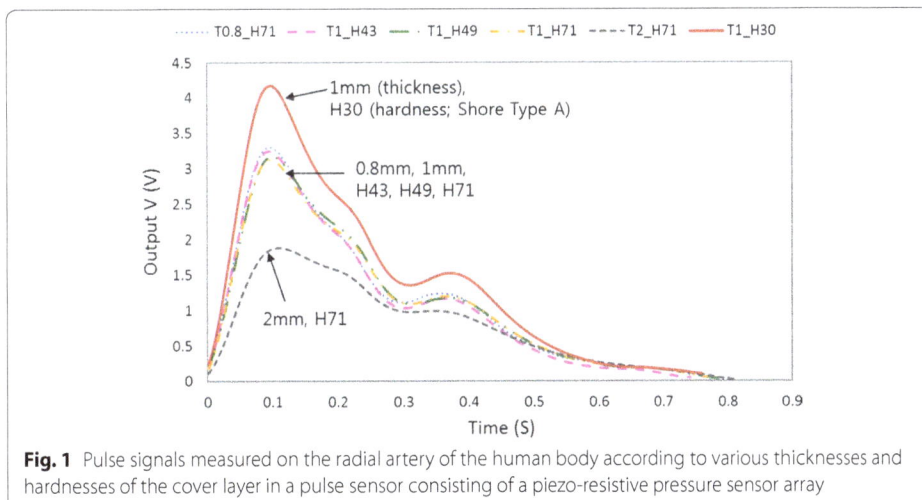

Fig. 1 Pulse signals measured on the radial artery of the human body according to various thicknesses and hardnesses of the cover layer in a pulse sensor consisting of a piezo-resistive pressure sensor array

fracture of the sensor cells due to contact with the wrist skin. Three types of pulse sensors coated with the hardest silicone as the cover layer were fabricated with cover layer thicknesses of 0.8 mm, 1.0 mm, and 2.0 mm. The fabrication sequence of the pulse sensors with the cover layer is shown in Fig. 2. A thermistor that measures skin temperature was soldered to the patterned PCB. Six pressure sensors were arranged in-line and fixed by soldering, and the pressure sensors were then connected to the patterned PCB by wire bonding. To protect the connected wires, underfill (HI-FILL 3085B, HI-TECH KOREA CO., LTD, South Korea) was coated on the wires. Barriers were placed on the edges of the patterned PCB, and silicone was poured in the barriers to form the cover layer on the pulse sensor. The thickness of the cover layer was controlled by the amount of silicone. The silicone was cured by baking the pulse sensor in an oven at 150 °C for 30 min. The individual pulse sensor with the cover layer was cut out of the construct using a dicing saw. A fabricated pulse sensor with the cover layer included 1 thermistor and 6 piezo-resistive pressure sensors and measured 10 mm × 8 mm in size, as shown in Fig. 3. The 3 kinds of silicone used to make the cover layer were XE14-C2042, IVS4546, and IVS4742 (MOMENTIVE, NY, USA); their hardnesses were 43, 49, and 71 (Shore type A durometer), and their tensile strengths were 6.0 MPa, 7.1 MPa, and 11 MPa, respectively [32]. Although IVS4312 (hardness: 29 and tensile strength: 0.8 MPa) was tested, it was too sticky to be used as a cover layer for the pulse sensors. Hardness was considered in evaluating the dynamic response of the pulse sensors because the mechanical properties, such as Young's modulus and Poisson's ratio, among others, were not provided by the manufacturer. A pulse sensor with a 1-mm-thick polydimethylsiloxane (PDMS) cover layer was used as a reference sensor in the evaluation of the dynamic response of the pulse sensors. The reference pulse sensor with the PDMS cover layer was made via the standard fabrication method; PDMS has a hardness of 30. The reference sensor was used for the comparative analysis of the pulse sensors with the silicone cover layers because of the difficulty in obtaining the simulator waveform.

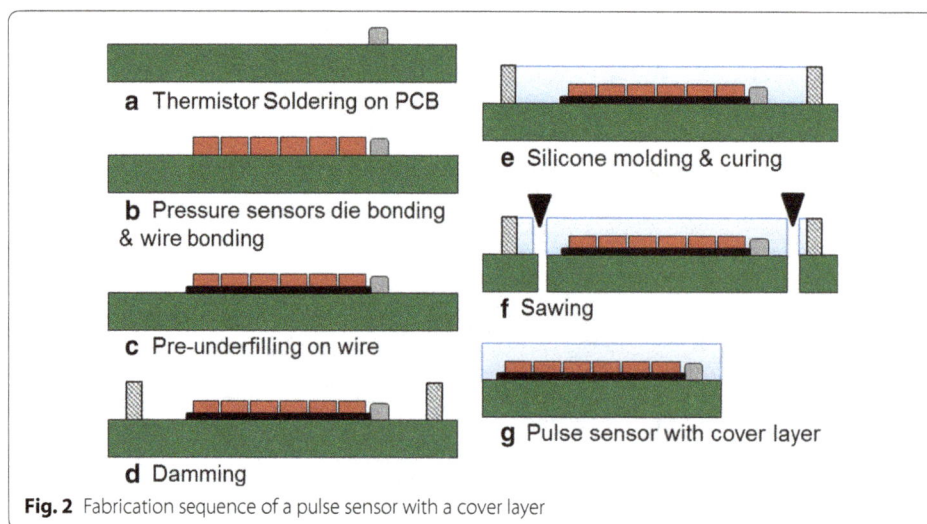

a Thermistor Soldering on PCB

b Pressure sensors die bonding & wire bonding

c Pre-underfilling on wire

d Damming

e Silicone molding & curing

f Sawing

g Pulse sensor with cover layer

Fig. 2 Fabrication sequence of a pulse sensor with a cover layer

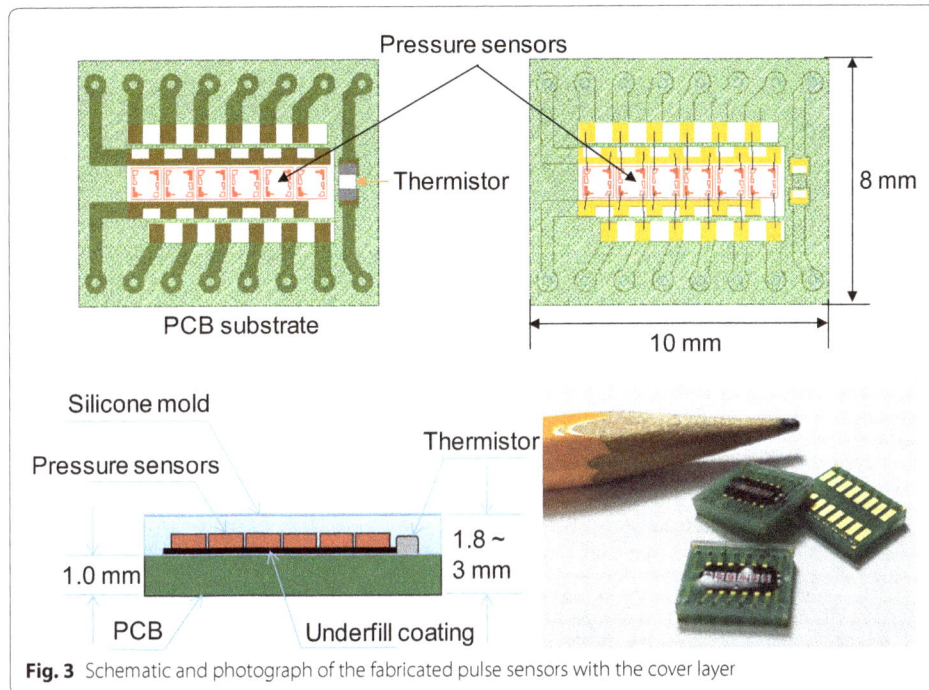

Fig. 3 Schematic and photograph of the fabricated pulse sensors with the cover layer

Experiment with the fabricated pulse sensors

To evaluate the dynamic responses of the fabricated pulse sensors, experiments with the pulse sensors were conducted using a simulator that artificially generated a constant pulse wave. The pulse wave simulator consisted of a motorized cam device that generated the artificial radial pulse waveform by adjusting the stroke of the cylindrical air pump and an air tube that conveyed the pulse to the artificial wrist, as shown in Fig. 4. The reference input signal of the radial pulse waveform was a typical pressure signal from a young adult acquired from clinical data [33–35]. The pulse pressure and heart rate of the simulated input pulse signals can be modulated by changing the length and volume of the air and the rotational speed of the cam. In the experiments, the pulse pressure values and the heart rate values were set to 50 mmHg and 75 bpm, respectively. To adjust the pressure and heart rate of the simulator precisely and accurately, pressure and heart rate are measured from the pressure sensor installed in front of syringe. While working for 5 min, the simulator showed repeatability of CV = 0.23% and CV = 0.82% for the heart rate and the pulse pressure, respectively.

The pulse sensor was attached to the end effector of a 6 DOF robotic tonometry device to maintain constant posture and contact force on the artificial radial artery. The 3 DOF motorized stage moved the center of the pulse sensor to the exact pulsation position. The contact direction between the pulse sensor and the skin surface was controlled by 2 harmonic-drive rotational actuators without gear backlash. A ball-screw type linear actuator was used to precisely control the contact force of the pulse sensor. Figure 5 shows the pulse sensor array attached to the end of the robotic tonometry device (KIOM PAS V3, KIOM, South Korea). The 2-axis tilting sensor was laid on the pulse sensor to measure the tilting angle values of the sensor surface along the gravitational axis. Additionally, the contact force direction of

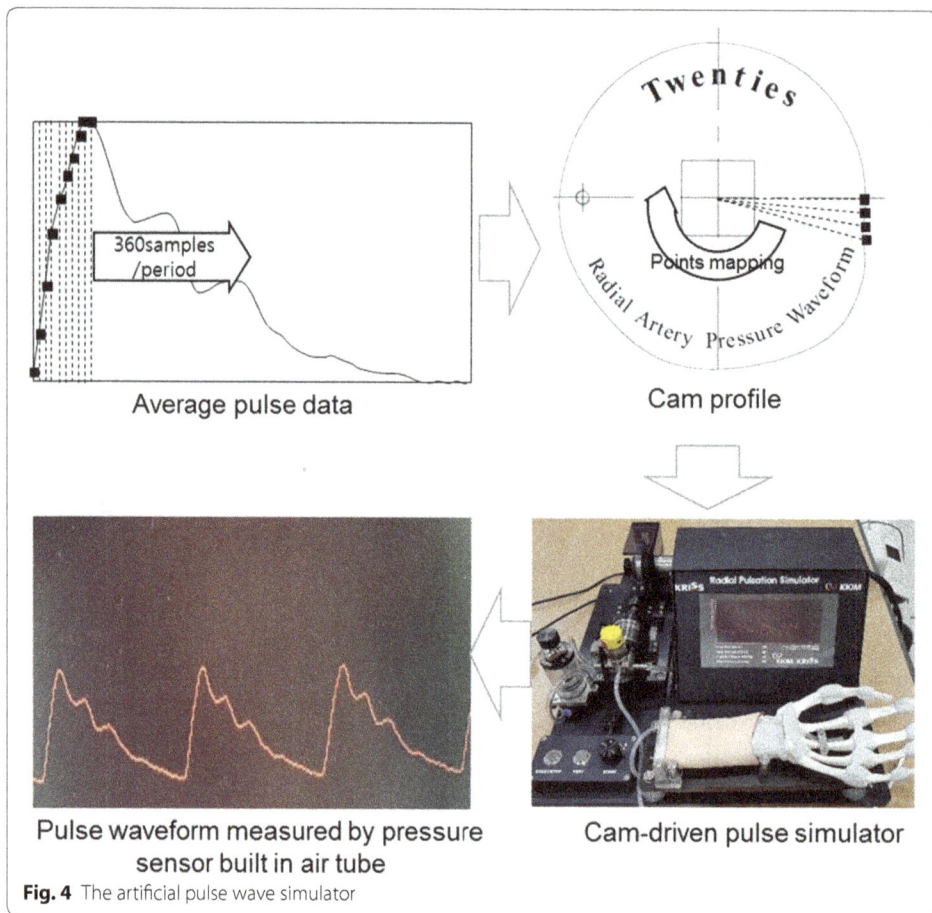

Fig. 4 The artificial pulse wave simulator

the pulse sensor could be kept constant when the pulse sensor surface angles with the gravitational axis were controlled by the constant target values, $\alpha = -5.0°$ and $\beta = 2.0°$ degrees, because the artificial arm of the simulator was fixed on the base of the robotic tonometry device, as shown in Fig. 5. In the experiment, the 2 contact angles between the pulse sensor surface and the base plane of the simulator were controlled with error bounds of $\pm 0.21°$ and $\pm 0.37°$ degrees, respectively.

Figure 6 shows the raw data of the pulse wave measured from the time that the pulse sensor reached the artificial wrist surface to pressurization of the artificial radial artery. The center of the pulse sensor was laid on the same contact point of the surface, and the radial artery was incrementally pressured until the maximum pulse pressure values were found. When the maximum pulse pressure was detected, the tonometry device maintained the contact force for 30 s to reliably record the raw signals of the maximum pulse pressure of the radial artery pulse. Approximately 30 pulse waveforms obtained in the reliable region were averaged to analyze the dynamic responses of the radial pulse for different thickness and hardness conditions. Figure 7 shows the experimental set-up to analyze the pulse responses of the pulse sensors using the simulator and the data acquisition for pulse signals on the artificial radial artery.

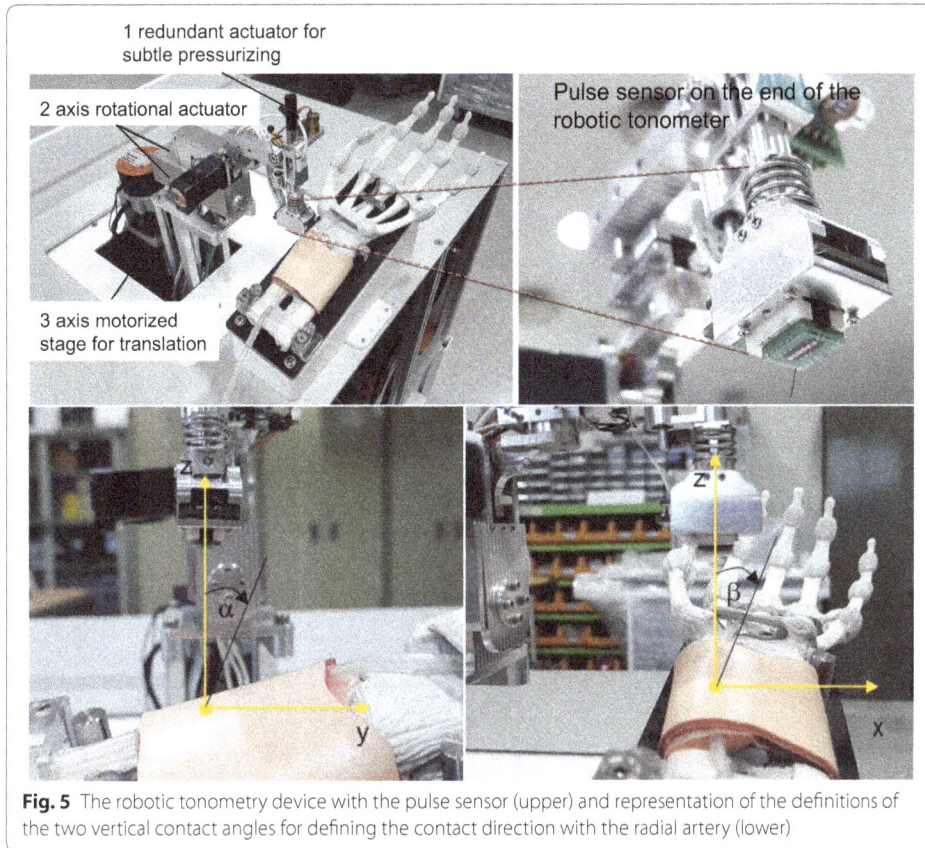

Fig. 5 The robotic tonometry device with the pulse sensor (upper) and representation of the definitions of the two vertical contact angles for defining the contact direction with the radial artery (lower)

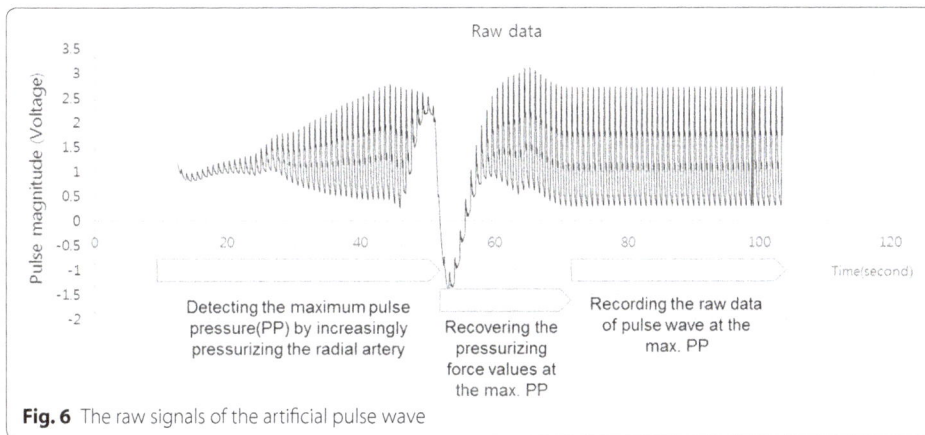

Fig. 6 The raw signals of the artificial pulse wave

Results

The dynamic responses of 5 kinds of fabricated pulse sensors and 1 pulse sensor with a PDMS cover layer were measured. The pulse sensor with the cover layer composed of PDMS, which is a much softer material than silicone and is often used as a mold given its good detachability property, was used to record the pulse waves as a reference due to its short lifetime and weak adhesion [36, 37]. The heart rate (HR) of the pulse wave simulator was set to 75 Hz, and the systolic/diastolic blood pressure difference was set to 50 mmHg. Each sensor was measured 3 times under the same

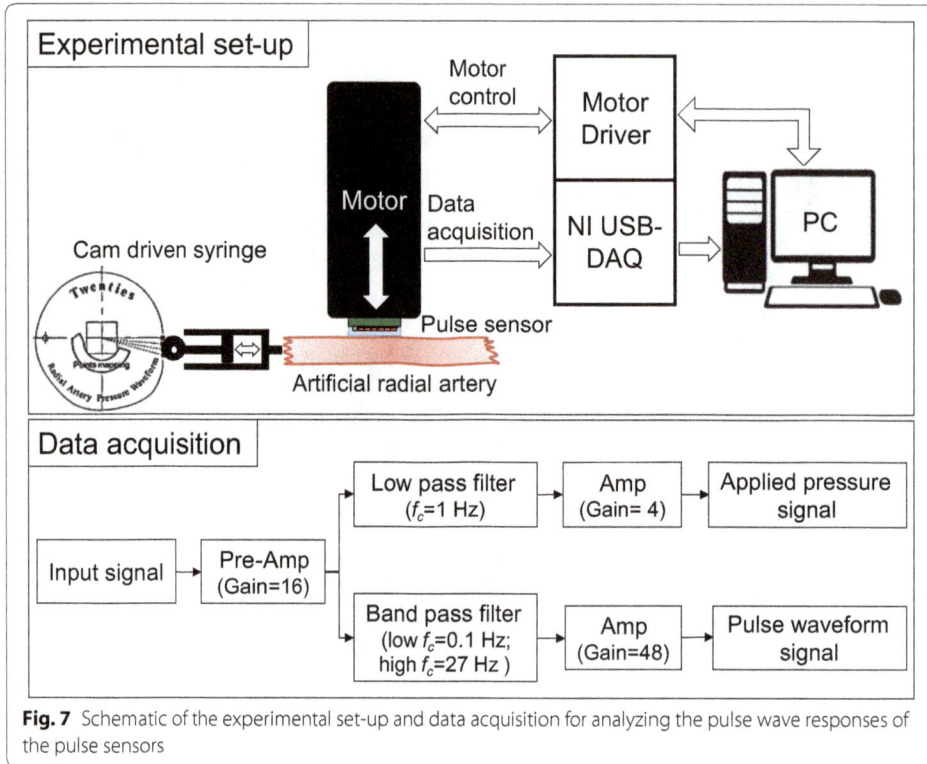

Fig. 7 Schematic of the experimental set-up and data acquisition for analyzing the pulse wave responses of the pulse sensors

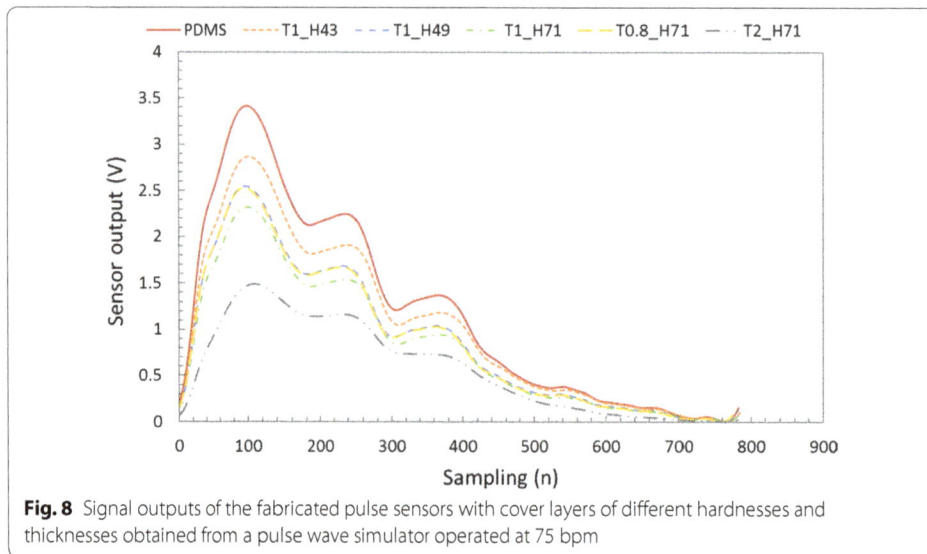

Fig. 8 Signal outputs of the fabricated pulse sensors with cover layers of different hardnesses and thicknesses obtained from a pulse wave simulator operated at 75 bpm

conditions, and the averages of the measured data are plotted in Fig. 8. The sampling rate of the measured signal was 1000 samples per second.

In Fig. 8, PDMS refers to the pulse sensor with a cover layer made of 1-mm-thick PDMS; T1_H43 refers to a cover layer made of 1-mm-thick XE14-C2042 (hardness: 43); T1_H49 refers to a cover layer made of 1-mm-thick IVS4546 (hardness: 49); and T0.8_H71, T1_H71, and T2_H71 refer to 0.8-mm-thick, 1-mm-thick, and 2-mm-thick IVS4742 (hardness: 71) cover layers, respectively.

The overall output signal of the pulse sensor with the PDMS cover layer showed the largest signal for the same applied pressure, and the amplitudes of the signals decreased according to the hardness in the order of H43 > H49 > H71 and according to the thickness in the order of 0.8 mm > 1 mm > 2 mm of the cover layers. The amplitude of the measured signal for each pulse sensor varied depending on the hardness and the thickness of the cover layer. The sensor with the cover layer made up of the soft material with a thin thickness had a higher pressure resolution than sensors with cover layers of other materials and thicknesses. The peak values of signal outputs were 3.42 ± 0.02 V in PDMS, 2.84 ± 0.06 V in T1_H43, 2.58 ± 0.05 V in T1_H49, 2.43 ± 0.08 V in T1_H71, 2.52 ± 0.06 V in T0.8_H71, and 1.50 ± 0.02 V in T2_H71 as shown in Table 1. The peak amplitude of the T1_H43 sensor was 83.08% of the peak amplitude of the reference sensor, while those of the T1_H49, T1_H71, T0.8_H71, and T2_H71 sensors were 75.59, 71.03, 73.77, and 43.85%, respectively. H71, the hardest of the tested materials, corresponded to a 28.97% decreased response compared with the reference, and the 2-mm-thick cover layer corresponded to a 27.18% decreased response compared with the 1-mm-thick cover layer of the same material. The rising time to reach the peak values of the pulse sensors was the 14.76% longer for T2_H71 compared to the reference sensor as shown in Table 1. These results showed that both the thickness and the hardness of the cover layer affected the output signals for the same input pulse wave. However, the sensitivity of the pulse sensors did not depend on the thickness of the cover layers and had constant values. As the results of static response for applied pressure, when the thicknesses of the cover layer were 1 mm, 1.5 mm, 2 mm, and 2.5 mm, the sensitivities of the pulse sensors were 6.48 ± 0.06 mV/mmHg (mean ± SD), 6.49 ± 0.08 mV/mmHg, 6.52 ± 0.08 mV/mmHg, and 6.54 ± 0.05 mV/mmHg, respectively. The sensitivities of static response of pulse sensor for applied pressure were measured according to the change in output signals for constant applied pressure differences to the pulse sensors in the sealed chamber.

To evaluate the pulse dynamic response of each sensor, the output amplitude of each pulse sensor was normalized, as shown in Fig. 9. The waveforms of the pulse sensors were almost the same except for the pulse sensor with the 2-mm-thick cover layer. For the quantitative analysis, the normalized waveforms were analyzed using the percent root-mean-square difference (PRD) method. The PRD method is a typical method used to measure the similarity between any 2 waveforms [38]. The PRD equation can be expressed as:

$$\mathrm{PRD} = \sqrt{\left. \sum_{i=1}^{N} |S(i) - S_c(i)|^2 \middle/ \sum_{i=1}^{N} S_c(i)^2 \right.}, \tag{1}$$

where S(i) and Sc(i) are the 2 waveforms, and N is the number of samples. In addition, the radial augmentation index (AIx) was also analyzed for each pulse sensor. The AIx for the radial artery waveform was calculated as the ratio of P2/P1 and is expressed as a percentage [39]. The PRD and AIx results for the tested cover layers compared with the PDMS cover layer are shown in Table 2. The waveform similarity of most of the sensors had an error of < 2% compared with the pulse sensor with the 1-mm-thick PDMS cover layer, but the pulse sensor with the 2-mm-thick H71 (IVS4732) cover layer had an

Table 1 Averaged peak values and rising time of measured pulse wave

Sensor type	PDMS	T1_H43	T1_H49	T1_H71	T0.8_H71	T2_H71
Peak values (V) (% ref)	3.42 ± 0.02 (0)	2.84 ± 0.06 (83.08)	2.58 ± 0.05 (75.59)	2.43 ± 0.08 (71.03)	2.52 ± 0.06 (73.77)	1.50 ± 0.02 (43.85)
Rising time (ms) (% ref)	97.75 ± 0.97 (0)	99.78 ± 1.99 (102.07)	96.11 ± 1.45 (96.33)	99.78 ± 0.92 (103.82)	96.33 ± 1.05 (96.55)	110.56 ± 2.22 (114.76)

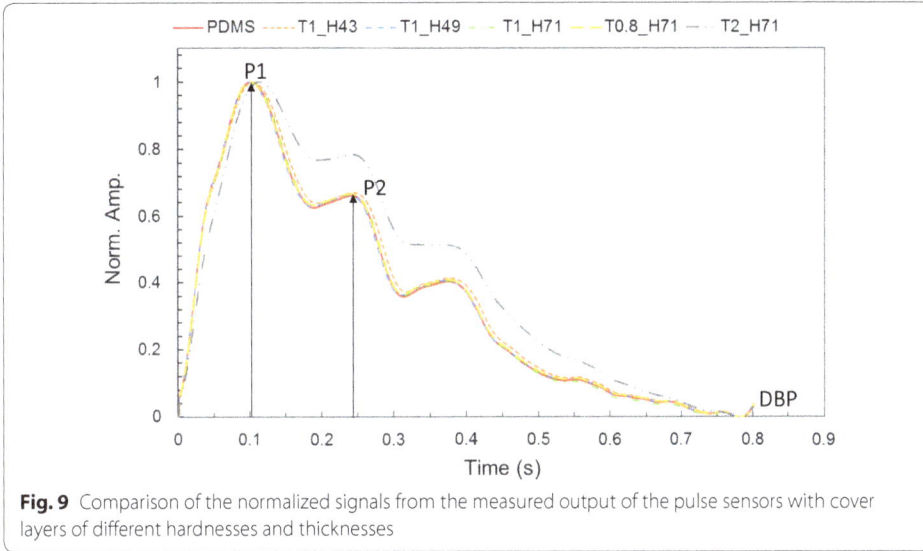

Fig. 9 Comparison of the normalized signals from the measured output of the pulse sensors with cover layers of different hardnesses and thicknesses

Table 2 Comparison of prd and aix for normalized signals of fabricated pulse sensors

Index	PDMS	T1_H43	T1_H49	T1_H71	T0.8_H71	T2_H71
PRD (%)	0	1.99	0.62	1.99	0.77	11.02
AIx (%)	66.11 (0)	66.06 (▼0.05)	66.37 (▲0.26)	66.51 (▲0.40)	66.63 (▲0.52)	78.38 (▲12.27)

error of more than 11%. The radial AIx of most of the sensors also showed a difference of less than 0.6% compared with the pulse sensor with the PDMS cover layer, but the pulse sensor with the T2_H71 cover layer showed a difference greater than 12%. As shown in Fig. 9 and Table 2, the waveform similarity and radial AIx did not show any significant differences among cover layers, except for the T2_H71 cover layer.

The amplitude was recorded from each channel of the pulse sensor with a simulated pulse input to quantitatively determine the force distribution of the applied dynamic pressure due to the cover layer in the pulse sensor. The maximum peak values of the channels correspond to the averages of the 3 measured values, and the averaged maximum peak value of each channel with the points connected by a dotted line are displayed in Fig. 10. The average pressure (AP) to the pulse sensor was evaluated by dividing the sum of the 6 peak values by the number of pressure sensors.

$$\text{AP} = \left(\sum_{x=1}^{n} \text{P}_{Chx} \right) \Big/ n, \quad (n = 6), \tag{2}$$

where P_{chx} is an averaged output peak value of x channel and n is the number of channels. The AP rates were calculated to evaluate the pulse sensors with different cover layers by comparing the amplitudes of the measured signals for delivery of a constant external pulse pressure. The AP was used as an index to evaluate the extent to which identical external pressure was transferred to the pulse sensor. The calculated AP/Ref for each sensor, the ratio of the AP for each sensor to the reference pulse sensor with the PDMS cover layer, and the AP error compared with the reference pulse sensor are shown

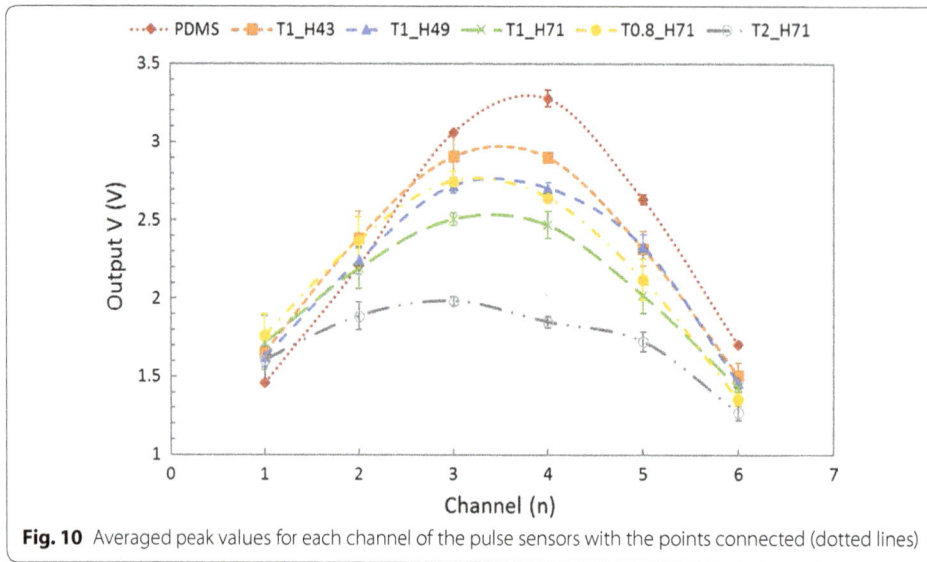

Fig. 10 Averaged peak values for each channel of the pulse sensors with the points connected (dotted lines)

Table 3 Comparison of average pressure according to different cover layers of pulse sensors

Index	PDMS	T1_H43	T1_H49	T1_H71	T0.8_H71	T2_H71
AP (V/Ch)	2.38	2.28	2.18	2.06	2.17	1.72
AP/Ref (%) (error (%))	100 (0)	95.90 (4.10)	91.76 (8.24)	86.51 (13.49)	91.21 (8.79)	72.37 (27.63)

in Table 3. The spacing of the pressure sensors corresponding to each channel was constant (1.1 mm), and the tube of the pulse simulator had an outer diameter of 4 mm and an inner diameter of 2 mm. In the analysis results, the ratio errors of the T1_H71 and T2_H71 sensors were more than 10%. The pulse sensors with thick and hard cover layers showed higher damping for input dynamic pressure than other sensors.

Discussion

The AP errors for calculating the AP rates of the T1_H71 and T2_H71 sensors are too large for reliable pulse wave measurements because the errors are larger than 10%. In addition, it was difficult to fabricate the pulse sensor with the 0.8-mm-thick cover layer covering the 0.5-mm-thick pressure sensor, wire bonding, and underfilling for protection of the wires on the sensor cells. The 0.8-mm-thick cover layer was fabricated to compare only the pulse sensor dynamic responses of other cover layers with the thinnest possible cover layer that could be implemented via our fabrication process. Although the thin cover layer showed good pressure resolution, high waveform reproducibility, and low pressure transfer losses in the pulse sensor dynamics analysis, fabrication of pulse sensors with a cover layer of 1 mm or greater thickness is suggested due to the difficulty in fabricating thinner layers, mass production considerations, and large thickness errors.

Although it has been experimentally shown that the dynamic response of pulse sensors for monitoring pulse waves depends on the cover layer, there are limits to determining the optimal thickness and hardness of the cover layer. In addition, further studies of the cover layer using materials with a wide variety of mechanical properties are needed to determine the optimal cover layer in the pulse sensor. Additionally, in order to analyze a dynamic frequency response for fabricated pulse sensors, it is necessary to trace the frequency response of the pulse sensor as the input frequencies were swept. However, the frequency range of the pulse wave simulator driven by cam is not wide enough to analyze the frequency response (from 50 to 90 bpm). We will develop a pulse generator to analyze the dynamic frequency response of the pulse sensors, and analyze the dynamic response as a further study.

To precisely evaluate the force transfer rate of the pulse sensor, the sensor output with respect to the contact area and the pulse wave can be analyzed through theoretical calculations of the stress and the strain due to the pressure between a cylinder and a flat plate of elastic bodies [40, 41]. However, it is difficult to calculate the pressure on the contact area because of the unknown mechanical properties. Therefore, the force transfer rates of the pulse sensors were evaluated by the AP values, which represent the entire pressure output divided into the number of sensor channels. In addition, it is necessary to understand the forced vibration system of periodic excitation to clarify the damping effect of the cover layers.

A good way to evaluate the reproducibility of the pulse wave by pulse sensors with respect to the external pressure is to compare the output signals of the fabricated pulse sensors with the external pressure as the input signal. However, a reference pulse sensor was used because it would be difficult to perform a direct comparison with the external pressure using our system. A more accurate external pressure on the artificial radial artery can be calculated by the modeling of the waveform transmission considering various properties of the artificial tube, such as elasticity, diameter, length, and thickness. More research on modeling of the artificial radial artery and waveform transmission are required to develop precise sensors for measuring the pulse wave. In addition, the lifetime of the pulse sensors was not verified because durability tests of the pulse sensors with the H43, H49, or H71 cover layers have not yet been conducted. It is also necessary to study whether the silicones used are the most suitable cover layer materials for pulse sensors.

Conclusions

In this study, in-line array pulse sensors with various cover layers were fabricated, the dynamic pulse wave responses according to the thickness and the hardness of the cover layer were analyzed, and an appropriate thickness and hardness for the cover layer were suggested. Pulse sensors with cover layers of 3 different thicknesses (0.8 mm, 1 mm, 2 mm) and 4 different hardnesses (Shore type A; 30, 43, 49, 71) were fabricated. Experiments to evaluate the dynamic responses of the fabricated sensors were performed using a pulse simulator, and 3 repeated measurements were made for each sensor. The averaged amplitudes of the measured pulse pressure were 3.42 V for the PDMS sensor, 2.84 V for the T1_H43 sensor, 2.58 V for the T1_H49 sensor, 2.43 V for the T1_H71 sensor, 2.52 V for the T0.8_H71 sensor, and 1.50 V for the T2_H71 sensor. The normalized waveform analysis using PRD showed that the waveform errors were 1.99% for T1_H43, 0.62% for T1_H49, 1.99% for T1_H71, 0.77% for T0.8_H71, and 11.02% for T2_H71 in relation to the sensor with the PDMS cover layer. AP analysis of the pulse sensors showed that the errors were 4.10% for T1_H43, 8.24% for T1_H49, 13.49% for T1_H71, 8.79% for T0.8_H71, and 27.63% for T2_H71. This study suggests that pulse sensors with a 1-mm-thick H43 cover layer or a 1-mm-thick H49 cover layer are suitable for measuring radial pulse waves, as fabricating pulse sensors with a 0.8-mm-thick cover layer is difficult. This study of dynamic pulse wave response will contribute to accurate and precise pulse wave measurements and analysis.

Authors' contributions
M-HJ wrote the manuscript and designed and conducted the experiments of the sensor characteristics with Y-MK, Y-JJ designed the pulse sensor and the electrical circuits for the experiments, J-HC installed the experimental setup and helped with the experiments, and Y-MK managed the experiments of the pulse sensor characteristics and contributed to writing and revising the manuscript. All authors read and approved the final manuscript.

Acknowledgements
This work was supported by a Grant (K18022) from the Korea Institute of Oriental Medicine (KIOM), funded by the Korean government.

Competing interests
The authors declare that they have no competing interests.

Funding
This work was supported by a Grant (K18022) from the Korea Institute of Oriental Medicine (KIOM), funded by the Korean government.

References
1. Gao Z, Li Y, Sun Y, Yang J, Xiong H, Zhang H, Liu X, Wu W, Liang D, Li S. Motion tracking of the carotid artery wall from ultrasound image sequences: a nonlinear state-space approach. IEEE Trans Med Imaging. 2018;37(1):273–83.
2. Zhao S, Gao Z, Zhang H, Xie Y, Luo J, Ghista D, Wei Z, Bi X, Xiong H, Xu C. Robust segmentation of intima-media borders with different morphologies and dynamics during the cardiac cycle. IEEE J Biomed Health. 2017. https://doi.org/10.1109/JBHI.2017.2776246.
3. Gao ZF, Xiong HH, Liu X, Zhang HY, Ghista DJ, Wu WQ, Li S. Robust estimation of carotid artery wall motion using the elasticity-based state-space approach. Med Image Anal. 2017;37:1–21.
4. Liu X, Gao Z, Xiong H, Ghista D, Ren L, Zhang H, Wu W, Huang W, Hau WK. Three-dimensional hemodynamics analysis of the circle of Willis in the patient-specific nonintegral arterial structures. Biomech Model Mechanobiol. 2016;15(6):1439–56.
5. Liu GY, Wu JH, Huang WH, Wu WD, Zhang HN, Wong KKL, Ghista DN. Numerical simulation of flow in curved coronary arteries with progressive amounts of stenosis using fluid-structure interaction modelling. J Med Imaging Health Inform. 2014;4(4):605–11.
6. Wong KKL, Tu J, Mazumdar J, Abbott D. Modelling of blood flow resistance for an atherosclerotic artery with multiple stenoses and poststenotic dilatations. ANZIAM J. 2010;51:66–82.
7. Wong K, Mazumdar J, Pincombe B, Worthley SG, Sanders P, Abbott D. Theoretical modeling of micro-scale biological phenomena in human coronary arteries. Med Biol Eng Comput. 2006;44(11):971–82.
8. Matskiv AS, Kravets SL, Tsemekhin BD. Treatment of purulent and necrotic lesions of the lower extremities in patients with diabetes mellitus. Klin Khir. 1993;9–10:37–40.
9. Liu SH, Tyan CC. Quantitative analysis of sensor for pressure waveform measurement. Biomed Eng Online. 2010;9:6.
10. Jun MH, Kim YM, Bae JH, Jung CJ, Cho JH, Jeon YJ. Development of a tonometric sensor with a decoupled circular array for precisely measuring radial artery pulse. Sensors (Basel). 2016;16(6):768. https://doi.org/10.3390/s16060768
11. Jeon Y-J, Kim JU, Kim Y-M, Bae J-H, Kim J-Y. Development of an array sensor for measuring radial pulse wave. In: 11th International conference on wearable and implantable body sensor networks, 16–19 June 2014, Zurich, Switzerland.
12. Chung C-Y, Chung Y-F, Chu Y-W, Luo C-H. Spatial feature extraction from wrist pulse signals. In: International conference on Orange Technologies (ICOT), 2013. New York: IEEE; 2013. p. 1–4.
13. Choi SD, Kim SW, Kim GW, Ahn MC, Kim MS, Hwang DG, Lee SS. Development of spatial pulse diagnostic apparatus with magnetic sensor array. J Magn Magn Mater. 2007;310(2):E983–5.
14. Kim SW, Hwang DG, Choi YK, Lee HS, Park DH, Lee SS, Kim GW, Lee SG, Lee SJ. Improvement of pulse diagnostic apparatus with array sensor of magnetic tunneling junctions. J Appl Phys. 2006;99(8):08R908.
15. Hu CS, Chung YF, Yeh CC, Luo CH. Temporal and spatial properties of arterial pulsation measurement using pressure sensor array. Evid Based Complement Altern Med. 2012;2012:745127. https://doi.org/10.1155/2012/745127.
16. Peng JY, Lu MSC. A flexible capacitive tactile sensor array with CMOS readout circuits for pulse diagnosis. IEEE Sens J. 2015;15(2):1170–7.
17. Chang H, Chen J-X. Piezoelectric pulse diagnosis transducer of 9×9 sensing arrays and pulse signal processing. In: International conference on applied informatics and communication. New York: Springer; 2011. p. 541–8.

18. Xu L, Meng MQ-H, Shi C, Wang K, Li N. Quantitative analyses of pulse images in traditional Chinese medicine. Med Acupunct. 2008;20(3):175–89.

19. Jun M-H, Jeon YJ, Kim Y-M. Interference effects on the thickness of a pulse pressure sensor array coated with silicone. J Sens Sci Technol. 2016;25(1):35–40.

20. Jun M-H, Jeon YJ, Kim Y-M. Signal change and compensation of pulse pressure sensor array due to wrist surface temperature. J Sens Sci Technol. 2017;26(2):141–7.

21. Yoo SK, Shin KY, Lee TB, Jin SO, Kim JU. Development of a radial pulse tonometric (RPT) sensor with a temperature compensation mechanism. Sensors (Basel). 2013;13(1):611–25.

22. Zhang YL, Zheng YY, Ma ZC, Sun YN. Radial pulse transit time is an index of arterial stiffness. Hypertens Res. 2011;34(7):884–7.

23. Filipovsky J, Ticha M, Cifkova R, Lanska V, Stastna V, Roucka P. Large artery stiffness and pulse wave reflection: results of a population-based study. Blood Press. 2005;14(1):45–52.

24. Chrysohoou C, Angelis A, Tsitsinakis G, Spetsioti S, Nasis I, Tsiachris D, Rapakoulias P, Pitsavos C, Koulouris NG, Vogiatzis I, Dimitris T. Cardiovascular effects of high-intensity interval aerobic training combined with strength exercise in patients with chronic heart failure. A randomized phase III clinical trial. Int J Cardiol. 2015;179:269–74.

25. Nelson MR, Stepanek J, Cevette M, Covalciuc M, Hurst RT, Tajik AJ. Noninvasive measurement of central vascular pressures with arterial tonometry: clinical revival of the pulse pressure waveform? Mayo Clin Proc. 2010;85(5):460–72.

26. Townsend RR. Analyzing the radial pulse waveform: narrowing the gap between blood pressure and outcomes. Curr Opin Nephrol Hypertens. 2007;16(3):261–6.

27. Takazawa K, Kobayashi H, Shindo N, Tanaka N, Yamashina A. Relationship between radial and central arterial pulse wave and evaluation of central aortic pressure using the radial arterial pulse wave. Hypertens Res. 2007;30(3):219–28.

28. Larsson M, Bjallmark A, Lind B, Balzano R, Peolsson M, Winter R, Brodin LA. Wave intensity wall analysis: a novel noninvasive method to measure wave intensity. Heart Vessels. 2009;24(5):357–65.

29. Vlachopoulos C, O'Rourke M, Nichols WW. McDonald's blood flow in arteries: theoretical, experimental and clinical principles. Boca Raton: CRC Press; 2011.

30. Choudhury MI, Singh P, Juneja R, Tuli S, Deepak KK, Prasad A, Roy S. A novel modular tonometry-based device to measure pulse pressure waveforms in radial artery. J Med Devices. 2018;12(1):011011. https://doi.org/10.1115/1.4039010

31. Jeon YJ, Kim JU, Lee HJ, Lee J, Ryu HH, Lee YJ, Kim JY. A clinical study of the pulse wave characteristics at the three pulse diagnosis positions of Chon, Gwan and Cheok. Evid Based Complement Altern Med. 2011. https://doi.org/10.1093/ecam/nep150.

32. Momentive Performance materials Inc., Silicone Material Solutions for LED Packages and Assemblies. https://www.momentive.com/en-us/search-results/?q=xe14-c2042. Accessed 7 Dec 2017.

33. Kim TH, Ku B, Bae JH, Shin JY, Jun MH, Kang JW, Kim J, Lee JH, Kim JU. Hemodynamic changes caused by acupuncture in healthy volunteers: a prospective, single-arm exploratory clinical study. BMC Complement Altern Med. 2017;17(1):274.

34. Bae JH, Ku B, Jeon YJ, Kim H, Kim J, Lee H, Kim JY, Kim JU. Radial pulse and electrocardiography modulation by mild thermal stresses applied to feet: an exploratory study with randomized, crossover design. Chin J Integr Med. 2017. https://doi.org/10.1007/s11655-017-2972-0.

35. Bae J-H, Jeon YJ, Lee S, Kim JU. A feasibility study on age-related factors of wrist pulse using principal component analysis. In: 2016 IEEE 38th annual international conference of the IEEE engineering in medicine and biology society (EMBC). New York: IEEE; 2016. p. 6202–5.

36. Baek JY, An JH, Choi JM, Park KS, Lee SH. Flexible polymeric dry electrodes for the long-term monitoring of ECG. Sens Actuators A Phys. 2008;143(2):423–9.

37. Chen CY, Chang CL, Chang CW, Lai SC, Chien TF, Huang HY, Chiou JC, Luo CH. A low-power bio-potential acquisition system with flexible PDMS dry electrodes for portable ubiquitous healthcare applications. Senors (Basel). 2013;13(3):3077–91.

38. Xu L, Yao Y, Wang H, He D, Wang L, Jiang Y. Morphology variability of radial pulse wave during exercise. Biomed Mater Eng. 2014;24(6):3605–11.

39. Duprez DA, Kaiser DR, Whitwam W, Finkelstein S, Belalcazar A, Patterson R, Glasser S, Cohn JN. Determinants of radial artery pulse wave analysis in asymptomatic individuals. Am J Hypertens. 2004;17(8):647–53.

40. Norden BN, Norden BN. On the compression of a cylinder in contact with a plane surface. Gaithersburg: U.S. Dept. of Commerce, National Institute of Standards and Technology; 1973.

41. Puttock M, Thwaite E. Elastic compression of spheres and cylinders at point and line contact. Australia: Commonwealth Scientific and Industrial Research Organization; 1969.

4

Combining multi-scale composite windows with hierarchical smoothing strategy for fingerprint orientation field computation

Haiyan Li[1], Tangyu Wang[1], Yiying Tang[2], Jun Wu[1], Pengfei Yu[1*], Lei Guo[1], Jianhua Chen[1] and Yufeng Zhang[1]

*Correspondence:
pfyu@ynu.edu.cn
[1] School of Information Science and Engineering, Electronic Engineering, Yunnan University, Chenggong District, Kunming 650000, China
Full list of author information is available at the end of the article

Abstract

Background: Orientation field (OF) plays a very significant role in automatic fingerprint recognition systems. Many algorithms have been proposed for the estimation of fingerprints' OF but it is hard to solve the dilemma of correcting spurious ridge structure and avoiding singularity location deviation, especially for poor images. So far, the following drawbacks still need to be solved for OF construction methods for practical application: (1) How to adaptively choose block scales to resolve the contradiction between accuracy and anti-noise, since small scale is beneficial to accuracy but is sensitive to noise, while large scale is more resistant to noise, but the accuracy is deteriorated. (2) How to construct the genuine OF in the areas close-by singular points and to evade singularity location deviation? Current block based methods give spurious OF estimates in the area near singular points because these areas have large curvature thus the detected singular points deviate from the genuine localizations. When these singular points are used as the anchor for referencing minutiae, it makes the average error of matching or recognition even larger. Therefore, it is essentials to construct the genuine OF in the areas close-by singular points and to evade singularity deviation.

Methods: To overcome the above-mentioned limitations, a novel method, combining a weighted multi-scale composite window (WMCM) with a hierarchical smoothing strategy has been proposed for the computation of fingerprint OF. This method mainly contains two procedures: the approximate OF estimation and the hierarchical OF smoothing. In the first procedure, a series of OFs are established under multiple scales of composite windows by using a gradient based method then a coarse OF is estimated using the weight of each scale determined by a squared gradient consistency. In the second procedure, the OF is first quantized into a two-digitized orientation zone and a two-orientation-zone filtering strategy is adapted to the OF blocks based on a filtering mask obtained after eliminating the isolated blocks. In the end a similar three-digitized orientation zone is performed to obtain an accurate and smooth OF. To validate the performance, the proposed method has been applied to OF computation using the FVC2004 databases and three experiments are designed. Experiment 1 aims to validate whether the weighted multi-scale composite window can balance the dilemma of accuracy and robustness more effectively than the previous works do. Experiment 2 is designed to examine whether the hierarchical smoothing method can correct the spurious ridge flow and preserve the genuine localization of singular points. The purpose of experiment 3 is to test the performance of the proposed method on OF reconstruction in low quality fingerprint images. The fingerprint databases FVC 2004 DB1–DB4 are employed in this study.

Results: The results of experiment I shows that the proposed method is capable to extract the information of OF reliably and it is more robust against singularity localization deviation in comparison with the other three gradient based methods. The results of experiment II indicates that the proposed smoothing method can balance the contradiction in correcting spurious ridge structures and preserving genuine singularity localization. The results of experiment III illustrates that our approach combing WMCW with the hierarchical smoothing method is capable to extract the information of OF ridge reliably and it is more robust against singularity deviation in comparison with the other three gradient based methods. In a word, the experiment results demonstrate that the proposed method can correct spurious ridge structure and meanwhile avoid singularity deviation compared with the previous works.

Conclusions: A novel gradient based algorithm has been proposed which is more reliable for the estimation of the ridge information for fingerprint OF and is more accurate in preserving the singularity localization. Compared with the previously proposed gradient based methods, the advantages of the proposed RBSF lie in three aspects. Firstly a weighted multi-scale composite window is put forward to replace the single window used by conventional gradient based methods and to adaptively choose the scales of the blocks. Secondly, a hierarchical smoothing strategy is proposed to enhance the OF by using the two-orientation-zone filtering and the three-orientation-zone filtering, aiming to correct the spurious ridges and preserving the genuine location of singular points. Finally, three experiments are designed to test the proposed algorithm together with other popular gradient based methods on real fingerprint images, which are selected from different categories and all are suffering from obvious noise effects. All the experiment results show that the proposed method is superior with respect to reliable OF construction and avoiding singularity localization deviation.

Keywords: Fingerprint orientation field (OF), Gradient based method, Weighted multi-scale composite window, Two-digitized orientation-zone filtering, Three-digitized orientation-zone filtering

Background

Ridge orientation pattern, representing the ridge flow directions on regularly spaced grids and revealing intrinsic features of ridge topologies, plays a critical role in singularity extraction, fingerprint classification, fingerprint recognition and so on. Therefore, a huge amount of research efforts have been made towards the reliable estimation of fingerprint orientation pattern from fingerprint images which can roughly be classified as global modeling [1–8] and local estimation [9–19].

Global fingerprint structure can be used for OF estimation in such a way that OF in bad quality areas can be accurately interpolated by using singular points as heuristic knowledge. Pioneered global modeling based method, a so-called zero-pole model, for orientation field computation based on singular points was proposed [1]. The approach modeled cores and deltas as zero and a pole in the complex plane and the orientation was computed by the summation of the influence of singularities. Since then this model receives a wide acceptance and improvements. Vizcaya and Gerhardt [2] used a piecewise linear approximation model around singular points to adjust the zero and pole's behavior. Gu et al. [3, 4] proposed a combination model for orientation field representation, in which the global orientation is firstly constructed by a polynomial model and subsequently a point-charge model was applied to correct regions near singular points. However, these global based methods mostly depend on the accurate detection of singular points, and the precise detection of singular points, in turn, depends on the correct computation of the OF. As a result, the problem turns into the paradoxical chicken-eeg problem. In order to solve the paradox, OF estimation are treated as data fitting problems [5, 6]. Weights are assigned based on foreground [7] and background pixel or probable location of singular point [8] to evade spurious OF estimation in bad quality areas or singular points close-by areas. Even though these algorithms exhibit better performance than the methods in [5, 6], it is obvious that the success of data-fitting based global OF modeling mainly relies on accurate local OF estimates and proper weight assignment.

Comparatively, local methods do not require any prior knowledge of singular points or use neighborhoods for OF estimation. These are mainly categorized as: filter-bank based algorithm [9, 10] and gradient-based algorithm [11–19]. The filter-based methods are resistant to noise but they almost completely rely on the limited number of filters. Therefore, the results of OF computation are often not very accurate. Meanwhile, the computation cost is very high because all of the filter's outputs need to be compared. Gradient-based methods are more accurate and subtle to characterize the OF information compared with the above mentioned methods, and thus has become one of the most popular methods for OF computation. However, the following drawbacks still need to be solved for practical application:

1. How to adaptively choose the block scales? The pioneering research into the gradient-based method was proposed in 1987 [11], where the OF was calculated by the block gradient vectors. Several improvements were made by researchers over the past years and have achieved more accurate OF, which can be found in [12–19]. However, no final conclusion has yet been reached on how to choose suitable scales for the block to resolve the contradiction between accuracy and anti-noise, since small scale is beneficial to accuracy but is sensitive to noise, while large scale is more resistant to noise, but the accuracy is deteriorated. Literature [16] proposed a composite window based method to make a balance between accuracy and anti-noise, where the composite window consists of an inner window and an outer window. Thereafter, the composite window was adopted by [17] and it is proved to be able to achieve better performance compared with the previous single window based methods. However, the composite window introduces a more complex issue on how to choose proper scales for the inner and outer windows? So far, this dilemma has not been solved and current methods rely on a huge amount of experiments to obtain experientially proper scales [16, 17]. Therefore, it is critical to find a theoretical approach on how to adaptively define the scales for composite windows.

2. How to construct a genuine OF in the areas close-by singular points and to evade singularity location deviation? In order to overcoming the defect that gradient-based methods are not robust against large scale noise, a hierarchical scheme was proposed to dynamically adjust the re-estimation resolution by using coherence, defined as the deviation between the target block's orientation and other block's orientation around

it. If the consistency level is above a certain threshold, then the target block's orientation is re-estimated at a lower resolution level until it is under a certain level [18]. A weighted averaging method was proposed in [19], which intended to construct redundant estimation for each target block. In addition, various improvements have been achieved to enhance the robustness against large scale noise [15–17]. However, these block-based methods give spurious OF estimates in the area near singular points because these areas have large curvature thus the singular points deviate from the genuine localization. When these singular points are used as the anchor for referencing minutiae, it makes the average error of matching or recognition even larger. Therefore, it is essentials to construct the genuine OF in the areas close-by singular points and to evade singularity deviation.

The goal of this paper is to provide an OF computation method combining a weighted multi-scale composite window with hierarchical smoothing strategy. The method basically consists of two phases: the preliminary estimation of the local region orientation by a series of weighted multi-scale composite windows, followed by a refined phase for reconstructing the ridge OF using a two-hierarchy smoothing strategy. Compared to the existing approach, our approach are some of advantages. (1) An adaptive method is proposed to determine the proper scales of the composite windows, which obviates the drawback that most of the existing algorithms depend critically on a wide range of experiments for relatively proper scales. (2) A hierarchical strategy is proposed to construct genuine OF in the areas close-by singular points and to evade singularity deviation, aiming to minimize the error when singular points are used as the reference to describe minutiae for fingerprint matching. (3) Three groups of experiments are designed to validate whether the proposed algorithm is effective in computing and refining OF. The experiment results show that the proposed method can correct spurious ridge structure and avoid singularity deviation compared with the previous works.

Methods

The flow-graph of the proposed fingerprint OF computation algorithm is shown in Fig. 1. It consists of two procedures. In the first procedure, a coarse OF is estimated by using a gradient based local algorithm combing a weighted multi-scale composite window. It may produce spurious ridge information for bad quality areas and singular points close-by area in the coarse OF. Subsequently, a hierarchical smoothing strategy, consisting of a two-digitized-level of orientation zone filtering and a three-digitized-level of orientation zone filtering, are performed to obtain the refined OF where OF in uniform flow area as well as the genuine singularity locations are preserved while OF in non-uniform flow areas is refined.

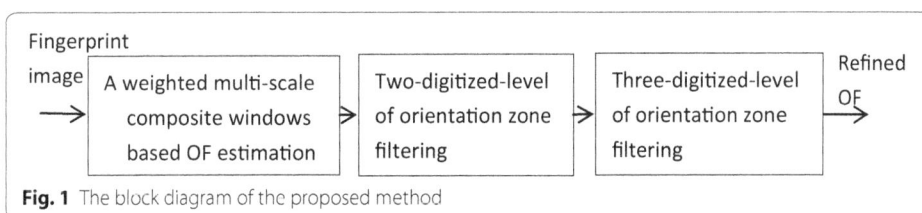

Fig. 1 The block diagram of the proposed method

Gradient based OF

In this section, we simply introduce the classical gradient-based method adopted by Kass et al. [11].

Assuming I (x, y) denotes the gray-scale values of point (x, y). The OF estimation based on the gradient method mainly contains the following steps.

1. In gradient based methods, the gradient vectors, denoted as $[G_x, G_y]^T$, are first calculated for a fingerprint image by taking the partial derivatives of gray intensity at each pixel according to Eq. (1):

$$G_x = \frac{\partial g(x,y,\sigma)}{\partial x} \times (x,y), G_y = \frac{\partial g(x,y,\sigma)}{\partial y} \times I(x,y) \tag{1}$$

where $g(x,y,\sigma)$ indicates a two dimensional Gaussian core with the variance σ. $\frac{\partial}{\partial x}$ and $\frac{\partial}{\partial y}$ represent the partial derivative in x and y directions, respectively. G_x and G_y denote the gradient vectors in x and y coordinates, respectively. T represents transpose.

A fingerprint orientation map is defined as a collection of two dimensional orientation fields. The magnitudes of these fields can be omitted. Only the angle information is of interest because it captures the dominant ridge direction in every regular spaced grid. An orientation map is commonly represented in the form of a matrix $\{\theta_{xy}\}$, where $\theta_{xy} \in [0, \pi]$, denoting the averaged gradient angle computed from the local gradient vectors in a grid as φ. In a fingerprint image, the gradient vectors always point to the directions of the highest variation of gray intensity, shown in Fig. 2, where the ridge orientation θ is orthogonal to the dominant gradient angle φ.

2. Since the ridge line has two edges, the gradient vectors at both sides of a ridge are opposite to each other. If φ is calculated by averaging the gradient angles directly, the opposite gradients at both sides of the ridge line are likely to cancel each other. To solve this problem, Kass et al. [11] proposed a simple yet effective idea of doubling the gradient angles before averaging. In this way, φ becomes 2φ and $(\varphi + \pi)$ becomes $(2\varphi + 2\pi)$ which is also equal to 2φ. In practice, 2φ is the angle of a squared gradient

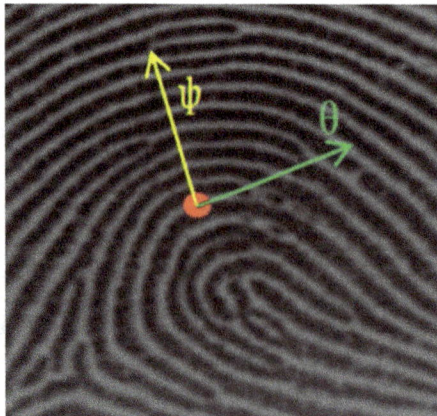

Fig. 2 The gradient direction and the ridge direction

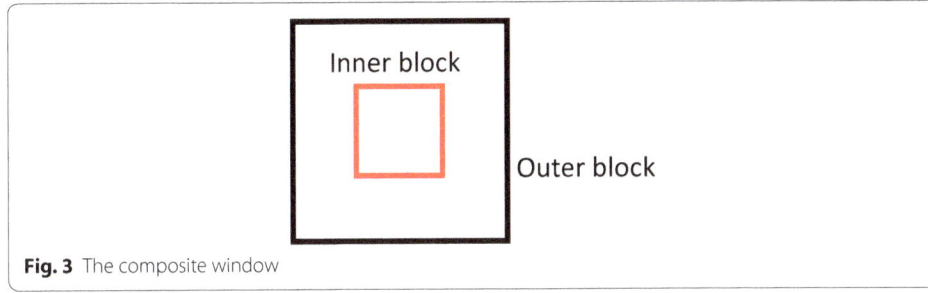

Fig. 3 The composite window

vector $[G_{sx}, G_{sy}]^T$ that has the following relation with $[G_x, G_y]^T$ according to trigonometric identities:

$$\begin{bmatrix} G_{sx} \\ G_{sy} \end{bmatrix} = \begin{bmatrix} G^2 \cos 2\varphi \\ G^2 \sin 2\varphi \end{bmatrix} = \begin{bmatrix} G^2 \left(\cos^2 \varphi - \sin^2 \varphi \right) \\ G^2 2 \sin \varphi \cos \varphi \end{bmatrix} = \begin{bmatrix} G_x^2 - G_y^2 \\ 2 G_x G_y \end{bmatrix} \tag{2}$$

3. The fingerprint image is then divided into equal-scaled and non-overlapping blocks, where $w \times w$ indicates the block scale.

4. The gradient over each block, denoted as $[G_{mx}, G_{my}]^T$, is calculated by performing average in each block independently.

$$\left[G_{mx}, G_{my} \right]^T = \left[\sum_{i=1}^{w} \sum_{j=1}^{w} G_{sx}, \sum_{i=1}^{w} \sum_{j=1}^{w} G_{sy} \right] \tag{3}$$

5. Subsequently, the averaged gradient angle of each block can be calculated by using Eq. (4).

$$\varphi = \frac{1}{2} \tan^{-1} \left(\frac{G_{my}}{G_{mx}} \right) \tag{4}$$

6. Since the ridge orientation is perpendicular to the gradient angle, therefore the ridge orientation can be obtained

$$\theta = \varphi + \frac{\pi}{2}. \tag{5}$$

The composite block

Since the ridge orientation is estimated in a block, there is a significant issue: how to choose a suitable scale for the block? As mentioned above, the gradient-based methods need to solve the contradiction between accuracy and robustness. If a small scale block is chosen, the orientation result is more accurate but is sensitive to noise. In contrast, if a large scale block is chosen, the robustness will be improved but the accuracy will degrade. A plain thought is choosing a suitable scale to make a balance between accuracy and robustness; however, it just makes a compromise [16].

A composite block was proposed in [16] to overcome the contradiction of accuracy and robustness. As shown in Fig. 3, the composite block consists of an inner block and an outer block, they both possess the same central point and $W_{in} \leq W_{out}$. In the case that $W_{in} = W_{out}$, the composite block is equal to the single window. Therefore, the single block is just a special case of the composite block [16].

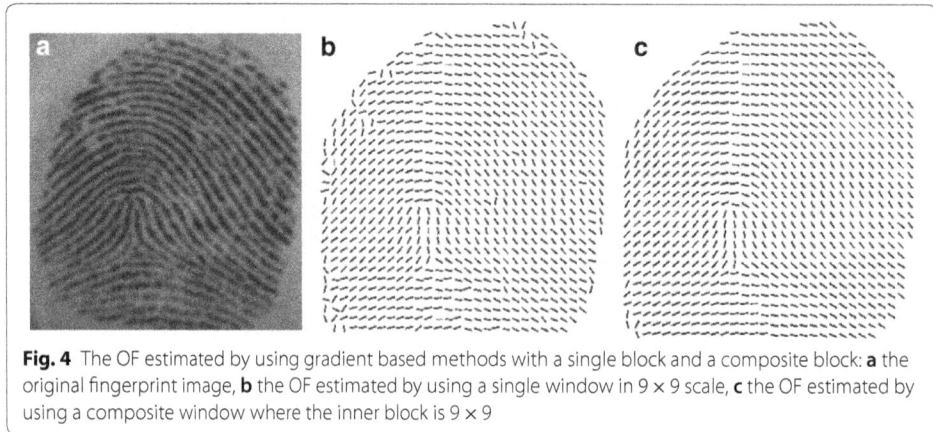

Fig. 4 The OF estimated by using gradient based methods with a single block and a composite block: **a** the original fingerprint image, **b** the OF estimated by using a single window in 9 × 9 scale, **c** the OF estimated by using a composite window where the inner block is 9 × 9

When the composite block is used for OF estimation, the orientation calculated by the outer block is set to the inner block. Since the small block is prone to accuracy and the large block is prone to robustness, the composite block combines the accuracy of a small inner block with the robustness of a large outer block. More detailed analysis can be found in [16].

Figure 4b presents the result calculated by using the gradient based method and the single block in 9 × 9 scale applied. Figure 4c illustrates the result calculate by using the same method and the composite block where the inner block is 9 × 9 and the outer block is 15 × 15. It is obvious that the result obtained by the composite block achieves better robustness and accuracy.

OF estimation by using the weighted multi-scale composite block

Though the composite block integrates the robustness of a large outer block and the accuracy of a small inner block, it incurs a more complex issue: how to choose the proper scales for the inner and outer blocks? So far, this issue has not yet been theoretically solved.

In order to choose proper scales for the composite block as well as balance accuracy and robustness, OF is computed by using the composite block with the same inner block in 9 × 9 scale and the outer block in various scales. Figure 5 shows one group results of our experiment. (a) Is the original fingerprint image with poor quality and low contrast. From (b) to (f), the scales of the outer block are 9 × 9, 15 × 15, 23 × 23, 31 × 31 and 37 × 37, respectively. In (b), since the inner block and outer block have the same scales, it is essentially a single window. By observation, we can find that the result of single block, shown in (b), can preserve genuine ridge structure in singularity close by area and accurate location for singular points, marked by the red circle, while it is seriously affected by noise. With the increase of the outer block, the influence of noise comes down. However, as the scale is larger, the singularity location deviation becomes larger, shown in (d)–(f) marked in the red circles. The reason is that when the scale of outer block increases, more low frequency information is used to compute the ridge orientation, which leads to the ridge orientation seriously deviate from the genuine direction. As mentioned before, if singular points are detected with localization error, then it indicates that the estimated OF in the neighborhood of the singular point deviates from the actual OF. Consequently,

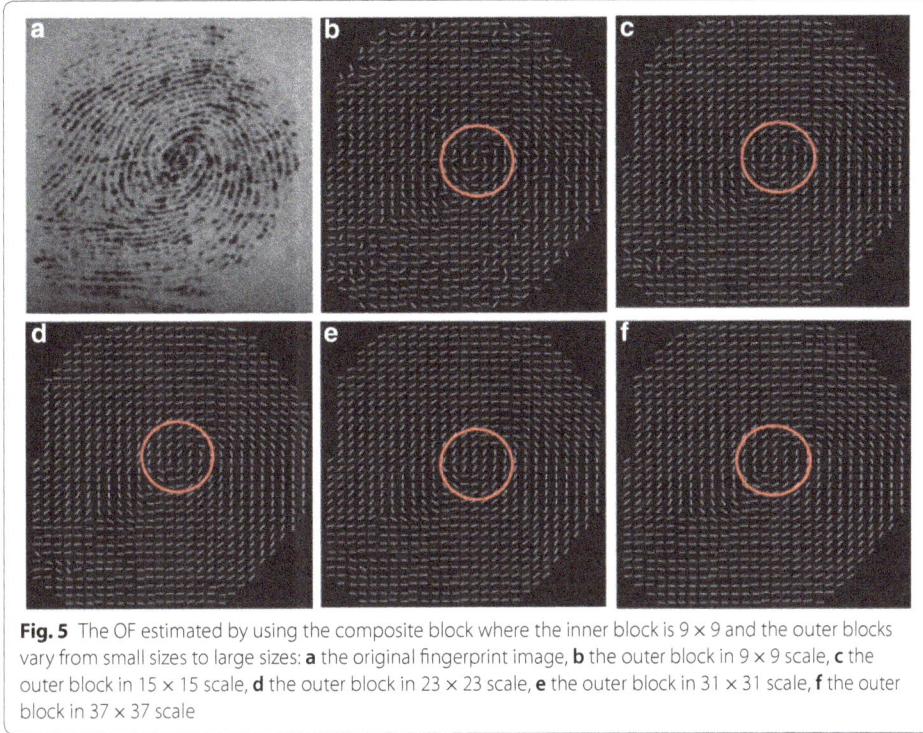

Fig. 5 The OF estimated by using the composite block where the inner block is 9 × 9 and the outer blocks vary from small sizes to large sizes: **a** the original fingerprint image, **b** the outer block in 9 × 9 scale, **c** the outer block in 15 × 15 scale, **d** the outer block in 23 × 23 scale, **e** the outer block in 31 × 31 scale, **f** the outer block in 37 × 37 scale

the localization error of singularities will lead to incorrect results when they are used as the reference to describe minutiae for fingerprint matching.

In order to adaptively define the scales of the composite block and meanwhile balance the robustness and accuracy of preserving singularity location, a weighted multi-scale composite block is proposed. The method first computes the local OF in multiple scales of outer blocks and then a weight is assigned to the local OF in each scale based on the squared gradient coherence. The final OF is obtained by integrating OFs in multiple scales. The proposed method is described with details in the following subsection.

In order to measure the reliability of estimation, Kass et al. [11] proposed a metric called coherence. The coherence metric calculates the strength of the average gradient in the distribution of local gradient vectors. Assuming $coh(p, q, l)$ denotes the squared gradient coherence and $\theta(p, q, l)$ represents the local OF for the block where the scale of inner block is $q * q$ and the scale of outer block is $p * p$. Then the coherence is given by:

$$coh(p, q, l) = \frac{\left| \sum_{i=-p/2}^{2/p} \sum_{j=-p/2}^{p/2} \left(G_{dx}(i,j), G_{dy}(i,j) \right) \right|}{\sum_{i=-p/2}^{2/p} \sum_{j=-p/2}^{p/2} \left| \left(G_{dx}(i,j), G_{dy}(i,j) \right) \right|} \tag{6}$$

where $p = q + l * k$ and $l = 1, 2, \ldots, L$. L is the number of the scales and k is the difference of scales between two adjacent blocks. In the following experiments, q is set as 9. If the coherence value is equal to 0, it indicates the gradients are equally distributed over all directions. On the contrary, if the coherence value is equal to 1, it means all squared gradient vectors share the same orientation angle. Therefore, a large weight is assigned to the block in the scale of large coherence; in contrast, a small weight is assigned to the block in the scale of small coherence. The orientation angle of the weighted multi-scale composite block is calculated by:

$$\theta_{final} = \tan^{-1}\frac{A}{B} \tag{7}$$

where,

$$A = \frac{coh(p,q,1)}{\sum_{l=1}^{L} coh(p,q,l)}\sin(2\theta(p,q,1)) + \frac{coh(p,q,2)}{\sum_{l=1}^{L} coh(p,q,l)}\sin(2\theta(p,q,2))$$
$$+ \cdots + \frac{coh(p,q,L)}{\sum_{l=1}^{L} coh(p,q,l)}\sin(2\theta(p,q,L)) \tag{8}$$

$$B = \frac{coh(p,q,1)}{\sum_{l=1}^{L} coh(p,q,l)}\cos(2\theta(p,q,1)) + \frac{coh(p,q,2)}{\sum_{l=1}^{L} coh(p,q,l)}\cos(2\theta(p,q,2))$$
$$+ \cdots + \frac{coh(p,q,L)}{\sum_{l=1}^{L} coh(p,q,l)}\cos(2\theta(p,q,L)) \tag{9}$$

The hierarchical OF smoothing

In general, the ridge flows are slowly varied across a fingerprint image except at several singular points while the ridge flows vary abruptly in the area close-by the singularities. Due to the difference of singularity close-by area and non-singularity area, in some extent, the method based on composite block can solve the contradiction compared with single block. The composite block performs well in the local region of small noise, but its performance deteriorates in the local region of some big fracture or holes. The reason is that the outer block is not large enough to cover the entire noise area. To compensate this issue, we can increase the scale of the outer block, however as mentioned above, the outer block in large scale leads to localization error of singular points. Another solution is to increase the number of the scales L or the difference of scales k, defined in Eqs. (6)–(9) while the computation cost is also magnified. Therefore, in order to smooth the estimated OF to obtain a refined OF, a hierarchical smoothing method is proposed which can correct the spurious ridge flow and preserve the genuine localization of singular points. The details of the proposed smoothing method are introduced as follows.

In order to visualize the directional image OF, the directions are colored in gray scales, black for 0 and white for $N-1$. The rest of the directions $n < N-1$ are represented by various gray scales. The resulting pseudo OF consists of a set of uniformly-colored regions, each called an orientation zone. Figure 6 illustrates how the number of directions applied can significantly affect the information of the orientation zones. Figure 6a is the original fingerprint image. (b) is the OF corresponding to (a). (c–e) Are the illustrated orientation zones digitized as two, three and six gray scales, respectively. A border line is defined as the separating boundary between two adjacent orientation zones. In other words, a border line is where the ridges' digitized orientation changes.

Several properties exist when an OF is quantified into two or more orientation zones.

Property 1 *Singular points only locate at the intersection of border lines.*

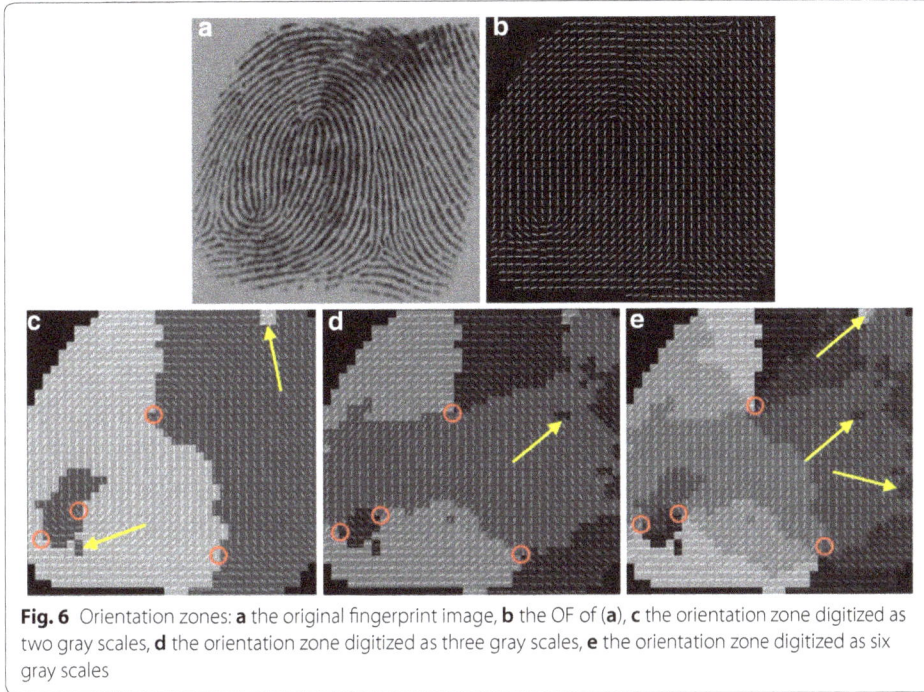

Fig. 6 Orientation zones: **a** the original fingerprint image, **b** the OF of (**a**), **c** the orientation zone digitized as two gray scales, **d** the orientation zone digitized as three gray scales, **e** the orientation zone digitized as six gray scales

Proof The intersections of border lines are marked by red circles in Fig. 6c–e. Let Ω_i and Ω_j be two distinct border lines, which Ω_i is the boundary between $\Omega(n)$ and $\Omega(n+1)$ as well as Ω_j is the boundary between $\Omega(n+2)$ and $\Omega(n+3)$, where $\Omega(n)$ is defined as the orientation zone of direction n. Assume that Ω_i intersects Ω_j at point X and X is not a singular point, since Ω_i and Ω_j are different border lines and intersect each other, there should be more than two distinct directions among $\{n, n+1, n+2, n+3\}$. If there are only two directions in the set, then Ω_i and Ω_j must be parallel. In addition, this set cannot have four distinct directions because it is assumed that Ω_i and Ω_j intersect. Therefore, there must be one region in common among $\Omega(n)$, $\Omega(n+1)$, $\Omega(n+2)$ and $\Omega(n+3)$, indicating that X is the center of a cycling domain. Therefore, X is a singular point according to the definition of singular points, which a core point is defined as a concentrate region where the region curvature is converging to a local maximum and a delta point is defined as a region where the ridge curvature is converging to a local minimum [20].

Property 2 *An island of discontinuities, called a hole, is noise if it is not on the border lines or close to the border lines.*

Proof Holes are marked by yellow arrows in Fig. 6c–e. According to the characteristics of noise and edge, edges and noise may be defined as sharp changes in intensity in gray level images [21]. That is, the intensity of noise and edge pixel varies sharply from the neighboring pixels' intensity. Therefore, an island of discontinuities is defined as noise if it is not on the border lines or close to the border lines because a border line is defined as the separating boundary between two adjacent gray scales.

Property 3 *When the number of gray levels, denoted as N, increases, the singular points become out of focus.*

Proof According to the definition of orientation zone, the number of gray level N is equivalent to the number of orientation. Compared Fig. 6e with (c) and (b), it is observed that the number of N greatly affect the number of border line presented. If$=1$, there will be no pattern at all. If $N=2$, then ridges will be bisected and no intersection can be formed. Therefore, we conclude that the orientation number N should be at least 3. However, when $N \geq 4$, the number of border lines becomes more and the singular points become out of focus with the increase of N.

Based on the three properties, we propose a hierarchical smoothing method, which the OF is digitized into two-orientation-zone in the first step and the OF is digitized into three-orientation-zone in the second step. The propose method does not digitize the OF into four or more direction zones since the singular points become out of focus with the increase of N. The details of the proposed method are introduced as follows.

The Gaussian low pass filter is applied for smoothing the fingerprint image, which is defined as:

$$g(x,y) = \exp\left(\frac{x^2 + y^2}{2\delta^2}\right). \tag{10}$$

When processing digital fingerprint images, Eq. (10) is digitized as:

$$g(x,y) = \exp\left(\frac{(x - i)^2 + (y - j)^2}{2\delta^2}\right). \tag{11}$$

where (i, j) is the center of the Gaussian kernel. Since 90% percent of the energy is located in $(-2\delta, 2\delta)$ in terms of the Gaussian low pass filter with the variance of δ. Therefore, $(-2\delta, 2\delta)$ is set as the dominant area of the Gaussian low pass filter. For example, when $\delta = 1.0$, the Gaussian kernel can be computed as:

$$\begin{bmatrix} 0.0183 & 0.0821 & 0.1353 & 0.0821 & 0.0183 \\ 0.0821 & 0.3679 & 0.6065 & 0.3679 & 0.0821 \\ 0.1353 & 0.6065 & 1 & 0.6065 & 0.1353 \\ 0.0821 & 0.3679 & 0.6065 & 0.3679 & 0.0821 \\ 0.0183 & 0.0821 & 0.1353 & 0.0821 & 0.0183 \end{bmatrix} \tag{12}$$

Two-orientation-zone filtering

The OF is first digitized as two equivalent directional zones where ω_1^1 and ω_2^1 represent the direction zones whose angle is $(0, 90°)$ and $[90°, 180°]$, respectively. Subsequently, holes are removed as noise if they are not on the border lines. The pixel number of the two-orientation-zone, denoted as ω_1^1 and ω_2^1 in the filtering window $S_n^1 \times S_n^1$ are counted and represented as num_{n1}^1 and num_{n2}^1, respectively. Assuming Num_n^1 is the maximal value of num_{n1}^1 and num_{n2}^1, then the variance of the Gaussian low pass filter for the two-orientation-zone filter, denoted as δ^1 can be defined as:

$$\delta^1 = \begin{cases} 1.5 & if & Num_1^1 \geq threshold_1^1 \\ 1.0 & else\ if & Num_2^1 \geq threshold_2^1 \\ 0.5 & else\ if & Num_3^1 \geq threshold_3^1 \\ empty & else \end{cases} \tag{13}$$

where $\delta^1 = empty$ represents no filtering is performed. For the two-orientation-zone filter, the filter window $S_n^1 \times S_n^1$ is set as $S_1^1 = 7$, $S_2^1 = 5$ and $S_3^1 = 3$ where $n = 1, 2, 3$, $threshold_n^1 = int(S_n^1 \times S_n^1 \times 0.8)$ and $int(\bullet)$ indicates the rounding.

Three-orientation-zone filtering

The OF is first digitized as three equivalent directional zones where ω_1^2, ω_2^2 and ω_3^2 represent the orientation zones whose angle is within $(0, 60°)$, $[60°, 120°)$ and $[120°, 180°]$, respectively. Subsequently, holes are removed as noise if they are not on the border lines. The pixel number of the three-direction-zone, denoted as ω_1^2, ω_2^2 and ω_3^2 in the filtering window $S_n^2 \times S_n^2$ are counted and represented as num_{n1}^2, num_{n2}^2 and num_{n3}^2, respectively. Thereafter, the sum of every two-orientation-zone are calculated, that is, $sum_{n1}^2 = num_{n1}^2 + num_{n2}^2$, $sum_{n2}^2 = num_{n1}^2 + num_{n3}^2$ and $sum_{n3}^2 = num_{n2}^2 + num_{n3}^2$. Assuming Num_n^2 is the maximal value of sum_{n1}^2, sum_{n2}^2 and sum_{n3}^2, then the variance of the Gaussian low pass filter for the three-orientation-zone filter, denoted as δ^2 can be defined as:

$$\delta^2 = \begin{cases} 1.0 & if & Num_1^2 \geq threshold_1^2 \\ 0.5 & else\ if & Num_2^2 \geq threshold_2^2 \\ empty & else \end{cases} \tag{14}$$

For the three-orientation-zone filter, the filter window $S_n^2 \times S_n^2$ is set as $S_1^2 = 5$ and $S_2^2 = 3$ where $n = 1, 2$ and $threshold_n^2 = int(S_n^2 \times S_n^2 \times 0.9)$.

Experimental results and discussions

To evaluate the performance of the proposed weighted multi-scale composite block and the hierarchical smoothing method for OF reconstruction, three experiments are designed. Experiment 1 aims to validate whether the weighted multi-scale composite block can balance the dilemma of accuracy and robustness more effectively than the previous work. Experiment 2 is designed to examine whether the hierarchical smoothing method can correct the spurious ridge flow and preserve the genuine localization of singular points? The purpose of experiment 3 is to test the performance of the proposed method combining the weighted multi-scale composite blocks with the hierarchical smoothing strategy for OF reconstruction in low quality fingerprint images. The fingerprint databases FVC 2004 DB1–DB4 are employed in this study. All the test images suffer from the different large noise effects caused by wounds, scars, dirtiness, creases, moisture or greasiness. The experiments are carried out in Visual Studio2010+OpenCV2.4.4 and Windows 7. Due to the limited space, a part of the experimental results are presented in this session for discussion.

OF reconstruction by using the weighted multi-scale composite block

In order to examine the performance of the weighted multi-scale composite block, the proposed approach is compared with three state-of-the-art methods according to their performance in gradient based OF computation. The fingerprint OF are reconstructed by the conventional gradient based method (CG) [13], the enhanced gradient based method (EG) [23], the gradient based voting method (GV) [22] and the proposed method. The results of the four methods are shown in Figs. 7, 8. The images scale of the original fingerprints are resized as 440×440 for all the experiments. For the methods using single block, the scale of the block is set as 9×9. For the proposed method using composite block, the scale of inner is set as 9×9 and the scale of the outer block is defined by Eqs. (7)–(9).

Figure 7 shows the experiment results of several OF computation methods. (a) Is the original fingerprint image affected by serious level of dryness. There are lots of breakpoints caused by dryness in the ridges and the ridge information in most regions of the original image is unclear, such as the left, bottom and right regions. (b) Shows the result obtained by using the conventional gradient based method [13], obviously, the result is seriously sensitive to noise. There is no reliable OF information in (b) and the ridge flows are completely incorrect in the left, right and bottom regions since these regions are too noisy. (c) Illustrates the result estimated by using the enhanced gradient based method [23]. It is observed that the reconstructed OF is more accurate than (b) in small scale noise, but for the low quality area containing large scale noise, it is very difficult to obtain the accurate OF, marked by the yellow circle. Furthermore, it produces spurious ridge in the singular point area, marked by the red circle. Although the estimation is improved by using the gradient based voting method, shown in (d), the OF contains

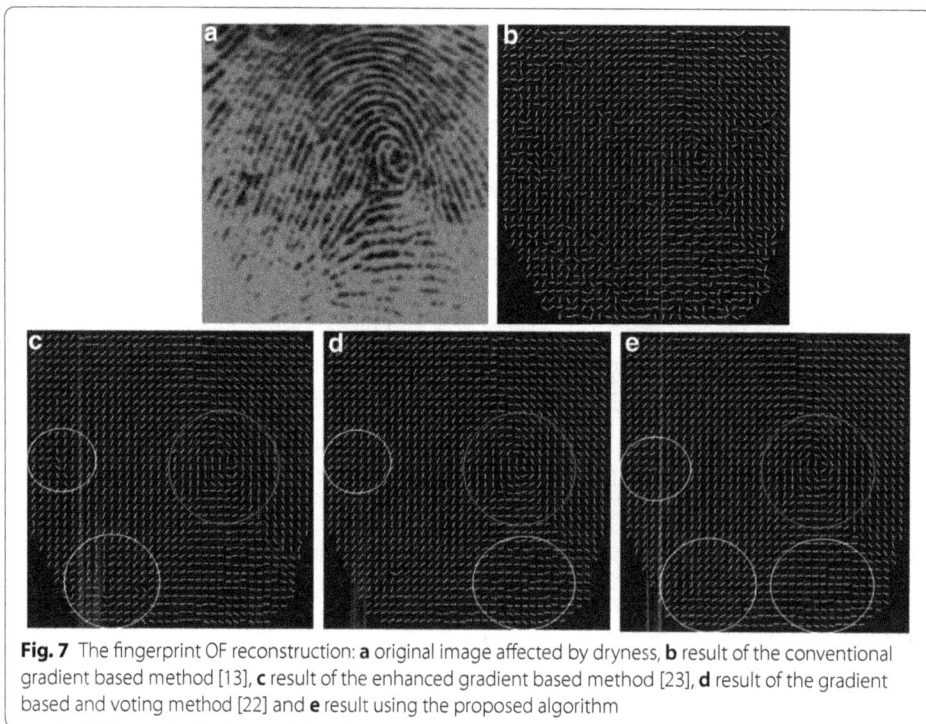

Fig. 7 The fingerprint OF reconstruction: **a** original image affected by dryness, **b** result of the conventional gradient based method [13], **c** result of the enhanced gradient based method [23], **d** result of the gradient based and voting method [22] and **e** result using the proposed algorithm

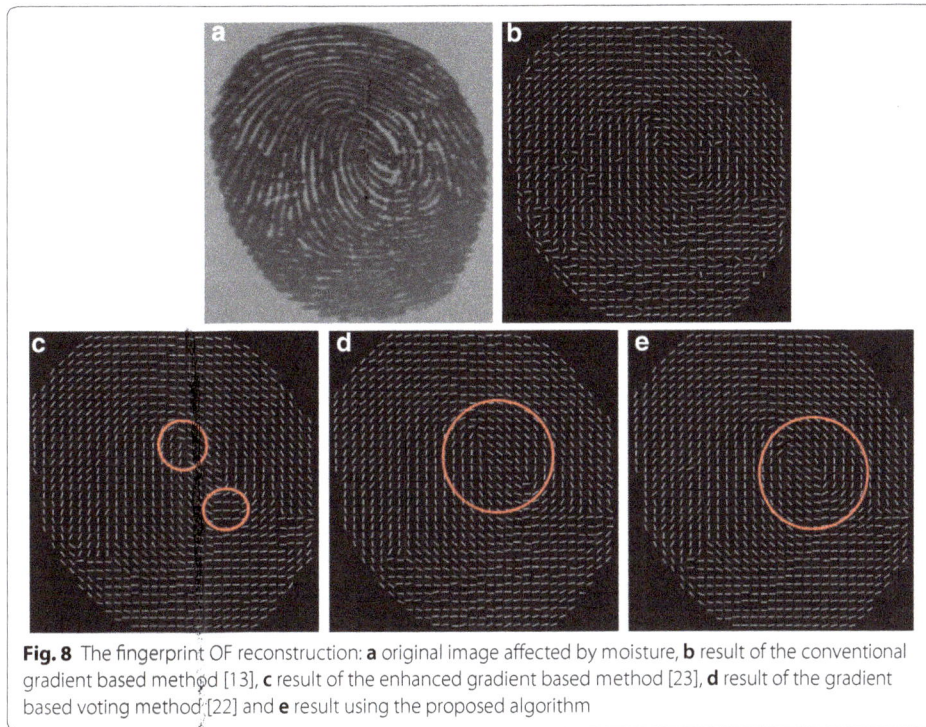

Fig. 8 The fingerprint OF reconstruction: **a** original image affected by moisture, **b** result of the conventional gradient based method [13], **c** result of the enhanced gradient based method [23], **d** result of the gradient based voting method [22] and **e** result using the proposed algorithm

incorrect ridge flow in the lower left part, marked by the yellow circle, and it is noted that the singularity localization is inevitably deviated from the genuine place, marked by the red circle. Among the four, the proposed method performs the best by recovering the ridge structures, marked by the yellow circle, and preserving genuine singularity localization, marked by the red circle, shown in (e).

Figure 8a is the original fingerprint image which presents a number of indistinguishable ridge structures in the lower part due to moisture. The conventional gradient based method can only give a coarse OF containing lots of spurious ridge structures, shown in (b). (c) Is the estimation result from the enhanced gradient based method, where incorrect ridge structures are generated in the area close by the singular points, marked by the circles. While the result of the gradient based voting method incurs serious deviation for the singularity localization, shown in (d). For this fingerprint example, the result shows that the proposed method is capable to extract the information of OF reliably and it is more robust against singularity localization deviation in comparison with the other three gradient based methods.

OF smoothing by using the two-hierarchical smoothing method

The proposed smoothing approach aims to reconstructing correct ridge structures and preserving genuine singularity localizations. We first introduce the proposed hierarchical smoothing approach in details.

Figure 9b illustrates the OF estimated by using the conventional gradient based method, where there are several horizontal creases of different lengths running through the original fingerprint image, shown in (a). It is observed that spurious ridge structures are generated in the area close to the singular points due to noise, marked by the yellow

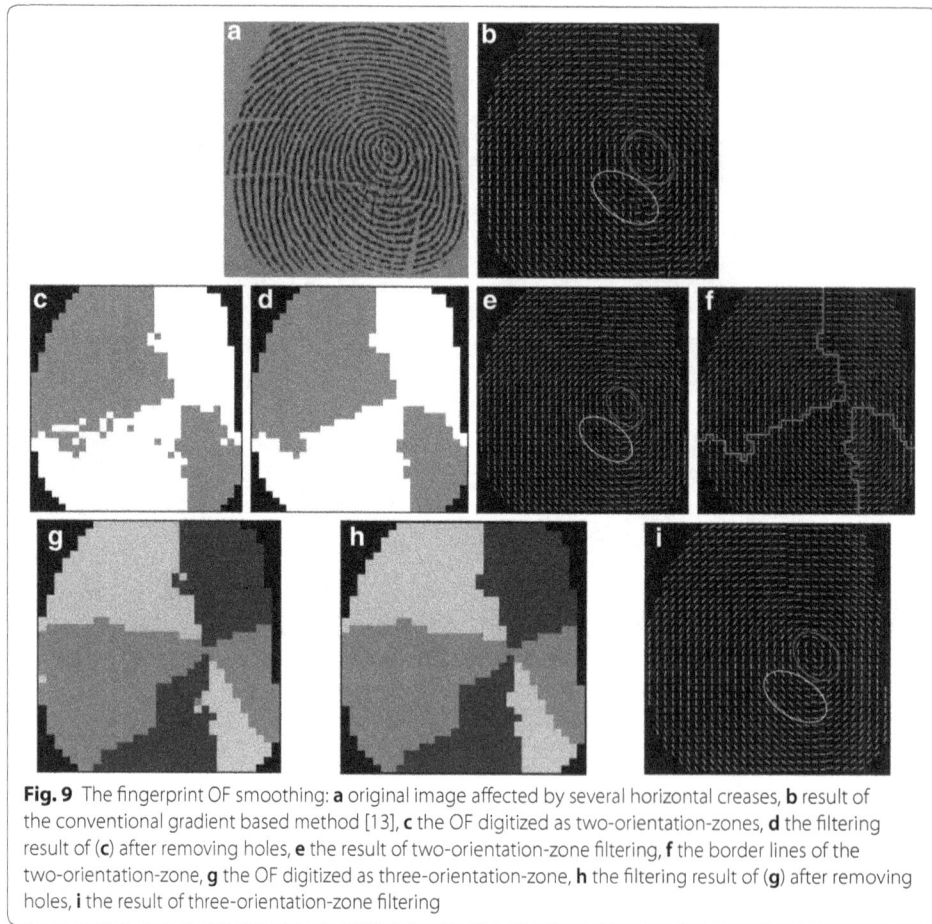

Fig. 9 The fingerprint OF smoothing: **a** original image affected by several horizontal creases, **b** result of the conventional gradient based method [13], **c** the OF digitized as two-orientation-zones, **d** the filtering result of (**c**) after removing holes, **e** the result of two-orientation-zone filtering, **f** the border lines of the two-orientation-zone, **g** the OF digitized as three-orientation-zone, **h** the filtering result of (**g**) after removing holes, **i** the result of three-orientation-zone filtering

circle. The purpose of our hierarchical smoothing method is to correct the spurious structures and to preserve the genuine singularity localization un-deviated, marked in the red circle. (c) Shows the OF digitized as two-orientation-zone and (d) is the corresponding result of (c) after removing holes. The result of two-orientation-zone filtering is demonstrated in (e). Comparing (b) and (e) from the figure, it can be observed that there are two significant improvements. One is that the entire OF smoothed by the two-orientation-zone filter is more consistent with respect to the actual flow of the fingerprint ridges. The other is that the spurious ridge structure is partially corrected in the area close to the singular points, marked in the yellow circle and the singular area is not affected, marked in the red circle.

According to the aforementioned Properties 1 and 2, it can be inferred that singular points must be localized along the border line when OF is digitized as two-orientation-zone. Therefore, in order to preserve the singular area unaffected, the two-orientation-zone filtering is not performed along the border lines, shown in (f). To reconstruct the actual ridge flow along the border lines, the three-orientation-zone filtering is subsequently applied.

Figure 9g shows the OF digitized as three directional zones and (h) is the corresponding result of (g) after removing holes. The result of the three-orientation-zone filtering is demonstrated in (i). Compared with the result of two-orientation-zone

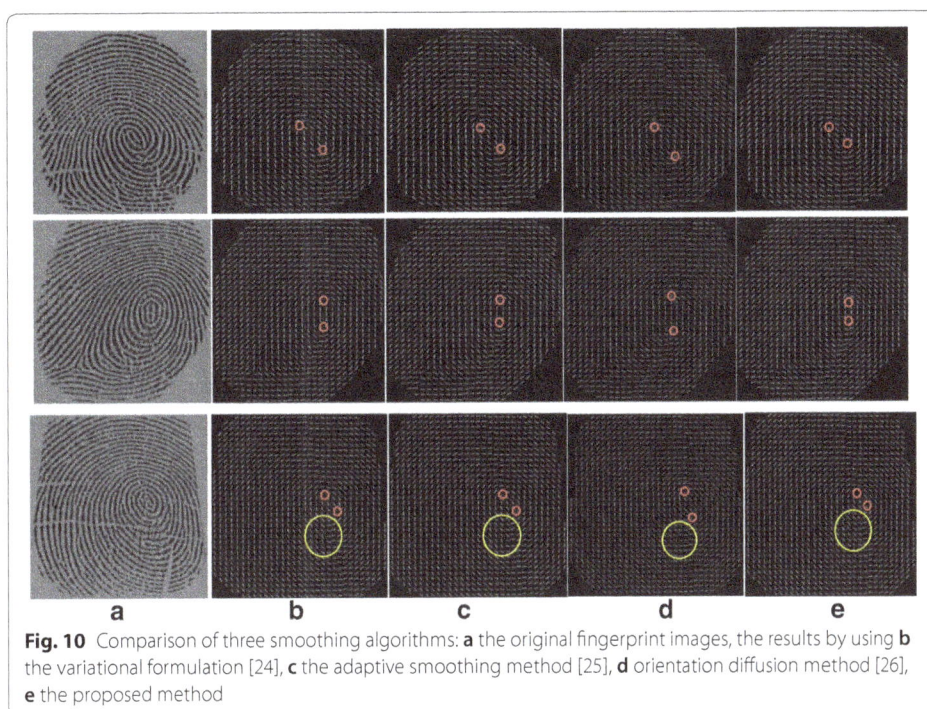

Fig. 10 Comparison of three smoothing algorithms: **a** the original fingerprint images, the results by using **b** the variational formulation [24], **c** the adaptive smoothing method [25], **d** orientation diffusion method [26], **e** the proposed method

filtering, the spurious ridge flow close by the singular points is completely corrected, marked in the yellow circle and the singular area is nearly perfect.

To test the performance of the hierarchical smoothing algorithm, the proposed method is compare with three state-of-the-art methods according to their performance in OF enhancement. The fingerprint OF are smoothed by the variational formulation by Hou [24], the adaptive smoothing method proposed by Liu [25], the orientation diffusion method proposed by Bian [26] and the proposed method. The results of the methods are shown in Fig. 10.

Figure 10a shows three examples of fingerprint images containing several vertical and horizontal scars of various lengths. The localizations of the singular points are highlighted by red circles in the estimated OFs. For these fingerprint examples, the results illustrate that the OFs estimated by using the variational formulation, shown in (b), and the adaptive smoothing method, shown in (c) produce distinct singularity localization deviation, while the proposed method can preserve genuine singularity localization. Furthermore, our proposed method is superior to the variational formulation method and the adaptive smoothing method in terms of correcting the perturbation, shown in the area marked by the yellow circles in the third line. The orientation diffusion method [26] can greatly improve the distortion shown in the area marked by the yellow circles in the third line. Unfortunately, the detected singularities obviously deviate away the genuine location due to the break line close to the singularity. Therefore, it is concluded that our proposed smoothing method can balance the contradiction in correcting spurious ridge structures and preserving genuine singularity localization.

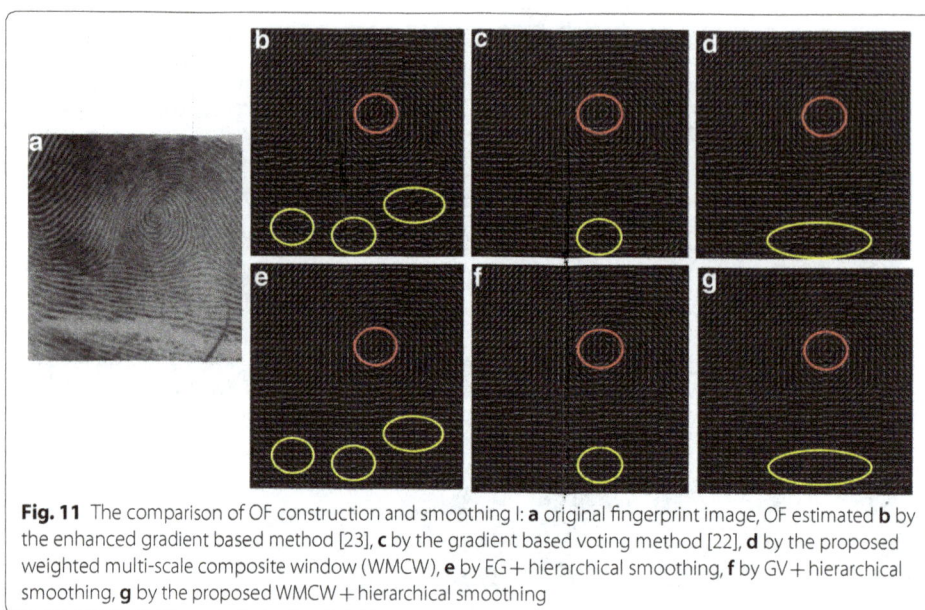

Fig. 11 The comparison of OF construction and smoothing I: **a** original fingerprint image, OF estimated **b** by the enhanced gradient based method [23], **c** by the gradient based voting method [22], **d** by the proposed weighted multi-scale composite window (WMCW), **e** by EG + hierarchical smoothing, **f** by GV + hierarchical smoothing, **g** by the proposed WMCW + hierarchical smoothing

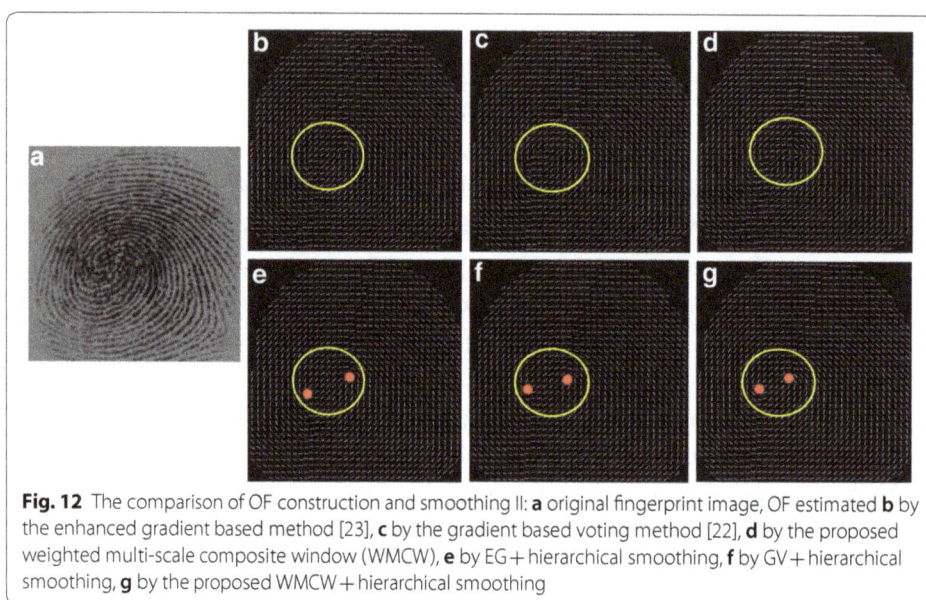

Fig. 12 The comparison of OF construction and smoothing II: **a** original fingerprint image, OF estimated **b** by the enhanced gradient based method [23], **c** by the gradient based voting method [22], **d** by the proposed weighted multi-scale composite window (WMCW), **e** by EG + hierarchical smoothing, **f** by GV + hierarchical smoothing, **g** by the proposed WMCW + hierarchical smoothing

OF reconstruction by combining the weighted multi-scale composite blocks with the hierarchical smoothing strategy

Different OF estimation methods based on gradient are applied to the sample fingerprint images of low quality that are selected from the database FVC2004. These fingerprint samples are affected by serious level of dirtiness, creases, moisture or dryness. Figures 11, 12, 13 and 14 show that the performance comparison for the OF reconstruction by using different OF extraction methods. For the purpose of comparison, the estimated spurious ridge structures are marked in yellow circles and the area of singular points are marked in red circles.

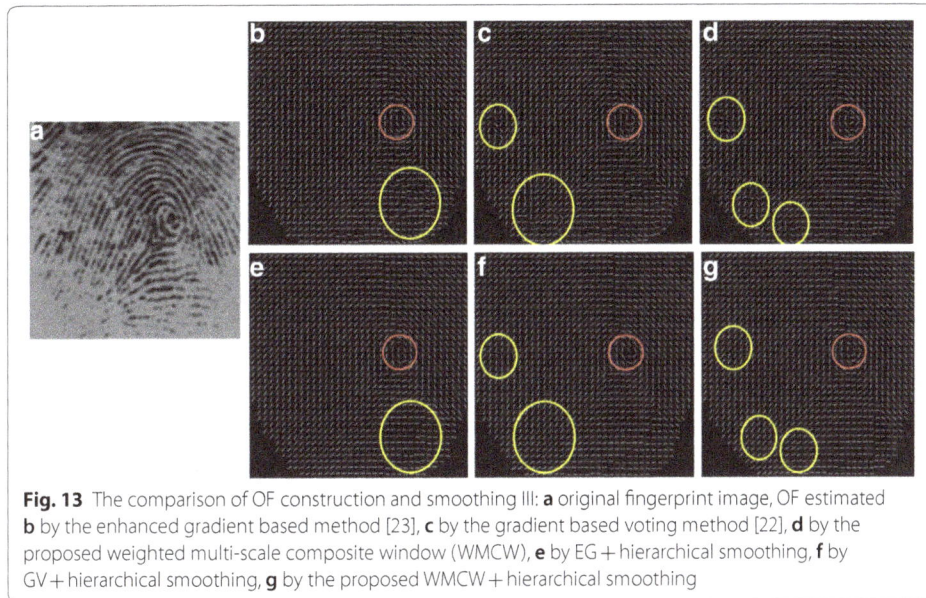

Fig. 13 The comparison of OF construction and smoothing III: **a** original fingerprint image, OF estimated **b** by the enhanced gradient based method [23], **c** by the gradient based voting method [22], **d** by the proposed weighted multi-scale composite window (WMCW), **e** by EG + hierarchical smoothing, **f** by GV + hierarchical smoothing, **g** by the proposed WMCW + hierarchical smoothing

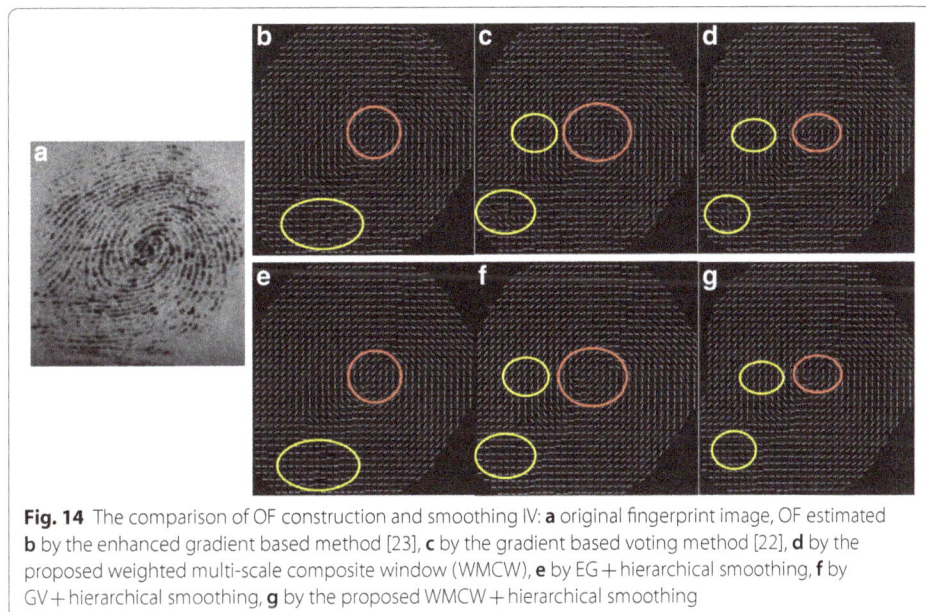

Fig. 14 The comparison of OF construction and smoothing IV: **a** original fingerprint image, OF estimated **b** by the enhanced gradient based method [23], **c** by the gradient based voting method [22], **d** by the proposed weighted multi-scale composite window (WMCW), **e** by EG + hierarchical smoothing, **f** by GV + hierarchical smoothing, **g** by the proposed WMCW + hierarchical smoothing

From Fig. 11a we can observe that a long and wide blotch, possibly caused by callus, appears at the lower site of the original image and the ridge pattern in this area is completely lost. We then superimpose the estimated OF from different gradient based methods on the original fingerprint image and display the results in Fig. 11b–d. (b) and (c) Are produced from the enhanced gradient based method (EG) and the gradient-based voting method (GV), where several spurious ridge structures are contained. Although the estimation of the proposed weighted multi-scale composite window (WMCW) generates incorrect ridge flow, the OF are the smoothest among the results of the three methods. From (e) to (g) we see the spurious ridge structures are correctly reconstructed by using the proposed hierarchical smoothing method. The red circles in the figure indicate the singular areas. One can see that the proposed WMCW combining the hierarchical smoothing method performs the best by preserving genuine singularity localization, shown in (d) and (g), while the other methods produce singularity localization deviations, shown in (b), (c), (e) and (f).

Figures 12, 13 and 14 are another three experiment results. For these fingerprint examples, the results show that our approach combing WMCW with the hierarchical smoothing method is capable to extract the information of ridge OF reliably and it is more robust against singularity deviation in comparison with the other two gradient based methods.

In order to objectively evaluate the singularity deviation of the proposed method compared with the state-of-arts approaches, 300 fingerprint images are randomly selected from the database of FVC2004DB1, which the image sizes are fixed to 480×480. The genuine singularities are marked by experts as the ground truth for measuring the distance of the detected singularities and the genuine ones. The images contain 358 cores and 135 deltas. We use the method of Ref. [27] for the sake of singularity detection. Euclidean distance is utilized to calculate the distance between the genuine singularities and the detected ones by using Eq. (15):

$$d = \sqrt{(x - C_X)^2 + (y - C_Y)^2} \tag{15}$$

where d indicates the distance, (x, y) is the center pixel of the detected singularity, (C_X, C_Y) is the center pixel of the singularity marked by experts.

Table 1 is the distance between the detected cores and the genuine ones. From Table 1 we can observe that the localization deviation incurred by the proposed method is much less than that produced by the enhanced gradient based method (EG) and the gradient-based voting method (GV). Most of the deviation distances of the test fingerprint images by using the proposed method are within 5 pixels. Table 2 is the distance between the detected deltas and the genuine ones. In most cases, deltas locate far away from the center of the image and furthermore the OF close to delta is more stable than that close to the core, so it is not prone to be distorted by smoothing. Table 2 illustrates that the deviation distance of most deltas from the enhanced gradient based method (EG) and the gradient-based voting method (GV) is within 5 pixels. However the proposed method can achieve better result that the deviation distances of most deltas are within 3 pixels. Therefore, we can conclude the proposed method obtains less singularity localization deviation compared to the state-of-arts algorithms.

Table 1 The distance between the detected cores and the genuine ones (pixels)

	EG + hierarchical smoothing	GV + hierarchical smoothing	The proposed WMCW + hierarchical smoothing
≥ 15	58	49	31
(10, 15]	132	147	72
(5, 10]	112	104	112
≤ 5	56	76	143

Table 2 The distance between the detected deltas and the genuine ones (pixels)

	EG + hierarchical smoothing	GV + hierarchical smoothing	The proposed WMCW + hierarchical smoothing
(5, 10]	17	12	15
(3, 5]	81	72	78
≤ 3	25	41	42

In a word, three experiments results show that the proposed gradient based algorithm is more reliable for the estimation of the ridge information for fingerprint OF and is more accurate in preserving the singularity localization.

Conclusions

In this paper, a gradient based algorithm, which uses a weighted multi-scale composite window to adapt the scales of the blocks, has been proposed. In order to correct the spurious ridges and preserve the genuine location of singular points, we refine the OF by using a hierarchical smoothing strategy. To verify the performance of the proposed method, three experiments are designed to test the proposed algorithm together with other popular gradient based methods on real fingerprint images,which are selected from different categories and all are suffered from obvious noise effects. The experiment results obtained show that the proposed method is superior with respect to reliable OF construction and avoiding singularity localization deviation.

Abbreviations
OF: orientation field; CG: conventional gradient based method; EG: enhanced gradient based method; GV: gradient based voting method; WMCW: weighted multi-scale composite window.

Authors' contributions
HL and YT implemented the experiments, draft and analyzed the data. JW and PY suggested the proposed algorithm. JC and YZ were responsible for theoretical guidance. All authors read and approved the final manuscript.

Author details
[1] School of Information Science and Engineering, Electronic Engineering, Yunnan University, Chenggong District, Kunming 650000, China. [2] Breast Surgery Department, The Third Hospital Affiliated to the Medical University of Kunming, Kunming 605118, Yunnan Province, China.

Acknowledgements
This work was supported by the Grant (61561050), (61462094) from the Natural Science Foundation of China.

Competing interests
The authors declare that they have no competing interests.

References
1. Sherlock B, Monro D. A model for interpreting fingerprint topology. Pattern Recognit. 1993;26(7):1047–55.
2. Vizcaya P, Gerhardt L. A nonlinear orientation model for global description of fingerprints. Pattern Recognit. 1996;29(7):1221–31.
3. Gu J, Zhou J, Zhang D. A combination model for orientation field of fingerprints. Pattern Recognit. 2004;37(3):s543–53.
4. Zhou J, Gu J. Modeling orientation fields of fingerprint with rational complex functions. Pattern Recognit. 2004;37(2):389–91.
5. Wang Y, Hu J, Phillips D. A fingerprint orientation model based on 2d fourier expansion (FOMFE) and its application to singular-point detection and fingerprint indexing. IEEE Trans Pattern Anal Mach Intell. 2007;29(4):573–85.
6. Zhou J, Chen F, Gu J. A novel algorithm for detecting singular points from fingerprint images. IEEE Trans Pattern Anal Mach Intell. 2009;31(7):1239–50.
7. Ram S, Bischof H, Birchbauer J. Modeling fingerprint ridge orientation using legendre polynomials. Pattern Recognit. 2010;43(1):342–57.

8. Liu M, Liu S, Zhao Q. Fingerprint orientation field reconstruction by weighted discrete cosine transform. Inf Sci. 2014;268:65–77.
9. Karu K, Jain A. Fingerprint classification. Pattern Recognit. 1996;29(3):389–404.
10. Jain AK, Prabhakar S, Hong I. A multichannel approach to fingerprint classification. IEEE Trans Pattern Anal Mach Intell. 1999;21:348–59.
11. Kass M, Witkin A. Analyzing oriented patterns. In: Computer vision. Graphics and image processing. 1987;37(3):362–385.
12. Rao AR, Jain RC. Computerized flow field analysis: oriented texture fields. IEEE Trans Pattern Anal Mach Intell. 1992;14(7):693–709.
13. Bazen A, Gerez S. Systematic methods for the computation of the directional fields and singular points of fingerprints. IEEE Trans Pattern Anal Mach Intell. 2002;24(7):905–19.
14. Jiang X. On orientation and anisotropy estimation for online fingerprint authentication. IEEE Trans Signal Process. 2005;53(10):4038–49.
15. Mei Y, Sun HJ, Xia DS. A gradient-based combined method for the computation of fingerprint's orientation field. Image Vis Comput. 2009;27(8):1169–77.
16. Mei Y, Cao G, Sun HJ, et al. A systematic gradient-based method for the computation of fingerprint's orientation field. Comput Electron Eng. 2012;38:1035–46.
17. Bian W, Luo Y, Xu D, et al. Fingerprint ridge orientation field reconstruction using the best quadratic approximation by orthogonal polynomial in two discrete variables. Pattern Recognit. 2014;47:3304–13.
18. Cavusoglu A, Gorgunoglu S. A fast fingerprint image enhancement algorithm using a parabolic mask. Comput Electron Eng. 2008;34(3):250–6.
19. Wang Y, Hu J, Schroder H. A gradient based weighted averaging method for estimation of fingerprint OF. In: Proceedings of digital imaging computing: techniques and applications. Queensland. 2005.
20. Srinivasan VS, Murthy NN. Detection of singular points in fingerprint images. Pattern Recognit. 1992;25:139–53.
21. Zhang WC, Zhao YL, Breckon TP, et al. Noise robust image edge detection based upon the automatic anisotropic Gaussian kernels. Pattern Recognit. 2017;63:193–205.
22. Jain A, Hong L, Bolle R. On-line fingerprint verification. IEEE Trans Pattern Anal Mach Intell. 1997;19(4):302–14.
23. Wang Y, Hu J, Han F. Enhanced gradient-based algorithm for the estimation of fingerprint orientation fields. Appl Math Comput. 2007;185(2):823–33.
24. Hou Z, Yau WY. A variational formulation for fingerprint orientation modeling. Pattern Recognit. 2010;45(5):1915–26.
25. Liu M, Jiang X, Kot AC. Fingerprint reference-point detection. EURASIP J Adv Signal Process. 2005;2005(4):1–12.
26. Bian WX, Ding SF, Xue Y. Combining weighted linear project analysis with orientation diffusion for fingerprint orientation field reconstruction. Inf Sci. 2017;396:55–71.
27. Wang L, Bhattacharjee N, Srinivasan B. A novel technique for singular point detection based on Poincaré index. In: International conference on advances in mobile computing and multimedia. ACM. 2011. pp. 12–18.

Non-orthogonal one-step calibration method for robotized transcranial magnetic stimulation

He Wang[1] ⓘ, Jingna Jin[1], Xin Wang[1], Ying Li[1], Zhipeng Liu[1] and Tao Yin[1,2]* ⓘ

*Correspondence:
bme500@163.com
[1] Institute of Biomedical
Engineering, Chinese
Academy of Medical Science
& Peking Union Medical
College, Tianjin 300192,
China
Full list of author information
is available at the end of the
article

Abstract

Background: Robotized transcranial magnetic stimulation (TMS) combines the benefits of neuro-navigation with automation and provides a precision brain stimulation method. Since the coil will normally remain unmounted between different clinical uses, hand/eye calibration and coil calibration are required before each experiment. Today, these two steps are still separate: hand/eye calibration is performed using methods proposed by Tsai/Lenz or Floris Ernst, and then the coil calibration is carried out based on the traditional TMS experimental step. The process is complex and time-consuming, and traditional coil calibration using a handheld probe is susceptible to greater calibration error.

Methods: A novel one-step calibration method has been developed to confirm hand/eye and coil calibration results by formulating a matrix equation system and estimating its solution. Hand/eye calibration and coil calibration are performed to confirm the pose relationships of the marker/end effector 'X', probe/end effector 'Y', and robot/world 'Z'. First, the coil is fixed on the end effector of the robot. During the one-step calibration process, a marker is mounted on the top of the coil and a calibration probe is fixed at the actual effective position of the coil. Next, the robot end effector is moved to a series of random positions 'A', the tracking data of marker 'B' and probe 'C' is obtained correspondingly. Then, a matrix equation system $AX = ZB$ and $AY = ZC$ can be acquired, and it is computed using a least-squares approach. Finally, the calibration probe is removed after calibration, while the marker remains fixed to the coil during the TMS experiment. The methods were evaluated based on simulation data and on experimental data from an optical tracking device. We compared our methods with two classical methods: the QR24 method proposed by Floris Ernst and the handheld coil calibration method.

Results: The new methods outperform the QR24 method in the aspect of translational accuracy and performs similarly in the aspect of rotational accuracy, the total translational error decreased more than fifty percent. The new approach also outperforms traditional handheld coil calibration of navigated TMS systems, the total translational error decreased three- to fourfold, and the rotational error decreased six- to eightfold. Furthermore, the convergence speed is improved 16- to 27-fold for the new algorithms.

Conclusion: These results suggest that the new method can be used for hand/eye and coil calibration of a robotized TMS system. Two complex steps can be simplified using a least-squares approach.

Keywords: Transcranial magnetic stimulation, TMS, Medical robots and systems, Hand/eye and coil calibration, Least-squares approach

Background

Transcranial magnetic stimulation (TMS) is a non-invasive and painless method for stimulating the cerebral cortex nerve [1–3]. Based on the principle of electromagnetic induction, an electric current is created on the cerebral cortex using a magnetic coil that is manually placed on top of the patient's head. Recently, single-pulse and repetitive TMS have been used in clinical study for the therapy of mental disease [4–9]. However, TMS is still not widely promoted because its therapeutic effect changes between subjects [10, 11].

The variability is partly because of how accurate the stimulus process is carried out [12–14]. Initially, the magnetic stimulation coil was located over the subject's head manually and imprecisely, without the help of any navigation system [10–12]. To solve this problem, neural-navigation technology has been developed to position the coil on the subject's head more accurately [13, 14]. Magnetic resonance imaging (MRI) of both the subject and the optical tracking device are used in the neuro-navigation system, and navigation software can indicate the actual stimulation point of the coil on the subjects' brain in real time. However, the stimulation coil, which usually weighs more than 2 kg (Fig. 1), is difficult for the operator to hold for more than 30 min in each procedure. Compared with handheld approach, a robot-assisted coil positioning method is more stable and repeatable, which is beneficial to the clinical or academic application of TMS.

Recently, two types of robotic TMS systems have been reported and shown to improve stimulation accuracy [15–18]. First, a robotic TMS system has been developed using a general industrial robot [15–17]. A six-jointed industrial robot (Adept Viper s850) was used in a robotic TMS system designed by Lars et al. [17]. For this kind of robotic system, the coil was mounted to the robot's end effector, a Polaris Spectra infrared tracking system was used for navigation. After calibration, the magnetic coil is placed quite precisely and directly over a selected target region by the robot. However, the safety of this kind of robotic system has been queried, because the system is equipped with actuators selected for high-speed motions like any other industrial robot [18]. Second, a dedicated robotic system for TMS has been designed based on mechanical architecture and a control strategy [18]. A seven-jointed dedicated robot is designed from the kinematic scheme of the mechanism, including the arm, the prismatic joint, and the wrist [19]. The design and control of the robotic system optimizes the safety of the procedure, and allows the force applied between the robot and the head to be controlled. Robot-guided neuro-navigated TMS has been used in some recent studies of psychiatric diseases [19–21].

Robotized TMS systems are designed and requested for high-precision stimulus, which is primarily determined by calibration [17–22]. Besides, in order to ensure the safety of the robotic TMS system, a marker is mounted on the top of the coil to detect the coil position in real time [17–22], but the actual effective position of the coil is at the

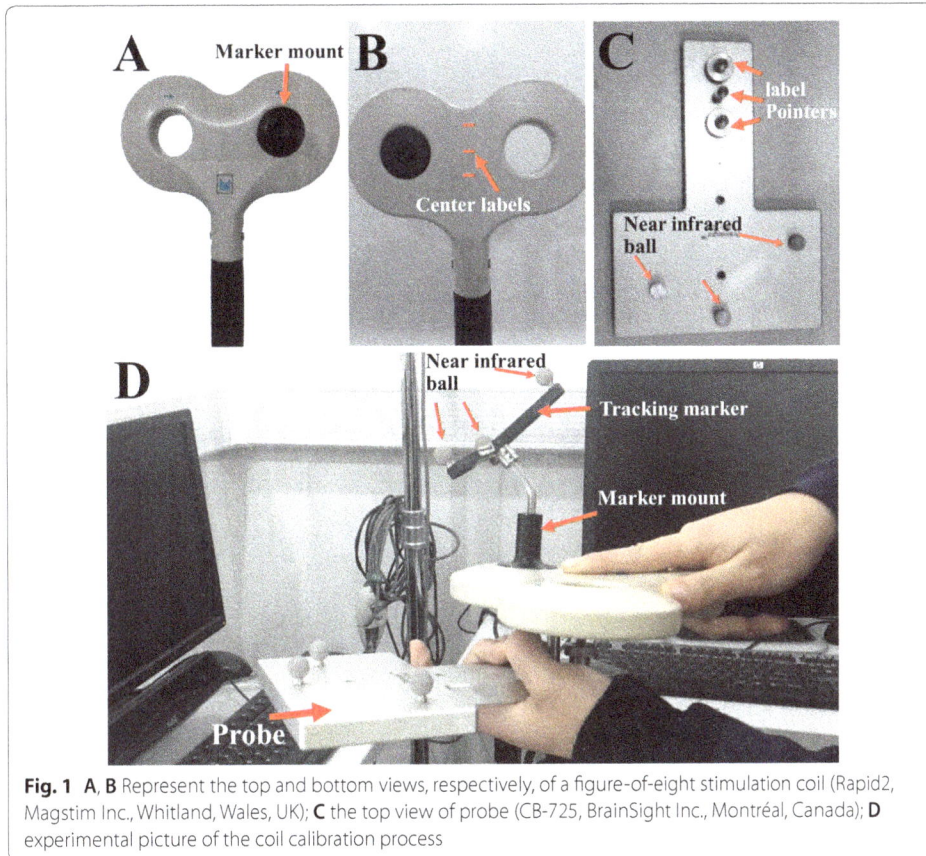

Fig. 1 A, B Represent the top and bottom views, respectively, of a figure-of-eight stimulation coil (Rapid2, Magstim Inc., Whitland, Wales, UK); **C** the top view of probe (CB-725, BrainSight Inc., Montréal, Canada); **D** experimental picture of the coil calibration process

bottom centre of the coil, as shown in Fig. 1. Thus, hand/eye calibration and coil calibration are the most important steps in the workflow of a robotized TMS, and these two steps directly determine the accuracy of the system. Coil calibration involves confirming the first unknown pose matrices between the actual effective positions of the coil and the marker. Traditionally, a handheld coil calibration method is used, as shown in Fig. 1. A marker is fixed on the top of the coil (Fig. 1A), then a probe (Fig. 1C) is placed under the coil and the pointer is aimed at the coil centre labels (Fig. 1B, D). Finally, the positions of the marker and the coil centre in the coordinates of the tracking device can be detected, and the first unknown pose matrices can be calculated. Hand/eye calibration involves accurately computing the other two unknown pose matrices, which are the end-effector/marker matrix and the robot/tracking-device matrix [22, 23]. Traditional hand/eye calibration involves solving an $AX = ZB$ form matrix equation, where the matrices 'A' and 'B' are known, and 'X' and 'Z' are unknown, as shown in Fig. 2D [22–26]. Typically, 'A' represents the transform from the robot's base to the robot's end effector, and 'B' represents the position and orientation data providing the full six degrees of freedom (DOF) of a marker obtained from a tracking device. 'X' represents the transforms from the robot's end effector to the marker, and 'Z' represents the transform from the robot's base to the tracking device.

The handheld coil calibration method typically suffers from larger calibration errors; a translational error of 3 mm is acceptable for handheld coil calibration [27–30]. In

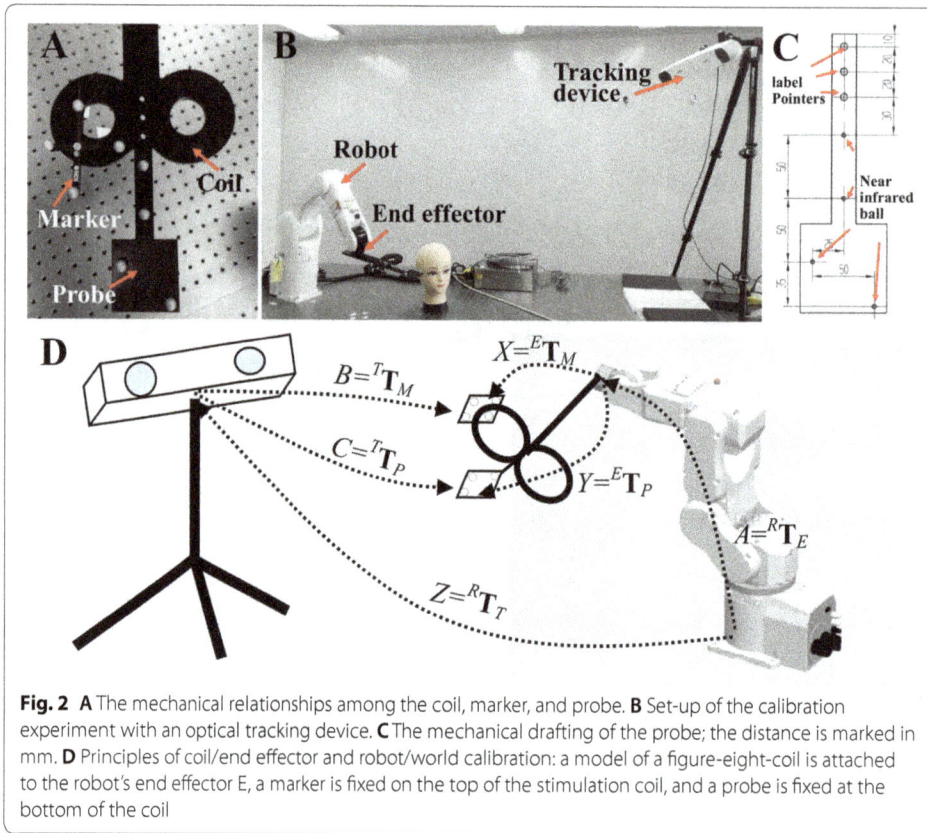

Fig. 2 **A** The mechanical relationships among the coil, marker, and probe. **B** Set-up of the calibration experiment with an optical tracking device. **C** The mechanical drafting of the probe; the distance is marked in mm. **D** Principles of coil/end effector and robot/world calibration: a model of a figure-eight-coil is attached to the robot's end effector E, a marker is fixed on the top of the stimulation coil, and a probe is fixed at the bottom of the coil

addition, the traditional hand/eye calibration method was reported separately by Shiu and Ahmad [23, 24] and Tsai and Lenz [25, 26]. Matrix algebra and the special properties of homogeneous matrices were used for determining the unknown matrices mentioned above. A review of these calibration algorithms was proposed by Wang et al. [31]. All these methods require that orthogonal homogeneous unknown matrices can be found. To overcome this limitation, the QR24 method, which uses three different variations on the basis of a naïve least-squares solution of the equation system, was developed for simultaneous hand/eye calibration [22]. However, all the hand/eye calibration methods mentioned above are used for determining two unknown pose matrices. For a robotized navigated TMS system, there are three unknown pose matrices, which need to be solved. The least-squares method can be applied to get unknown data and to minimize the error sum of squares between the gotten data and the actual data. For this reason, it is not unreasonable to assume that the simultaneous calibration of two steps in a naïve least-squares solution of the equation system might result in improved accuracy and efficiency.

In order to simplify the two complex calibration steps and improve the accuracy, this report presents an innovative approach for simultaneous hand/eye and coil calibration. A matrix equation system $AX = ZB$ and $AY = ZC$ was acquired at different robot positions, as shown in Fig. 2D. A linear equation system was obtained based on the matrix equation system, and a least-squares solution was calculated for determining 'X', 'Y', and 'Z'. The feasibility and effectiveness of the method were demonstrated by comparing with two classical calibration methods: the QR24 method and the handheld coil calibration method.

Methods

Synchronous hand/eye and coil calibration

In this paper, we present a calibration method for robotized TMS with six DOF, which can be applied to any situation where there are three unknown pose matrices that need to be solved in the robotic system. As shown in Fig. 2A, B, typically, a coil is fixed to the end effector $\mathbf{E}$ of the robot $\mathbf{R}$. A marker $\mathbf{M}$ with four near-infrared reflector balls is attached to the top of the coil and, synchronously, a probe $\mathbf{P}$ is fixed at the bottom centre of the coil. The mechanical drafting of the probe is shown in Fig. 2C. The tracking device $\mathbf{T}$ is placed so that the coil can be adjusted to the centre of the view of $\mathbf{T}$. The design of this new approach to synchronous hand/eye and coil calibration is based on the following relationships:

$$
\begin{aligned}
{}^{R}\mathbf{T}_{E}\,{}^{E}\mathbf{T}_{M} &= {}^{R}\mathbf{T}_{T}\,{}^{T}\mathbf{T}_{M} \\
{}^{R}\mathbf{T}_{E}\,{}^{E}\mathbf{T}_{P} &= {}^{R}\mathbf{T}_{T}\,{}^{T}\mathbf{T}_{P}
\end{aligned}
\tag{1}
$$

where the matrix ${}^{R}\mathbf{T}_{E}'$ which is obtained by forward kinematic represents the transform from the robot's base to the robot's end effector, and the matrices ${}^{T}\mathbf{T}_{M}'$ and ${}^{T}\mathbf{T}_{P}'$ which are obtained directly from the tracking system represent the tracking data providing full six DOF of the marker and the probe, respectively. These three pose matrices are known parameters. The matrices ${}^{E}\mathbf{T}_{M}'$ and ${}^{E}\mathbf{T}_{P}'$ represent the transforms from the robot's end effector to the marker and the probe, respectively, and ${}^{R}\mathbf{T}_{T}'$ represents the transform from the robot's base to the tracking device. These three pose matrices are unknown parameters. To obtain the unknown parameters in Eq. 1, the end effector of the robot is moved to n different positions to obtain Eq. 2.

$$
\begin{aligned}
\left({}^{R}\mathbf{T}_{E}\right)_{i}{}^{E}\mathbf{T}_{M} &= {}^{R}\mathbf{T}_{T}\left({}^{T}\mathbf{T}_{M}\right)_{i} \\
\left({}^{R}\mathbf{T}_{E}\right)_{i}{}^{E}\mathbf{T}_{P} &= {}^{R}\mathbf{T}_{T}\left({}^{T}\mathbf{T}_{P}\right)_{i}
\end{aligned}
\quad i = 1,\ldots,n.
\tag{2}
$$

Generally, the position of the robot's end effector in Eq. 2 is chosen randomly in a sphere of radius r. The rotation angles of yaw, pitch, and roll for each position are also selected randomly between $\pm d$ degrees.

Let $A_i = \left({}^{R}\mathbf{T}_{E}\right)_i$, $X = {}^{E}\mathbf{T}_{M}$, $B_i = \left({}^{T}\mathbf{T}_{M}\right)_i$, $Y = {}^{E}\mathbf{T}_{P}$, $C_i = \left({}^{T}\mathbf{T}_{P}\right)_i$, and $Z = {}^{R}\mathbf{T}_{T}$. Then we obtain Eq. 3:

$$
\begin{aligned}
ZB_i - A_iX &= 0 \\
ZC_i - A_iY &= 0
\end{aligned}
\quad i = 1,\ldots,n.
\tag{3}
$$

Let

$$
w = [z_{1,1}, z_{2,1}, \ldots z_{3,4}, x_{1,1}, x_{2,1}, \ldots x_{3,4}, y_{1,1}, y_{2,1}, \ldots y_{3,4}]^{T} \in \mathbb{R}^{36}.
$$

Thus, a linear system of equations can be derived from Eq. 3:

$$
Dw = b,
\tag{4}
$$

where $D \in \mathbb{R}^{24n \times 36}$ and $b \in \mathbb{R}^{24n}$. To be more specific,

$$D = \begin{bmatrix} D_1 \\ D_2 \\ \vdots \\ D_n \end{bmatrix} \quad \text{and} \quad b = \begin{bmatrix} b_1 \\ b_2 \\ \vdots \\ b_n \end{bmatrix} \tag{5}$$

where

$$D_i = \begin{bmatrix} R[A_i^{-1}](B_i)_{1,1} & R[A_i^{-1}](B_i)_{2,1} & R[A_i^{-1}](B_i)_{3,1} & Zero_{3*3} & & \\ R[A_i^{-1}](B_i)_{1,2} & R[A_i^{-1}](B_i)_{2,2} & R[A_i^{-1}](B_i)_{3,2} & Zero_{3*3} & & \\ R[A_i^{-1}](B_i)_{1,3} & R[A_i^{-1}](B_i)_{2,3} & R[A_i^{-1}](B_i)_{3,3} & Zero_{3*3} & -Eye_{12} & Zero_{12} \\ R[A_i^{-1}](B_i)_{1,4} & R[A_i^{-1}](B_i)_{2,4} & R[A_i^{-1}](B_i)_{3,4} & R[A_i^{-1}] & & \\ R[A_i^{-1}](C_i)_{1,1} & R[A_i^{-1}](C_i)_{2,1} & R[A_i^{-1}](C_i)_{3,1} & Zero_{3*3} & & \\ R[A_i^{-1}](C_i)_{1,2} & R[A_i^{-1}](C_i)_{2,2} & R[A_i^{-1}](C_i)_{3,2} & Zero_{3*3} & & \\ R[A_i^{-1}](C_i)_{1,3} & R[A_i^{-1}](C_i)_{2,3} & R[A_i^{-1}](C_i)_{3,3} & Zero_{3*3} & Zero_{12} & -Eye_{12} \\ R[A_i^{-1}](C_i)_{1,4} & R[A_i^{-1}](C_i)_{2,4} & R[A_i^{-1}](C_i)_{3,4} & R[A_i^{-1}] & & \end{bmatrix}. \tag{6}$$

$$\text{and,} \quad b_i = \begin{bmatrix} Zero_{9*1} \\ -T[A_i^{-1}] \\ Zero_{9*1} \\ -T[A_i^{-1}] \end{bmatrix}$$

The general form of the homogenous transformation matrix 'A_i^{-1}' can be presented as

$$A_i^{-1} = \begin{bmatrix} R[A_i^{-1}] & T[A_i^{-1}] \\ 0 & 1 \end{bmatrix}$$

where $R[A_i^{-1}] \in \mathbb{R}^{3 \times 3}$ represents the rotation matrix of 'A_i^{-1}' and $T[A_i^{-1}] \in \mathbb{R}^3$ represents the translation vector of 'A_i^{-1}'. '$Zero_{m \times n}$' is a zero matrix of $m \times n$ size, and 'Eye_k' is the identity matrix of $k \times k$ size. A least-squares method with QR-factorization can be used to solve a linear system of equations such as Eq. 4 [22]. In the definition of linear algebra, QR-factorization is a decomposition of the matrix 'D' into a result '$D = QR$', where 'Q' is an orthogonal matrix and 'R' is an upper triangular matrix. All elements of unknown matrix 'X', 'Y' and 'Z' can be determined from the solution vector w of Eq. 4 based on the decomposed matrix 'Q' and 'R'. All elements of unknown matrices 'X', 'Y' and 'Z' can be determined from the solution vector w of Eq. 4. It is important to point out that this method gives the optimum solutions for 'X', 'Y' and 'Z'; thus, they minimize Eq. 7

$$\sum_{i=1}^{n} (\|A_i X - Z B_i\|_F + \|A_i Y - Z C_i\|_F) \tag{7}$$

where $\|*\|_F$ is the Frobenius norm. The method presented above is called the QR36 calibration method, because a linear system of equations with 36 unknown parameters can be solved by QR-factorization.

To use these matrices, orthonormalization is the final and most important step, because 'X', 'Y' and 'Z' are not orthogonally solved in the QR36 method. Without this step, the homogeneous coordinate transformations obtained from 'X', 'Y' and 'Z' would be incorrect. In this paper, the singular value decomposition (SVD) method was used for orthonormalization. For a position N obtained from a tracking device, for which we want to calculate the position in robotic coordinates, we compute ZN to obtain a non-orthogonal matrix; next, the SVD of ZN is calculated as $U\Sigma V^T = ZN$. Finally, the ortho-normalized ZN can be calculated with $(ZN)^{\perp} = UV^T$.

Calibration errors

With the calibration matrices 'X', 'Y' and 'Z' calculated by the QR36 method, we can obtain the calibration error using Eq. 1:

$$
\begin{aligned}
E_M &= \left(^E T_M\right)^{-1} \left(^R T_E\right)^{-1} {}^R T_T {}^T T_M \\
E_p &= \left(^E T_P\right)^{-1} \left(^R T_E\right)^{-1} {}^R T_T {}^T T_P
\end{aligned}
\tag{8}
$$

where 'E_M' is defined as the calibration error matrix of the marker, and 'E_p' is defined as the calibration error matrix of the probe. To obtain the calibration quality, we define the translational error and rotational error based on Eqs. 9–13. Let $O_M = E_M^{\perp}$ and $O_p = E_p^{\perp}$. Thus, the translational error of the system for QR36 method is defined as

$$
\begin{aligned}
e_t &= e_{trans}[O_p] + e_{trans}[O_M] \\
e_{trans}[O_p] &= \sqrt{O_{p(1,4)}{}^2 + O_{p(2,4)}{}^2 + O_{p(3,4)}{}^2} \qquad for \quad QR36 \\
e_{trans}[O_M] &= \sqrt{O_{M(1,4)}{}^2 + O_{M(2,4)}{}^2 + O_{M(3,4)}{}^2}
\end{aligned}
\tag{9}
$$

where e_t is defined as the total translational error of the system, $e_{trans}[O_p]$ is defined as the translational error of the probe, $e_{trans}[O_M]$ is defined as the translational error of the marker.

Let (k_M, θ_M) and (k_p, θ_p) are the axis-angle representations of $R(O_M)$ and $R(O_p)$, respectively. Thus, the total rotational error of the system for QR36 method is defined as

$$
\theta_t = |\theta_p| + |\theta_M| \quad for \quad QR36 .
\tag{10}
$$

where θ_t is defined as the total rotational error of the system, $|\theta_p|$ is defined as the rotational error of the probe, $|\theta_M|$ is defined as the rotational error of the marker.

Finally, if the marker and probe are calibrated separately using the QR24 method [22], then the calibration matrix Z will not be identical for both optimal calibration procedures. Thus, another translational error e_z is defined for the system:

$$
e_z = \sqrt{\left(Z_{p(1,4)} - Z_{M(1,4)}\right)^2 + \left(Z_{p(2,4)} - Z_{M(2,4)}\right)^2 + \left(Z_{p(3,4)} - Z_{M(3,4)}\right)^2}
\tag{11}
$$

where 'Z_p' and 'Z_M' are the calibration results of the probe and the marker calculated using the QR24 method. Thus, the translational error of the system for QR24 method is defined as

$$e_t = e_{trans}[O_p] + e_{trans}[O_M] + e_z$$
$$e_{trans}[O_p] = \sqrt{O_{p(1,4)}^2 + O_{p(2,4)}^2 + O_{p(3,4)}^2} \quad for \ QR24 \ .$$
$$e_{trans}[O_M] = \sqrt{O_{M(1,4)}^2 + O_{M(2,4)}^2 + O_{M(3,4)}^2}$$

$$(12)$$

where e_t is defined as the total translational error of the system, $e_{trans}[O_p]$ is defined as the translational error of the probe, $e_{trans}[O_M]$ is defined as the translational error of the marker.

Let (k_z, θ_z) be the axis-angle representation of $R(Z_p - Z_M)$; thus, the rotational error is defined as $|\theta_z|$. Thus, the rotational error of the system for QR24 method is defined as

$$\theta_t = |\theta_p| + |\theta_M| + |\theta_z| \quad for \quad QR24 \ .$$

$$(13)$$

where θ_t is defined as the total rotational error of the system, $|\theta_p|$ is defined as the rotational error of the probe, $|\theta_M|$ is defined as the rotational error of the marker.

Data acquisition

Simulated and optical tracking data were obtained to estimate the accuracy and robustness of the QR36 method. To compare the quality of our method with the QR24 and handheld methods, the QR24 method and the QR36 method were performed on the same dataset. The results are presented in the following section.

Simulated data

A set of simulated data was generated to test the proposed method [22]. Random orthogonal matrices X, Y, and Z with same definition in Eq. 3 were created first. To generate realistic values, the elements of T[X] and T[Y] were chosen randomly from [−50, 50] mm, and the elements of T[Z] were chosen randomly from ±[500, 2000] mm. Subsequently, five hundred totally random orthogonal robot positions matrix A_i were created and used to calculate the corresponding tracking matrices B_i and C_i. To get the simulated data of an optical tracking device, a marker and a probe with four markers was defined as

$$marker = \begin{bmatrix} 0 & 50 & 50 & 0 \\ 0 & 0 & 50 & 50 \\ 0 & 0 & 0 & 0 \\ 1 & 1 & 1 & 1 \end{bmatrix}, \quad probe = \begin{bmatrix} 0 & 25 & 25 & 0 \\ 0 & 0 & 25 & 25 \\ 0 & 0 & 0 & 0 \\ 1 & 1 & 1 & 1 \end{bmatrix}.$$

In order to distort the tracking matrices $\tilde{B}_i$ and $\tilde{C}_i$, the marker and the probe were moved to the pose described by B_i and C_i, and according to Horn's algorithm, the matrix was computed without scaling [26], that is,

$$\tilde{B}_i = horn\left(marker, \ B_i * marker + \sum_i \right)$$

$$\tilde{C}_i = horn\left(probe, \ C_i * probe + \sum_i \right)$$

Here,

$$\sum_i = \begin{bmatrix} s_{i,1,1} & s_{i,1,2} & s_{i,1,3} & s_{i,1,4} \\ s_{i,2,1} & s_{i,2,2} & s_{i,2,3} & s_{i,2,4} \\ s_{i,3,1} & s_{i,3,2} & s_{i,3,3} & s_{i,3,4} \\ 0 & 0 & 0 & 0 \end{bmatrix}$$

where $s_{i,j,k}$ was obtained from the Gaussian distribution with standard deviation equal to 0.05 mm.

A different method was used to distort the robot matrices: the rotational matrix was distorted by multiplication with a rotation matrix around a random axis a_i. The rotation angles θ_i was drawn from the Gaussian process with standard deviation equal to 0.05 mm. Gaussian noise was used to distort the translational vector:

$$\tilde{A}_i = \begin{bmatrix} R[A_i] * rot_{a_i}(\theta_i) & T[A_i] + \sum_i \\ 0 & 1 \end{bmatrix}$$

where

$$\sum_i = \begin{bmatrix} s_{i,1} & s_{i,2} & s_{i,3} \end{bmatrix}^T$$

$s_{i,k}$ was obtained from the Gaussian distribution with standard deviation equal to 0.05 mm.

Data from optical tracking device

As shown in Fig. 2B, C, an optical tracking device (Polaris Spectra, Northern Digital Inc., Waterloo, Ontario, Canada) was mounted on tripods, and a coil model was fixed at the end effector of a 6-axis robot (VS060, Denso Inc., Aichi, Japan). The marker was attached to the top of the coil, and the probe was fixed at the bottom centre of the coil. Then, the position of the end effector and the tracking device were adjusted so that the positions of the marker and probe were near the centre of the tracking device's work volume (that is, $|x| < 100$ mm, $|Y| < 100$ mm, and $Z = -1500 \pm 100$ mm for both the marker and the probe), and the position of the end effector was defined as first position P_0. Next, the end effector was moved to 500 random positions around the first position P_0. The positions were confirmed by

$$P_i = P_0 * T(t) * R_x(\theta_1) * R_y(\theta_2) * R_z(\theta_3), \tag{14}$$

where $R_{x, y \text{ and } z}(\theta)$ is a rotational matrix around the x, y and z-axis, in which θ was randomly selected from $-10°$ to $10°$, and T is a translation matrix with translation vector t that was randomly selected in the range from -100 to 100 mm. 100 measurements (A, B, C) were averaged to reduce the errors from the robot and optical tracking device.

Results

Evaluation and implementation

The algorithms were evaluated on a standard personal computer (Intel core E7500 CPU with 4 GB of RAM, running 64-bit Windows 10 OS). The algorithms were all implemented in MATLAB 2010a. The performance period was less than 0.1 s for the QR24 and QR36 algorithms.

Calibration errors

The QR36 algorithm was performed using poses $P_1,...,P_n$ for $n = 5,...,250$. The rotational and translational errors were computed for each remaining testing pose $P_{251},...,P_{500}$ using Eqs. 9 and 10, and the average of all 250 testing poses was determined. The QR24 algorithm was calculated on the marker and the probe separately using poses $P_1,...,P_n$ for $n = 5,...,250$. The rotational and translational errors of the marker and the probe were computed for each remaining testing pose $P_{251},...,P_{500}$ using Eqs. 12 and 13, and the average of all 250 testing poses was determined. Then, error E_z and $|\theta_z|$ were determined for each remaining testing pose $P_{251},...,P_{500}$, and the average of all 250 testing poses was determined.

Simulated results

Figure 3 presents the average errors for the simulated data using the QR24 method (top) and the QR36 method (bottom).

Clearly, using more than 20 poses hardly influences the translational and rotational errors of the QR36 method. However, using the QR24 method to calibrate the marker and probe separately could create two different calibration matrices Z. This can create a new calibration error e_z, which significantly influences the translational error of the QR24 algorithm. In the simulated data, the resulting error came very close to the error obtained using the correct matrices, and remained within 1%. The minimum total translational error decreased 59.54% for the simulated data. The minimum total

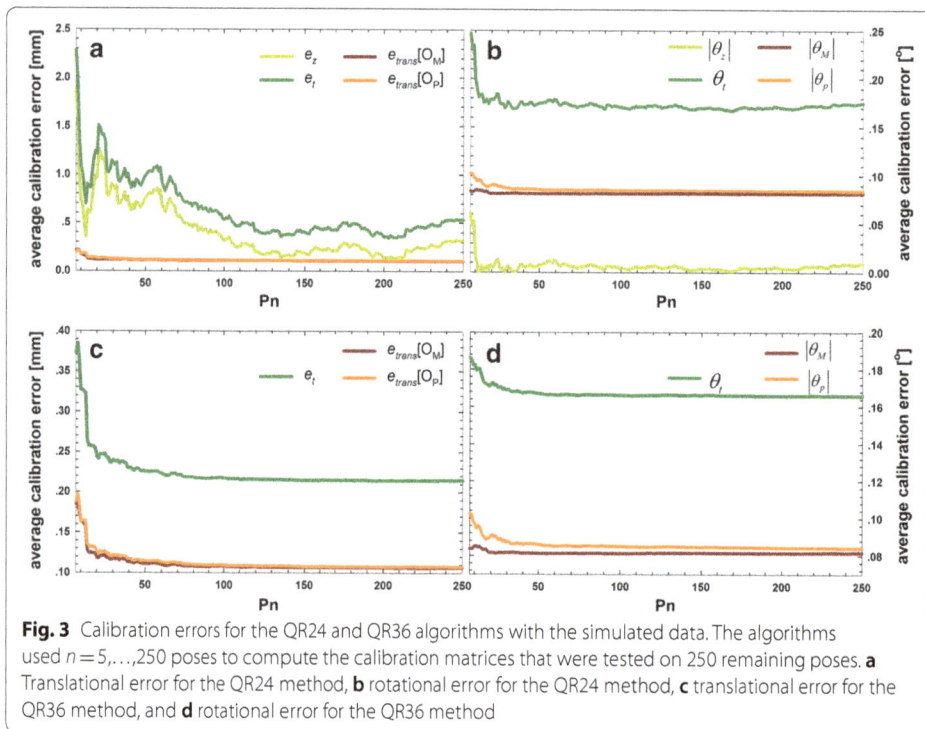

Fig. 3 Calibration errors for the QR24 and QR36 algorithms with the simulated data. The algorithms used $n = 5,...,250$ poses to compute the calibration matrices that were tested on 250 remaining poses. **a** Translational error for the QR24 method, **b** rotational error for the QR24 method, **c** translational error for the QR36 method, and **d** rotational error for the QR36 method

translational error e_t of the QR24 method was 0.3419 mm, with an e_z of 0.1267 mm, while the minimum total error of the QR36 method was 0.2143 mm. In other words, without error e_z, the minimum total errors of the QR24 method and the QR36 method are approximately equal; the same result was obtained from the data for the 25th, median, and 75th errors, but not the maximum total error. Insufficient sample data produced an extremely large error e_z; therefore, our method is more useful for a small data set. Overall, both convergence speed and precision of the translational error were improved for the QR36 method compared with the QR24 method. However, although the QR36 method outperformed the QR24 method regarding the rotational error of θ_t, the effect of θ_z was very slight. The θ_z of the QR24 method only had a 0.78% influence on the minimum total rotational error.

Additionally, Table 1 presents the statistics of the errors for the simulated data shown in Fig. 3. The minimum, 25th, median, 75th, and maximum total translational error e_t and total rotational error θ_t and corresponding parameters are summarized in Table 1.

Experimental results

Figure 4 presents the average errors for the optical tracking device. We obtained similar results from the experimental data. The translational errors of the QR24 and QR36 methods were hardly changed by using more than 30 poses. In the experimental data,

Table 1 Error statistics of the calibration algorithms using matrices on the simulated data

Alg	Para	Min	25th	Med	75th	Max
QR24	Translational error (mm)					
	$e_{trans}[O_p]$	0.1071	0.1075	0.1074	0.1088	0.5134
	$e_{trans}[O_M]$	0.1081	0.1085	0.1081	0.1138	0.7211
	e_z	0.1267	0.2059	0.2735	0.5078	5.9305
	e_t	0.3419	0.4219	0.4889	0.7305	7.1649
	P_n	203	147	182	73	6
	Rotational error (°)					
	θ_p	0.0814	0.0814	0.0812	0.0818	0.0839
	θ_m	0.0847	0.0847	0.0851	0.0842	0.1024
	θ_z	0.0013	0.0043	0.0056	0.0079	0.0609
	θ_t	0.1675	0.1704	0.1719	0.1739	0.2472
	P_n	168	173	97	245	7
QR36	Translational error (mm)					
	$e_{trans}[O_p]$	0.1066	0.1068	0.1071	0.1088	0.3004
	$e_{trans}[O_M]$	0.1076	0.1079	0.1084	0.1118	0.3203
	e_z	0	0	0	0	0
	e_t	0.2143	0.2148	0.2155	0.2206	0.6207
	P_n	202	158	133	64	6
	Rotational error (°)					
	θ_p	0.0815	0.0812	0.0815	0.0813	0.0861
	θ_m	0.0844	0.0848	0.0848	0.0853	0.1064
	θ_z	0	0	0	0	0
	θ_t	0.1659	0.1661	0.1663	0.1665	0.1925
	P_n	203	107	142	69	6

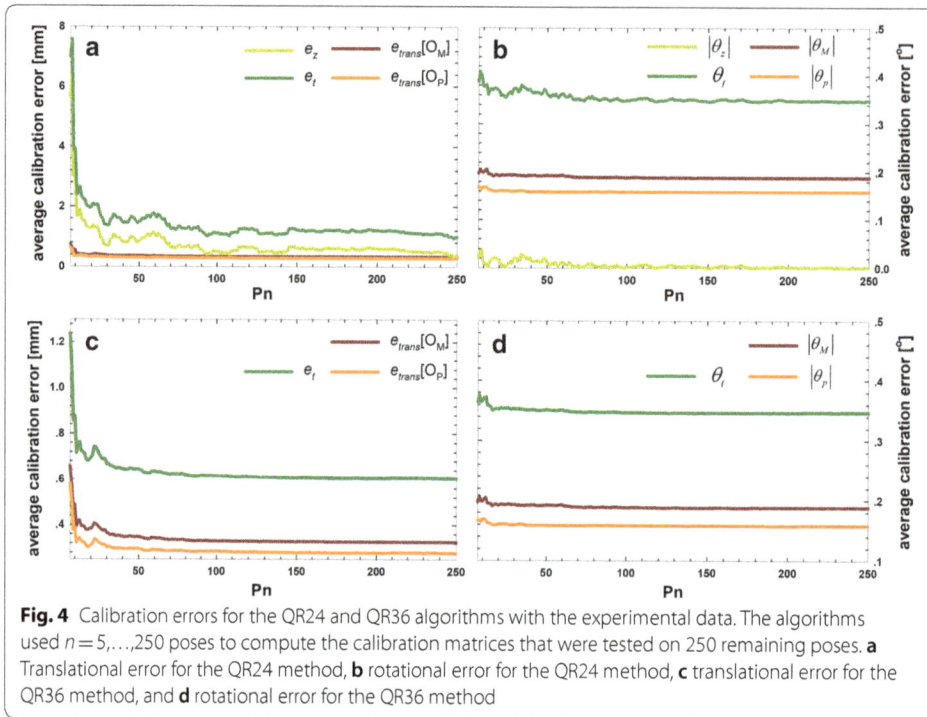

Fig. 4 Calibration errors for the QR24 and QR36 algorithms with the experimental data. The algorithms used $n=5,\ldots,250$ poses to compute the calibration matrices that were tested on 250 remaining poses. **a** Translational error for the QR24 method, **b** rotational error for the QR24 method, **c** translational error for the QR36 method, and **d** rotational error for the QR36 method

the resulting error became much larger than the error obtained from simulation. We also found that the error e_z had a greater influence on the translational error of the experimental data. The minimum total translational errors were decreased by 54.93%, as 0.9353 mm and 0.6037 mm for QR24 and QR36, respectively. e_z had a minimum total translational error of 0.3301 mm. The effect of θ_z on the rotational error of θ_t was very small, according to the experimental data. The maximum total translational errors were 8.3719 and 1.2384 mm for QR24 ($P_n=5$) and QR36 ($P_n=6$), respectively. e_z had a translational error of 5.7679 mm, indicating that fewer tracking data will lead to a larger calibration error e_z.

Additionally, Table 2 presents the statistics of the errors for the experimental data shown in Fig. 4. The minimum, 25th, median, 75th, and maximum total translational error e_t and total rotational error θ_t and corresponding parameters are summarized in Table 2.

Convergence properties. To acquire the convergence properties of the two algorithms, the simulated and experimental data were assessed again. It is clear from Figs. 3 and 4 that the QR36 method converged faster than the QR24 method. Additionally, it was possible to determine whether and how fast the algorithms converged to set values, which were presented in this paragraph. The results are shown in Table 3. Again, we can see that the convergence speed is improved 16- to 27-fold for the QR36 algorithms, stabilizing at a translational error of less than 0.4 mm and 0.3 mm at $P_n=7$ and 14, respectively. For the QR24 method, translational errors of <0.4 and 0.3 mm were found at $P_n=118$ and none, respectively. The translational errors of the experimental data followed a similar pattern: the convergence properties of the QR36 method were much better than

Table 2 Error statistics of the calibration algorithms using matrices on the experimental data

Alg	Para	Min	25th	Med	75th	Max
QR24	Translational error (mm)					
	$e_{trans}[O_p]$	0.3263	0.3296	0.3272	0.3331	1.4110
	$e_{trans}[O_M]$	0.2787	0.2845	0.2787	0.2854	1.1929
	e_z	0.3301	0.5024	0.5935	0.6995	5.7679
	e_t	0.9353	1.1166	1.1995	1.3181	8.3719
	P_n	247	109	200	81	5
	Rotational error (°)					
	θ_p	0.1882	0.1889	0.1886	0.1883	0.2157
	θ_m	0.1584	0.1590	0.1583	0.1583	0.1825
	θ_z	0.0005	0.0003	0.0030	0.0069	0.0832
	θ_t	0.3466	0.3482	0.3500	0.3536	0.4815
	P_n	176	88	190	127	5
QR36	Translational error (mm)					
	$e_{trans}[O_p]$	0.3256	0.3261	0.3277	0.3353	0.6580
	$e_{trans}[O_M]$	0.2781	0.2797	0.2807	0.2873	0.5803
	e_z	0	0	0	0	0
	e_t	0.6037	0.6059	0.6084	0.6226	1.2384
	P_n	250	227	132	73	6
	Rotational error (°)					
	θ_p	0.1880	0.1883	0.1885	0.1893	0.2074
	θ_m	0.1583	0.1584	0.1584	0.1586	0.1674
	θ_z	0	0	0	0	0
	θ_t	0.3464	0.3467	0.3470	0.3480	0.3749
	P_n	141	128	217	68	7

Table 3 Convergence properties of the calibration algorithms using matrices on the simulated and experimental data

Data group	Algorithm	Error type	Total errors	P_n
Simulation	QR24	Translation	< 0.4 mm	118
			< 0.3 mm	None
		Rotation	< 0.2°	10
			< 0.17°	28
	QR36	Translation	< 0.4 mm	7
			< 0.3 mm	14
		Rotation	< 0.2°	6
			< 0.17°	22
Experiment	QR24	Translation	< 1 mm	247
			< 0.7 mm	None
		Rotation	< 0.4°	10
			< 0.35°	66
	QR36	Translation	< 1 mm	9
			< 0.7 mm	27
		Rotation	< 0.4°	6
			< 0.35°	57

those of the QR24 method. However, the rotational errors showed a different picture: the accuracy of less than some set values is achieved on both algorithms similarly.

Workspace sizes

Workspace sizes have a substantial influence on the calibration error in many medical robotic applications. To demonstrate the quality of our method in this regard, additional calibration experiments with the Polaris Spectra system and increased workspace sizes were performed. Five different workspace sizes were chosen for the experiments (100 mm, 10°; 200 mm, 10°; 300 mm, 10°; 300 mm, 15°; and 300 mm, 20°). In all cases, 500 samples were successfully recorded. The minimum errors for different workspace sizes are presented in Fig. 5. Interestingly, the total translational error of the QR36 method increased with workspace size. However, the total translational error of the QR24 method was influenced by error e_z. e_z did not increase linearly with increasing workspace size; however, it did create a smaller calibration error for a large workspace ($e_z = 1.04$ mm for a workspace of 300 mm, 20°). The total rotational error of the QR24 and QR36 methods all increased with increasing workspace size, but error e_z had a small influence.

Handheld coil calibration

To compare the quality of our method with the handheld coil calibration method, two operators with similar experience of TMS participated in this experiment. The QR36 method and handheld calibration method were performed by the two operators. The translational error $e_{trans}[O_p]$ and rotational error θ_p of the probe, which were defined in Eqs. 9 and 10, were estimated to evaluate the calibration accuracy.

A set of five additional data acquisition experiments using the robotic TMS system in a workspace of 300 mm (10°) were performed by each operator. For the experimental procedure, refer to the experimental data acquisition paragraph in the Methods section. For each experiment, the probe was remounted on the coil by the operator, and 500 measurements (A_i, B_i, $C_{i\ i=1,...,500}$) were collected, with the end effector moving to 500 random positions around the first position P_0. Next, the robot was moved to the first position

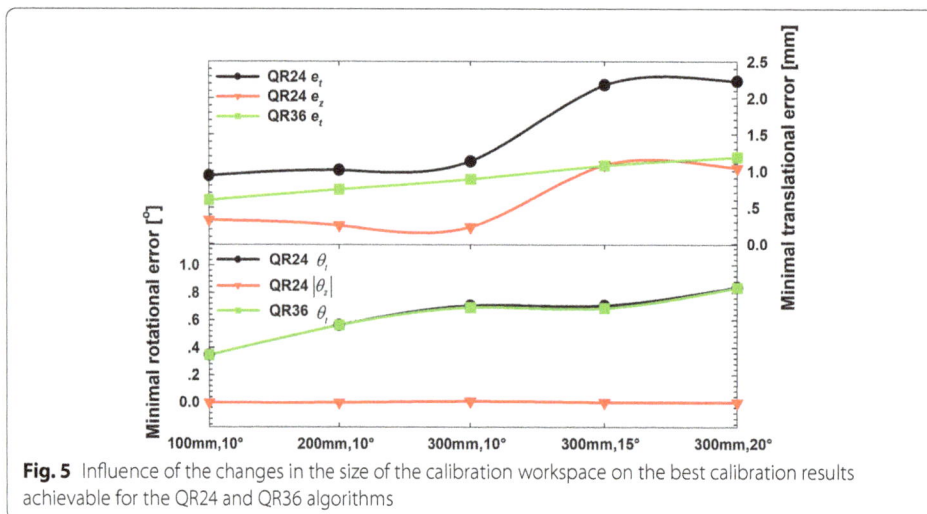

Fig. 5 Influence of the changes in the size of the calibration workspace on the best calibration results achievable for the QR24 and QR36 algorithms

P_0, and the probe was unloaded from the coil and held at the bottom centre labels of the coil by the operator. An additional measurement (B_a, C_a) was acquired with an optical tracking device. The acquired 500 measurements and the additional measurement were defined as the dataset of each experiment.

The QR36 algorithm was performed using poses $P_1,...,P_n$ for n = 5,...,250 of each experiment. The rotational and translational errors of the probe were computed for each remaining testing pose $P_{251},...,P_{500}$ using Eqs. 9 and 10, and the average of all 250 testing poses was determined. Then, the optimal translational and rotational calibration errors (MIN ($e_{trans}[O_p]$)$_n$ and MIN(θ_p)$_n$ for n = 5,...,250) were determined for each experiment. Finally, the mean value and standard deviation of the optimal calibration errors for five experiments per operator were determined. The results are presented in Fig. 6.

The traditional separate two-step calibration with QR24 and the handheld method were carried out for comparison with the QR36 method. Based on the dataset of marker (A_i, C_i $_{i=1,...,500}$), the optimal calibration result (X, Z) was determined by the QR24 method with the minimal translational error. Subsequently, Y was calculated using Eq. 13 with the additional position pair (B_a, C_a) [24]. For each experiment, the translational error $e_{trans}[O_p]$ and rotational error θ_p of the probe were computed for each testing pose (B_i, C_i $_{i=251,...,500}$) with the calibration result (Y, Z), and the average of all 250 testing poses was determined as the handheld calibration error of the experiment. Finally, the mean values and standard deviations of calibration errors of five experiments for each operator were determined. The results are presented in Fig. 6. All the symbols (A, B, C, X, Y, Z) in this section are as defined in Eq. 3.

$$Y = XB_a^{-1}C_a. \tag{15}$$

As shown in Fig. 6, both the mean values and the standard deviations of the rotational and translational errors significantly decreased when the QR36 method was used; the total translational error decreased three- to fourfold, and the rotational error decreased six- to eightfold. For the two operators, the minimum total translational errors for QR36 method were decreased to 0.7931 and 0.8361 mm, respectively, as compared with 2.4997 and 2.5354 mm for handheld method. The minimum total rotational errors were decreased to 0.5509° and 0.5624° using QR36 method,

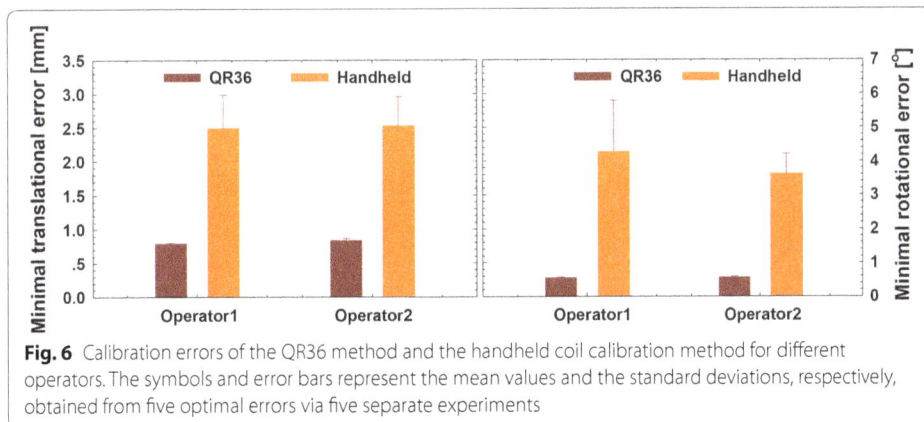

Fig. 6 Calibration errors of the QR36 method and the handheld coil calibration method for different operators. The symbols and error bars represent the mean values and the standard deviations, respectively, obtained from five optimal errors via five separate experiments

as compared with 4.2808° and 3.6202° using handheld method. The rotational error showed greater improvement, probably because there was still a positioning label on the coil for the position probe, whereas there is no method for reducing the perspective error of the operator. Moreover, because the probe was fixed on the coil mechanically and an optimization algorithm was used, no significant differences in either the translational or the rotational errors of the QR36 method were found for different operators. Finally, the QR36 method showed better robustness to choice of operator than the handheld coil calibration method.

Discussion

An innovative approach for synchronous hand/eye and coil calibration has been presented and evaluated. A least-squares method that has been widely used in medical imaging and robotics was used to estimate the optimal calibration matrix [32, 33]. The approach has been validated with synthetic data and experimental data from an optical tracking device. The calibration effect of our method was compared with that of traditional methods such as QR24 and the handheld coil calibration method.

These results show that QR36 calibration method is advisable for use in the robotized TMS system. We have shown that both the QR24 method and our new QR36 method perform very well, with errors below 1 mm and 0.2°, using both simulated and experimental data. This shows that the two methods can both be used for typical calibration. However, whereas more than 200 different robot positions were required to achieve a calibration error below 1 mm for the QR24 method, we found that fewer than 20 positions were typically required by our method. This means that the QR36 method is especially suitable to handle cases where fewer tracking data are available.

Traditional hand-eye calibration method proposed by Shiu and Ahmad [23, 24] and Tsai and Lenz [25, 26] calculates the rotational and translational parts of the unknown matrices separately using matrix algebra. Li and Betsis applied a geometric approach and least-squares solution for hand-eye calibration [34]. Dual quaternion approach was also used for hand-eye calibration [35]. All those methods expect that orthogonal homogeneous matrices can be found. Moreover, a robust real-time hand-eye calibration method was proposed by Lars and Floris present [17]. A marker is attached to the robot's third link for real-time hand-eye calibration. However, this method is not as precise as the QR24 algorithm. The total translational errors were 0.88 mm and 1.36 mm for QR24 method and real-time calibration method, respectively [17]. A robot system will not be calibrated perfectly, so orthogonality is not necessary. It is accepted or requested to permit non-orthogonality of the matrices in our calibration method. Our results demonstrated that the calibration method used in this study is more accurate than the classical hand-eye calibration approaches [23, 25]. In terms of translational accuracy, our method also outperforms the QR24 method. Therefore, the calibration method proposed in this study is more suitable for robotized transcranial magnetic stimulation.

We should point out that the maximum optimal translation errors for five handheld experiments per operator can reach 3 mm, as shown in Fig. 6a. This result is acceptable for a handheld navigated TMS experiment. During the robot-assisted TMS stimulation, the head motion is tracked using the marker on the subject's head and compensated by the robot [18]. But, without robotic assistance and head motion compensation, the relative motion between the subject's head and the handheld stimulation coil is greater than 3 mm during handheld TMS experiments [30]. Moreover, the calibration error of registration of the subject's MRI and optical tracking device is also controlled at around 3 mm for handheld navigated TMS experiments [36]. However, for a robotic TMS system, which is designed for high-precision stimulation, the handheld coil calibration method is not suitable [37–39].

It is important to note that the new method has three limitations. It can only be applied when:

- It is accepted or requested to permit non-orthogonality of the matrices in our calibration method;
- The calibrated robotic TMS system is used in the same space where calibration was carried out;
- There are three unknown matrices that need to be solved in the robotic system.

Conclusion

We have developed a new one-step calibration method to acquire three pose relationships from a navigated robotic system with a least-squares approach. The new method can significantly improve the accuracy of the robotic TMS system. Besides, the convergence speed is improved for the new method, which means that our method is particularly suited to handle fewer tracking data. The capability of the new method has been demonstrated for synchronous calibration and determination of the pose relationship of marker/end effector, probe/end effector, and robot/world for a robotic TMS system, which can be used to perform precision TMS experiment. Finally, for many robotic applications, where three unknown matrices need to be solved, the method presented in this paper provides an alternative solution to the classical approaches. In the future, it will be interesting to quantitatively compare the stimulation effect of robotic and manual techniques. Investigation on patients will then be pursued to evaluate the medical benefits of robotic TMS system.

Authors' contributions
HW and JNJ conceived and designed the study. YL and XW performed the experiments. HW wrote the paper. HW and TY reviewed and edited the manuscript. All authors read and approved the final manuscript.

Author details
[1] Institute of Biomedical Engineering, Chinese Academy of Medical Science & Peking Union Medical College, Tianjin 300192, China. [2] Neuroscience Center, Chinese Academy of Medical Science & Peking Union Medical College, Beijing 100730, China.

Acknowledgements
H.W. acknowledges the grant support from PUMC Youth Fund and the Fundamental Research Funds for the Central Universities (2017320025). T.Y. acknowledges grant support from CAMS Initiative for Innovative Medicine (CAMS-I2M 2016-I2M-1004) and the Natural Science Foundation of China (No. 81772003).

Competing interests
The authors declare that they have no competing interests.

Funding
CAMS Initiative for Innovative Medicine (CAMS-I2M 2016-I2M-1004). Natural Science Foundation of China NSFC (No. 81772003). PUMC Youth Fund and the Fundamental Research Funds for the Central Universities (2017320025).

References

1. Barker AT, Jalinous R, Freeston IL. Non-invasive magnetic stimulation of human motor cortex. Lancet. 1985;1:1106.
2. Hallett M. Transcranial magnetic stimulation. Nature. 2000;406:147–50.
3. Hallett M. Transcranial magnetic stimulation: a primer. Neuron. 2007;55:187–99.
4. Avenanti A, Bueti D, Galati G, et al. Transcranial magnetic stimulation highlights the sensorimotor side of empathy for pain. Nat Neurosci. 2005;8:955–60.
5. Fitzgerald PB, Brown TL, Marston NAU, et al. Transcranial magnetic stimulation in the treatment of depression: a double-blind, placebo-controlled trial. Am J Psychiatry. 2003;160:835.
6. Hirayama A, Saitoh Y, Kishima H, et al. Reduction of intractable deafferentation pain by navigation-guided repetitive transcranial magnetic stimulation of the primary motor cortex. Pain. 2006;122:22–7.
7. Hoffman RE, Cavus I. Slow transcranial magnetic stimulation, long-term depotentiation, and brain hyperexcitability disorders. Am J Psychiatry. 2002;159:1093.
8. Mcnamara B, Ray JL, Arthurs OJ, et al. Transcranial magnetic stimulation for depression and other psychiatric disorders. Psychol Med. 2001;31:1141–6.
9. Pascual-Leone A, Rubio B, Pallardó F, et al. Rapid-rate transcranial magnetic stimulation of left dorsolateral prefrontal cortex in drug-resistant depression. Lancet. 1996;348:233.
10. Herwig U, Padberg F, Unger J, et al. Transcranial magnetic stimulation in therapy studies: examination of the reliability of "standard" coil positioning by neuronavigation. Biol Psychiatry. 2001;50:58–61.
11. Lisanby SH, Kinnunen LH, Crupain MJ. Applications of TMS to therapy in psychiatry. J Clin Neurophysiol. 2002;19:344–60.
12. Sparing R, Buelte D, Meister IG, et al. Transcranial magnetic stimulation and the challenge of coil placement: a comparison of conventional and stereotaxic neuronavigational strategies. Hum Brain Mapp. 2008;29:82.
13. Herwig U, Schönfeldtlecuona C, Wunderlich AP, et al. The navigation of transcranial magnetic stimulation. Psychiatry Res Neuroimaging. 2001;108:123–31.
14. Neggers SFW, Langerak TR, Schutter DJLG, et al. A stereotactic method for image-guided transcranial magnetic stimulation validated with fMRI and motor-evoked potentials. Neuroimage. 2004;21:1805.
15. Kantelhardt SR, Fadini T, Finke M, et al. Robot-assisted image-guided transcranial magnetic stimulation for somatotopic mapping of the motor cortex: a clinical pilot study. Acta Neurochir (Wien). 2010;152:333–43.
16. Lancaster JL, Narayana S, Wenzel D, et al. Evaluation of an image-guided, robotically positioned transcranial magnetic stimulation system. Hum Brain Mapp. 2004;22:329.
17. Lars R, Floris E, Alexer S, et al. Robust real-time robot-world calibration for robotized transcranial magnetic stimulation. Int J Med Robot. 2011;7:414.
18. Zorn L, Renaud P, Bayle B, et al. Design and evaluation of a robotic system for transcranial magnetic stimulation. IEEE Trans Biomed Eng. 2012;59:805.
19. Yi X, Bicker R. Design of a robotic transcranial magnetic stimulation system. Robotics automation and mechatronics. IEEE, 2018. p. 78–83.
20. Hartmann A, Kalis R, Matthäus L, et al. P 226. Robot guided positioning of a magnetic coil for rTMS over the cortex in patients with cerebral lesions. Clin Neurophysiol. 2013;124(10):e173–4.
21. Quesada C, et al. Robot-guided neuronavigated repetitive transcranial magnetic stimulation (rTMS) in central neuropathic pain. An update of long-term follow-up. Arch Phys Med Rehabilit. 2018. https://doi.org/10.1016/j.apmr.2018.04.013.
22. Floris E, Lars R, Lars M, et al. Non-orthogonal tool/end effector and robot/world calibration. Int J Med Robot. 2012;8:407–20.
23. Shiu Y, Ahmad S. Finding the mounting position of a sensor by solving a homogeneous transform equation of the form AX = XB. In: IEEE international conference on robotics and automation proceedings. 1987. p. 1666–71.
24. Shiu YC, Ahmad S. Calibration of wrist-mounted robotic sensors by solving homogeneous transform equations of the form AX = XB. IEEE Trans Robot Autom. 1989;5:16–29.
25. Tsai RY, Lenz RK. A new technique for fully autonomous and efficient 3D robotics hand/eye calibration. In: International symposium on robotics research. 1988. p. 287–97.
26. Tsai RY, Lenz RK. A new technique for fully autonomous and efficient 3D robotics hand/eye calibration. IEEE Trans Robot Autom. 1989;5:345–58.
27. Matthaus L. A robotic assistance system for transcranial magnetic stimulation and its application to motor cortex mapping. Ph.D. thesis, Universität zu Lübeck; 2008.
28. Richter L. Robotized transcranial magnetic stimulation. Dodrecht: Springer; 2013. p. 1794.
29. Horn BKP. Closed-form solution of absolute orientation using unit quaternions. J Opt Soc Am A. 1987;4(4):629–42.
30. Richter L, Trillenberg P, Schweikard A, et al. Stimulus intensity for hand held and robotic transcranial magnetic stimulation. Brain Stimul. 2013;6(3):315–21.
31. Wang CC. Extrinsic calibration of a vision sensor mounted on a robot. IEEE Trans Robot Autom. 2002;8(2):161–75.
32. Lu H, Li X, Hsiao IT, et al. Analytical noise treatment for low-dose CT projection data by penalized weighted least-square smoothing in the K-L domain. Chin J Med Phys. 2002;4682(4):146–52.
33. Wang T, Nakamoto K, Zhang H, et al. Reweighted anisotropic total variation minimization for limited-angle CT reconstruction. IEEE Trans Nuclear Sci. 2017;64(10):2742–60.
34. Li M, Betsis D. Head-eye calibration. In: Proceedings of the 5th international conference on computer vision (ICCV'95). 1995. p. 40–5.
35. Daniilidis K. Hand-eye calibration using dual quaternions. Int J Robot Res. 1999;18(3):286–98.
36. Herwig U, Schönfeldtlecuona C, Wunderlich AP, et al. The navigation of transcranial magnetic stimulation. Psychiatry Res Neuroimaging. 2001;108(2):123–31.
37. Kantelhardt SR, Fadini T, Finke M, et al. Robot-assisted image-guided transcranial magnetic stimulation for somatotopic mapping of the motor cortex: a clinical pilot study. Acta Neurochir. 2010;152(2):333.
38. Wan Z, Wan N. Force-controlled transcranial magnetic stimulation (TMS) robotic system. In: 2010 8th world congress on intelligent control and automation (WCICA). IEEE. 2012. p. 3777–81.

6

Solving the general inter-ring distances optimization problem for concentric ring electrodes to improve Laplacian estimation

Oleksandr Makeyev[*] ⒾⒹ

*Correspondence:
omakeyev@dinecollege.edu
Department of Mathematics,
Diné College, 1 Circle Dr,
Tsaile, AZ 86556, USA

Abstract

Background: Superiority of noninvasive tripolar concentric ring electrodes over conventional disc electrodes in accuracy of surface Laplacian estimation has been demonstrated in a range of electrophysiological measurement applications. Recently, a general approach to Laplacian estimation for an $(n+1)$-polar electrode with n rings using the $(4n+1)$-point method has been proposed and used to introduce novel multipolar and variable inter-ring distances electrode configurations. While only linearly increasing and linearly decreasing inter-ring distances have been considered previously, this paper defines and solves the general inter-ring distances optimization problem for the $(4n+1)$-point method.

Results: General inter-ring distances optimization problem is solved for tripolar ($n=2$) and quadripolar ($n=3$) concentric ring electrode configurations through minimizing the truncation error of Laplacian estimation. For tripolar configuration with middle ring radius ar and outer ring radius r the optimal range of values for a was determined to be $0 < a \leq 0.22$ while for quadripolar configuration with an additional middle ring with radius βr the optimal range of values for a and β was determined by inequalities $0 < a < \beta < 1$ and $a\beta \leq 0.21$. Finite element method modeling and full factorial analysis of variance were used to confirm statistical significance of Laplacian estimation accuracy improvement due to optimization of inter-ring distances ($p < 0.0001$).

Conclusions: Obtained results suggest the potential of using optimization of inter-ring distances to improve the accuracy of surface Laplacian estimation via concentric ring electrodes. Identical approach can be applied to solving corresponding inter-ring distances optimization problems for electrode configurations with higher numbers of concentric rings. Solutions of the proposed inter-ring distances optimization problem define the class of the optimized inter-ring distances electrode designs. These designs may result in improved noninvasive sensors for measurement systems that use concentric ring electrodes to acquire electrical signals such as from the brain, intestines, heart or uterus for diagnostic purposes.

Keywords: Electrophysiology, Electroencephalography, Wearable sensors, Concentric ring electrodes, Laplacian, Optimization, Inter-ring distances, Finite element method, Modeling

Background

Noninvasive concentric ring electrodes (CREs) have been shown to estimate the surface Laplacian, the second spatial derivative of the potentials on the scalp surface for the case of electroencephalogram (EEG), directly at each electrode instead of combining the data from an array of conventional, single pole, disc electrodes (Fig. 1a). In particular, tripolar CREs (TCREs; Fig. 1b) estimate the surface Laplacian using the nine-point method, an extension of the five-point method (FPM) used for bipolar CREs, and significantly better than other electrode systems including bipolar and quasi-bipolar CRE configurations [1, 2]. Compared to EEG via disc electrodes Laplacian EEG via TCREs (tEEG) has been demonstrated to have significantly better spatial selectivity (approximately 2.5 times higher), signal-to-noise ratio (approximately 3.7 times higher), and mutual information (approximately 12 times lower) [3]. Thanks to these properties TCREs found numerous applications in a wide range of areas where electrical signals from the brain are measured including brain–computer interface [4, 5], seizure onset detection [6, 7], detection of high-frequency oscillations and seizure onset zones [8], etc. Review of recent advances in high-frequency oscillations and seizure onset detection based on tEEG via TCREs is available in [9]. These EEG related applications of TCREs along with recent CRE applications related to electroenterograms [10, 11], electrocardiograms (ECG) [12–15], and electrohysterograms [16] suggest the potential of CRE technology in noninvasive electrophysiological measurement.

In order to further improve the CRE design several approaches were proposed including printing disposable CREs on flexible substrates to increase the electrode's ability to adjust to body contours for better contact and to provide higher signal amplitude and signal-to-noise ratio [11, 13, 15, 16]. Other approaches concentrate on assessing the effect of ring dimensions [14, 15] and electrode position [14] on recorded signal and making the measurement system wireless [15]. However, the signal recorded from CREs in [11, 13–16] is either a surface Laplacian estimated for the case of the outer ring and the central disc of the TCRE being shorted together (quasi-bipolar CRE configuration) or a set of bipolar signals representing differences between potentials recorded from the rings and the central disc. Alternatively, signals from all the recording surfaces of each TCRE can be combined into a surface Laplacian estimate signal similar to tEEG. Previously, this approach has resulted in significantly higher Laplacian estimation accuracy

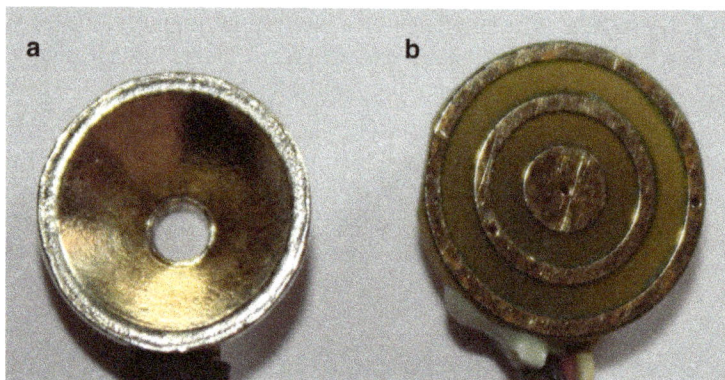

Fig. 1 Conventional disc electrode (**a**) and tripolar concentric ring electrode (**b**)

and radial attenuation for TCREs compared to bipolar and quasi-bipolar CRE configurations [1, 2]. This inspired the recent efforts to further improve the Laplacian estimation accuracy via CREs by increasing the number of concentric rings [17] and varying the inter-ring distances (distances between consecutive rings) [18] described below.

In [17] a general approach to estimation of the Laplacian for an $(n+1)$-polar electrode with n rings using the $(4n+1)$-point method for $n \geq 2$ has been proposed. This method allows cancellation of all the Taylor series truncation terms up to the order of $2n$ which has been shown to be the highest order achievable for a CRE with n rings [17]. In [17] $(4n+1)$-point method was used to demonstrate that accuracy of Laplacian estimation can be improved with an increase of the number of rings, n, by proposing multipolar CRE configurations. Such configurations with n equal to up to 6 rings (septapolar electrode configuration) were compared using finite element method (FEM) modeling and the obtained results suggested statistical significance ($p < 0.0001$) of the increase in Laplacian accuracy due to an increase of n [17]. In [18] $(4n+1)$-point method was used to demonstrate that accuracy of the Laplacian estimation can be improved with transitioning from the previously used constant inter-ring distances by proposing novel variable inter-ring distances CRE configurations. Laplacian estimates for linearly increasing and linearly decreasing inter-ring distances TCRE ($n=2$) and quadripolar CRE (QCRE; $n=3$) configurations were directly compared to their constant inter-ring distances counterparts using analytic analysis and FEM modeling. The main results included establishing a connection between the analytic truncation term coefficient ratios from the Taylor series used in $(4n+1)$-point method and respective ratios of Laplacian estimation errors computed using the FEM model [18]. Both analytic and FEM results were consistent in suggesting that CRE configurations with linearly increasing inter-ring distances may offer more accurate Laplacian estimates compared to CRE configurations with constant inter-ring distances. In particular, for TCREs the Laplacian estimation error may be decreased more than twofold while for QCREs more than a sixfold decrease in estimation error is expected [18]. First physical TCRE prototypes closely resembling the proposed increasing inter-ring distances TCRE design (physical TCRE prototype has a 4:7 ratio of inter-ring distances compared to the 1:2 ratio in the increasing inter-ring distances design proposed in [18]) were assessed in [19] on human EEG, ECG, and electromyogram (EMG) data with promising results.

One of the limitations of [18] was that only linearly variable inter-ring distances were considered while it was hypothesized that optimal inter-ring distances are likely to have a nonlinear relationship. In this paper, the general inter-ring distances optimization problem for the $(4n+1)$-point method of Laplacian estimation is proposed and solved for TCRE and QCRE configurations. The main results include determining the ranges of optimal distances between the central disc and the concentric rings that allow minimizing the truncation error of Laplacian estimation through minimizing the absolute values of truncation term coefficients to be within the 5th percentile. For TCRE with middle ring radius αr and outer ring radius r the optimal range of values for coefficient α was determined to be $0 < \alpha \leq 0.22$ while for QCRE with the first middle ring radius αr, the second middle ring radius βr, and the outer ring radius r the optimal range of values for coefficients α and β was determined to be defined by inequalities $0 < \alpha < \beta < 1$ and $\alpha \beta \leq 0.21$. Truncation term coefficient functions used to solve the general inter-ring distances optimization problem have been validated using ratios of truncation term coefficients for constant and linearly variable inter-ring distances TCRE and QCRE configurations from [18].

Moreover, while in [17] the analysis of variance (ANOVA) has been performed for multipolar CREs to confirm the statistical significance of obtained FEM results, no such analysis has been performed in [18] for variable inter-ring distances CREs. Even after it was added in [20] it lacked factor levels corresponding to optimized inter-ring distances CREs. In this paper, a full factorial design of ANOVA is performed on FEM data that included optimized inter-ring distances CRE configurations to assess statistical significance of the effect of optimization of inter-ring distances on accuracy of Laplacian estimation.

This paper is organized as follows: notations and preliminaries including basic case of FPM as well as the general $(4n+1)$-point method of surface Laplacian estimation for $(n+1)$-polar CRE with n rings are presented in "Methods" section. This section also contains derivation of the truncation term coefficient functions for TCRE and QCRE configurations and defines the general inter-ring distances optimization problem as a constrained optimization problem to minimize the absolute values of truncation term coefficients using the derived truncation term coefficient functions. Finally, FEM model and full factorial ANOVA design are presented. Main results including validation of the proposed truncation term coefficient functions using the ratios of truncation term coefficients for constant and linearly variable inter-ring distances TCRE and QCRE configurations from [18] and solving the proposed general inter-ring distances optimization problem for TCRE and QCRE configurations are presented in "Results" section along with FEM modeling and ANOVA results. Discussion of the obtained results and directions of future work are presented in "Discussion" section followed by the overall conclusions.

Methods

Notations and preliminaries

In [17] the general $(4n+1)$-point method for constant inter-ring distances $(n+1)$-polar CRE with n rings was proposed. It was derived using a regular plane square grid with all inter-point distances equal to r presented in Fig. 2.

First, FPM was applied to the points with potentials v_0, $v_{r,1}$, $v_{r,2}$, $v_{r,3}$, and $v_{r,4}$ (Fig. 2) following Huiskamp's calculation of the Laplacian potential Δv_0 using Taylor series [21]:

$$\Delta v_0 = \frac{d^2 v}{dx^2} + \frac{d^2 v}{dy^2} = \frac{1}{r^2}\left(\sum_{i=1}^{4} v_{r,i} - 4v_0\right) + O\left(r^2\right) \tag{1}$$

where $O\left(r^2\right) = \frac{r^2}{4!}\left(\frac{d^4 v}{dx^4} + \frac{d^4 v}{dy^4}\right) + \frac{r^4}{6!}\left(\frac{d^6 v}{dx^6} + \frac{d^6 v}{dy^6}\right) + \cdots$ is the truncation error.

Equation (1) can be generalized by taking the integral along the circle of radius r around the point with potential v_0. Defining $x = r\cos(\theta)$ and $y = r\sin(\theta)$ as in Huiskamp [21] we obtain:

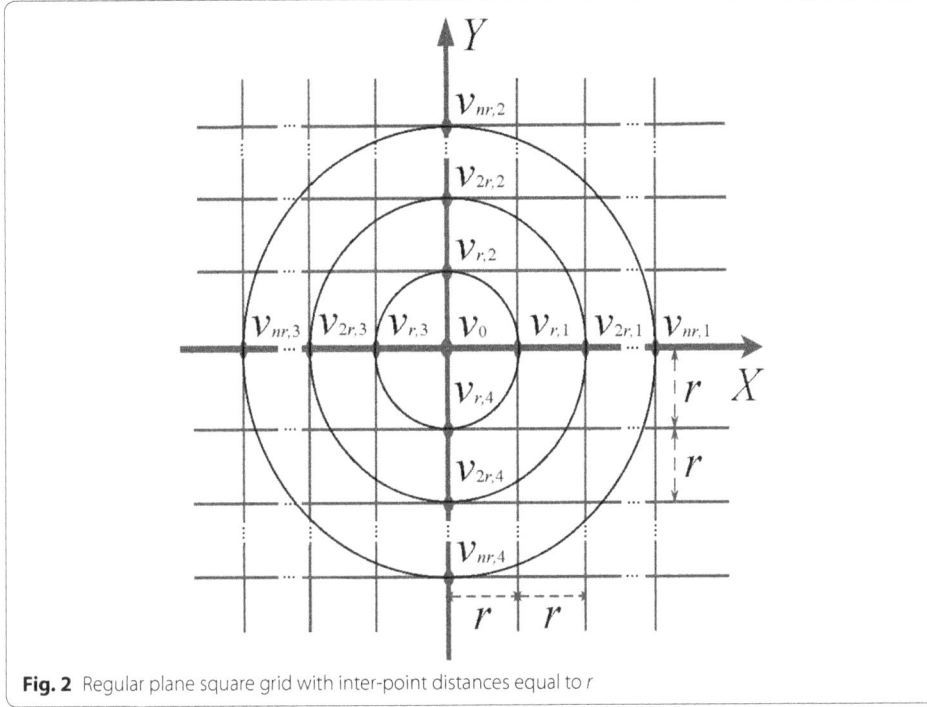

Fig. 2 Regular plane square grid with inter-point distances equal to r

$$\frac{1}{2\pi} \int\limits_0^{2\pi} v(r,\theta)d\theta - v_0 = \frac{r^2}{4}\Delta v_0 + \frac{r^4}{4!} \int\limits_0^{2\pi} \sum_{j=0}^{4} \sin^{4-j}(\theta)\cos^j(\theta)d\theta \left(\frac{d^4 v}{dx^{4-j}dy^j}\right) + \cdots$$

(2)

where $\frac{1}{2\pi}\int_0^{2\pi} v(r,\theta)d\theta$ is the average potential on the ring of radius r and v_0 is the potential on the central disc of the CRE.

Next, for the case of multipolar CRE with n rings ($n \geq 2$), we consider a set of n FPM equations. Each equation corresponds to one of the n rings with ring radii ranging from r to nr. These equations are derived in a manner identical to the way the FPM equation for the ring of radius r has been derived in Eq. (2). For example, we obtain the FPM equation for the ring of radius nr (points with potentials v_0, $v_{nr,1}$, $v_{nr,2}$, $v_{nr,3}$, and $v_{nr,4}$ in Fig. 2) as follows:

$$\frac{1}{2\pi} \int\limits_0^{2\pi} v(nr,\theta)d\theta - v_0 = \frac{(nr)^2}{4}\Delta v_0 + \frac{(nr)^4}{4!} \int\limits_0^{2\pi} \sum_{j=0}^{4} \sin^{4-j}(\theta)\cos^j(\theta)\,d\theta \left(\frac{d^4 v}{dx^{4-j}dy^j}\right)$$

$$+ \frac{(nr)^6}{6!} \int\limits_0^{2\pi} \sum_{j=0}^{6} \sin^{6-j}(\theta)\,\cos^j(\theta)d\theta \left(\frac{d^6 v}{dx^{6-j}dy^j}\right) + \cdots$$

(3)

where $\frac{1}{2\pi}\int_0^{2\pi} v(nr,\theta)d\theta$ is the average potential on the ring of radius nr and v_0 is the potential on the central disc of the CRE.

Finally, to estimate the Laplacian, the n equations, representing differences between average potentials on the n rings and the potential on the central disc of the CRE, are linearly combined in a way that cancels all the Taylor series truncation terms up to the order of $2n$. To obtain such linear combination, the coefficients l^k of the truncation terms with the general form $\frac{(lr)^k}{k!} \int_0^{2\pi} \sum_{j=0}^{k} \sin^{k-j}(\theta) \cos^j(\theta) d\theta \left(\frac{d^k v}{dx^{k-j} dy^j} \right)$ for even order k ranging from 4 to $2n$ and ring radius multiplier l ranging from 1 [Eq. (2)] to n [Eq. (3)] are arranged into an $n-1$ by n matrix A that is a function only of the number of the rings n:

$$
A = \begin{pmatrix} 1^4 & 2^4 & \cdots & n^4 \\ 1^6 & 2^6 & \cdots & n^6 \\ \vdots & \vdots & \ddots & \vdots \\ 1^{2n} & 2^{2n} & \cdots & n^{2n} \end{pmatrix} = \begin{pmatrix} 1 & 2^4 & \cdots & n^4 \\ 1 & 2^6 & \cdots & n^6 \\ \vdots & \vdots & \ddots & \vdots \\ 1 & 2^{2n} & \cdots & n^{2n} \end{pmatrix} \tag{4}
$$

The null space (or kernel) of matrix A is an n-dimensional vector $\bar{x} = (x_1, x_2, \ldots, x_n)$ that is a nontrivial solution of a matrix equation $A\bar{x} = \bar{0}$. The dot product of $\bar{x}$ and a vector consisting of n coefficients l^k corresponding to all the ring radii [i.e. $\left(1, 2^k, \ldots, n^k\right)$] for all even orders k ranging from 4 to $2n$ is equal to 0:

$$
x_1 + 2^k x_2 + \cdots + n^k x_n = 0 \tag{5}
$$

This allows cancellation of all the truncation terms up to the order of $2n$ when the Laplacian estimate is calculated as the linear combination of equations representing differences of potentials from each of the n rings and the central disc ranging from Eq. (2) for the first, innermost concentric ring and up to Eq. (3) for the n-th, outermost concentric ring. The null space vector $\bar{x}$ is used as coefficients and the linear combination is solved for the Laplacian Δv_0:

$$
\Delta v_0 \cong \frac{4}{r^2 \left(x_1 + \cdots + n^2 x_n\right)} \left[x_1 \left(\frac{1}{2\pi} \int_0^{2\pi} v(r, \theta) d\theta - v_0 \right) + \cdots + x_n \left(\frac{1}{2\pi} \int_0^{2\pi} v(nr, \theta) d\theta - v_0 \right) \right] \tag{6}
$$

This Laplacian estimate signal is calculated using a custom preamplifier board and is the only signal sent to the clinical amplifier for each CRE.

Finally, in [18] $(4n+1)$-point method from [17] has been modified to accommodate CRE configurations with variable inter-ring distances that increase or decrease linearly the further the concentric ring lies from the central disc. In both cases sums of all the inter-ring distances to the outermost, n-th, ring were calculated using the formula for the n-th term of the triangular number sequence equal to $n(n+1)/2$ [22]. Consequently, matrix A of truncation term coefficients l^k from Eq. (4) has been modified for linearly increasing (A') and linearly decreasing (A'') inter-ring distances CREs respectively [18]:

$$
A' = \begin{pmatrix} 1 & 3^4 & \cdots & \left(\frac{n(n+1)}{2} \right)^4 \\ 1 & 3^6 & \cdots & \left(\frac{n(n+1)}{2} \right)^6 \\ \vdots & \vdots & \ddots & \vdots \\ 1 & 3^{2n} & \cdots & \left(\frac{n(n+1)}{2} \right)^{2n} \end{pmatrix} \tag{7}
$$

$$A'' = \begin{pmatrix} n^4 & (2n-1)^4 & \cdots & \left(\frac{n(n+1)}{2}\right)^4 \\ n^6 & (2n-1)^6 & \cdots & \left(\frac{n(n+1)}{2}\right)^6 \\ \vdots & \vdots & \ddots & \vdots \\ n^{2n} & (2n-1)^{2n} & \cdots & \left(\frac{n(n+1)}{2}\right)^{2n} \end{pmatrix} \tag{8}$$

Instead of continuing to modify matrix A to assess any additional modalities of variable inter-ring distances CREs (including nonlinear ones) the way it was done in [18] resulting in Eqs. (7) and (8), in this paper the general inter-ring distances optimization problem for the $(4n+1)$-point method of Laplacian estimation is solved for TCRE and QCRE configurations.

Truncation term coefficient function for the TCRE configuration

Assuming that our TCRE $(n=2)$ has two rings with radii αr and r where coefficient α satisfies $0<\alpha<1$ (Fig. 3a), for each ring the integral of the Taylor series is taken along the circle with the corresponding radius. For the ring with radius r we obtain Eq. (2) while for the ring with radius αr we obtain:

$$\frac{1}{2\pi}\int_0^{2\pi} v(\alpha r, \theta)d\theta = v_0 + \frac{(\alpha r)^2}{4}\Delta v_0 + \frac{(\alpha r)^4}{4!}\int_0^{2\pi}\sum_{j=0}^{4}\sin^{4-j}(\theta)\cos^j(\theta)d\theta\left(\frac{d^4 v}{dx^{4-j}dy^j}\right)$$
$$+ \frac{(\alpha r)^6}{6!}\int_0^{2\pi}\sum_{j=0}^{6}\sin^{6-j}(\theta)\cos^j(\theta)d\theta\left(\frac{d^6 v}{dx^{6-j}dy^j}\right) + \cdots \tag{9}$$

For this generalized TCRE setup, modified matrix A of truncation term coefficients l^k from Eq. (4) becomes:

$$A^{TCRE} = \begin{pmatrix} \alpha^4 & 1^4 \end{pmatrix} = \begin{pmatrix} \alpha^4 & 1 \end{pmatrix} \tag{10}$$

The null space of A^{TCRE}, $\bar{x}^{TCRE}$, is equal up to (multiplication by) a constant factor to:

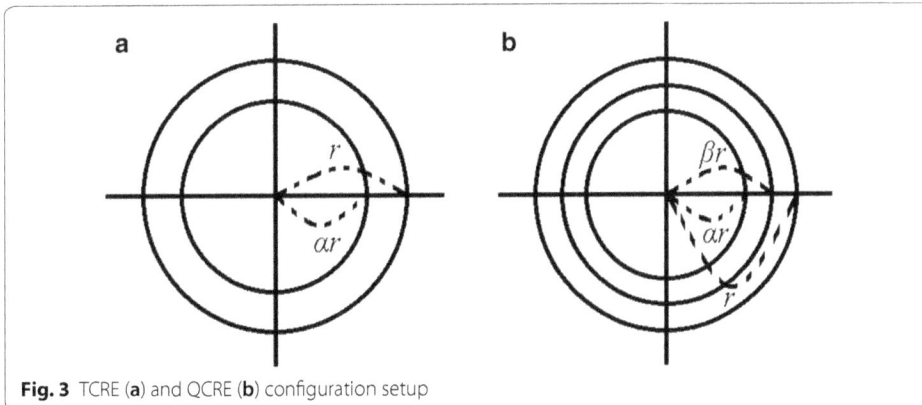

Fig. 3 TCRE (**a**) and QCRE (**b**) configuration setup

$$\bar{x}^{TCRE} = \left(-\frac{1}{\alpha^4}, \ 1 \right) \tag{11}$$

Null space vectors such as $\bar{x}^{TCRE}$ from (11) are not unique. From the properties of matrix multiplication it follows that for any vector $\bar{x}^{TCRE}$ that belongs to the null space of matrix A^{TCRE} and a constant factor c the scaled vector $c\bar{x}^{TCRE}$ also belongs to the null space of matrix A^{TCRE} since $A^{TCRE}(c\bar{x}^{TCRE}) = c(A^{TCRE}\bar{x}^{TCRE}) = c\bar{0} = \bar{0}$.

We combine Eqs. (9) and (2) using the null space vector $\bar{x}^{TCRE}$ from Eq. (11) as coefficients by multiplying Eq. (9) by $-1/\alpha^4$, multiplying Eq. (2) by 1, and adding the two resulting products together with the sum being solved for the Laplacian Δv_0:

$$\Delta v_0 = \frac{4}{r^2\left(1 - \frac{1}{\alpha^2}\right)} \left[-\frac{1}{\alpha^4}(v_{MR} - v_0) + (v_{OR} - v_0) \right.$$
$$\left. + \sum_{k=6,8,\dots}^{\infty} \frac{(1 - \alpha^{k-4})r^k}{k!} \int_0^{2\pi} \sum_{j=0}^{k} \sin^{k-j}(\theta)\cos^j(\theta)d\theta \left(\frac{\partial^k v}{\partial x^{k-j}\partial y^j} \right) \right] \tag{12}$$

where $v_{MR} = \frac{1}{2\pi}\int_0^{2\pi} v(\alpha r, \theta)d\theta$ is the potential on the middle ring of the radius αr and $v_{OR} = \frac{1}{2\pi}\int_0^{2\pi} v(r, \theta)d\theta$ is the potential on the outer ring of the radius r.

The Laplacian estimate from Eq. (12) allows cancellation of the fourth ($2n=4$ for $n=2$) order truncation term. After simplification, the coefficients $c^{TCRE}(\alpha, k)$ of truncation terms with the general form $\frac{c^{TCRE}(\alpha,k)r^{k-2}}{k!}\int_0^{2\pi}\sum_{j=0}^{k}\sin^{k-j}(\theta)\cos^j(\theta)d\theta\left(\frac{\partial^k v}{\partial x^{k-j}\partial y^j}\right)$

can be expressed as the function of coefficient α and the truncation term order k for even $k \geq 6$:

$$c^{TCRE}(\alpha, k) = \frac{4\left(\alpha^4 - \alpha^k\right)}{\alpha^2\left(\alpha^2 - 1\right)} \tag{13}$$

Truncation term coefficient function for the QCRE configuration

Assuming that our QCRE ($n=3$) has three rings with radii αr, βr, and r where coefficients α and β satisfy $0 < \alpha < \beta < 1$ (Fig. 3b), for each ring the integral of the Taylor series is taken along the circle with the corresponding radius. For the ring with radius r we obtain Eq. (2), for the ring with radius αr we obtain Eq. (9), and for the ring with radius βr we obtain:

$$\frac{1}{2\pi}\int_0^{2\pi} v(\beta r, \theta)d\theta = v_0 + \frac{(\beta r)^2}{4}\Delta v_0 + \frac{(\beta r)^4}{4!}\int_0^{2\pi}\sum_{j=0}^{4}\sin^{4-j}(\theta)\cos^j(\theta)d\theta\left(\frac{d^4 v}{dx^{4-j}dy^j}\right)$$
$$+ \frac{(\beta r)^6}{6!}\int_0^{2\pi}\sum_{j=0}^{6}\sin^{6-j}(\theta)\cos^j(\theta)d\theta\left(\frac{d^6 v}{dx^{6-j}dy^j}\right) + \cdots \tag{14}$$

For this generalized QCRE setup, modified matrix A of truncation term coefficients l^k from Eq. (4) becomes:

$$A^{QCRE} = \begin{pmatrix} \alpha^4 & \beta^4 & 1^4 \\ \alpha^6 & \beta^6 & 1^6 \end{pmatrix} = \begin{pmatrix} \alpha^4 & \beta^4 & 1 \\ \alpha^6 & \beta^6 & 1 \end{pmatrix} \tag{15}$$

The null space of A^{QCRE}, $\bar{x}^{QCRE}$, is equal up to a (multiplication by) a constant factor to:

$$\bar{x}^{QCRE} = \left(-\frac{1 - \beta^2}{\alpha^4 (\alpha^2 - \beta^2)}, -\frac{\alpha^2 - 1}{\beta^4 (\alpha^2 - \beta^2)}, 1 \right) \tag{16}$$

We combine Eqs. (2), (9), and (14) using the null space vector $\bar{x}^{QCRE}$ from Eq. (16) as coefficients by multiplying Eq. (9) by $-\frac{1-\beta^2}{\alpha^4(\alpha^2-\beta^2)}$, multiplying Eq. (14) by $-\frac{\alpha^2-1}{\beta^4(\alpha^2-\beta^2)}$,

multiplying Eq. (2) by 1, and adding the three resulting products together with the sum being solved for the Laplacian Δv_0. Such a Laplacian estimate allows cancellation of the fourth and the sixth ($2n=6$ for $n=3$) order truncation terms. It can be shown that, after simplification, the coefficients $c^{QCRE}(\alpha, \beta, k)$ of truncation terms with the general form $\frac{c^{QCRE}(\alpha,\beta,k)r^{k-2}}{k!} \int_0^{2\pi} \sum_{j=0}^{k} \sin^{k-j}(\theta) \cos^j(\theta) d\theta \left(\frac{\partial^k v}{\partial x^{k-j} \partial y^j} \right)$ can be expressed as the function

of coefficients α and β and the truncation term order k for even $k \geq 8$:

$$c^{QCRE}(\alpha, \beta, k) = \frac{4\left[\alpha^k \beta^4 (\beta^2 - 1) + \alpha^6 (\beta^4 - \beta^k) + \alpha^4 (\beta^k - \beta^6)\right]}{\alpha^2 \beta^2 (\alpha^2 - 1)(\beta^2 - 1)(\alpha^2 - \beta^2)} \tag{17}$$

General inter-ring distances optimization problem and its constraints

A constrained optimization problem is proposed to minimize the absolute values of truncation term coefficients for TCRE and QCRE configurations using functions $c^{TCRE}(\alpha, k)$ and $c^{QCRE}(\alpha, \beta, k)$ from Eqs. (13) and (17) respectively. Solving this problem will result in optimized inter-ring distances TCRE and QCRE designs that minimize the truncation error and, therefore, maximize the accuracy of surface Laplacian estimates. Absolute values of truncation term coefficients are used since the signs of the truncation term coefficients have been shown in [18] to be consistent for both constant and variable inter-ring distances CRE configurations: all negative for TCREs and all positive for QCREs. Therefore, for both configurations larger absolute values of truncation term coefficients will translate into larger truncation error. The optimization problem is solved for the lowest nonzero truncation term order equal to 6 and 8 for TCRE and QCRE configurations respectively as the ones that contribute the most to the truncation error since according to [23] for Taylor series "higher-order terms usually contribute negligibly to the final sum and can be justifiably discarded." Formal definitions of the optimization problem for TCRE and QCRE configurations are $\min\limits_{0<\alpha<1} \left| c^{TCRE}(\alpha, 6) \right|$ and $\min\limits_{0<\alpha<\beta<1} \left| c^{QCRE}(\alpha, \beta, 8) \right|$ respectively.

The algorithm of finding global solution to this constrained optimization problem is based on using the 5th percentile to determine the boundary values separating the lowest 5% from the highest 95% of the absolute values of truncation term coefficients. Absolute values of truncation term coefficients within the 5th percentile determine the range of optimal distances between the central disc and the concentric rings to be used in the optimized inter-ring distances TCRE and QCRE designs.

FEM modeling

To directly compare the surface Laplacian estimates for constant inter-ring distances TCRE and QCRE configurations to their counterparts with variable (including optimized) inter-ring distances a FEM model from [17, 18] was used. Evenly spaced square mesh size of 5000×5000 was located in the first quadrant of the $X-Y$ plane above a unit charge dipole projected to the center of the mesh and oriented towards the positive direction of the Z axis. Comparisons to the linearly increasing [18] and novel quadratically increasing inter-ring distances TCRE and QCRE configurations respectively were drawn. In the novel quadratically increasing CRE configurations the inter-ring distances are increasing as a quadratic function $f(s) = s^2$ rather than as a linear identity function $f(s) = s$ of the concentric ring number s counting from the central disc. Bipolar CRE configuration ($n = 1$) was also included into the FEM model. Matlab (Mathworks, Natick, MA, USA) was used for all the FEM modeling.

At each point of the mesh, the electric potential was generated by a unity dipole at depth equal to 3 cm. The medium was assumed to be homogeneous with the conductivity of 7.14 mS/cm to emulate biological tissue [24]. The analytical Laplacian was then calculated at each point of the mesh, by taking the second derivative of the electric potential [17, 18]. Laplacian estimates for different CRE configurations were computed at each point of the mesh where appropriate boundary conditions could be applied for different CRE diameters. Laplacian estimate coefficients for constant inter-ring distances CRE configurations were previously derived using the null space of matrix A from Eq. (4): (16, -1) for TCRE and (270, -27, 2) for QCRE [17]. Coefficients for linearly increasing inter-ring distances CRE configurations were previously derived using the null space of matrix A' from Eq. (7): (81, -1) for TCRE and (4374, -70, 1) for QCRE [18]. Derivation of Laplacian estimate coefficients for novel quadratically increasing inter-ring distances CRE configurations was performed using generalized null space equations proposed in this paper. For the TCRE configuration Eq. (11) was used for $\alpha = 1/5$ to obtain coefficients (625, -1) while for the QCRE configuration (16) was used for $\alpha = 1/14$ and $\beta = 5/14$ to obtain coefficients (34,214,250, $-62,426$, 125). These seven Laplacian estimates including three for TCREs (with constant, linearly increasing, and quadratically increasing inter-ring distances respectively), three for QCREs, and one for the bipolar CRE configuration were then compared with the calculated analytical Laplacian for each point of the mesh where corresponding Laplacian estimates were computed using Relative Error and Maximum Error measures [17, 18]:

$$\text{Relative Error}^i = \sqrt{\frac{\sum (\Delta v - \Delta^i v)^2}{\sum (\Delta v)^2}} \tag{18}$$

$$\text{Maximum Error}^i = \max \left| \Delta v - \Delta^i v \right| \tag{19}$$

where i represents seven CRE configurations, $\Delta^i v$ represents their corresponding Laplacian estimates, and Δv represents the analytical Laplacian potential. More detail on the FEM model used can be found in [17, 18].

Design-Expert (Stat-Ease Inc., Minneapolis, MN, USA) was used for all the statistical analysis of FEM modeling results. Full factorial ANOVA was used with one categorical and two numerical factors [25]. The categorical factor (A) was the inter-ring distances of the CRE presented at three levels corresponding to electrodes with constant inter-ring distances, linearly increasing inter-ring distances, and novel quadratically increasing inter-ring distances respectively. The first numerical factor (B) was the number of concentric rings in the CRE presented at two levels corresponding to TCRE (two concentric rings) and QCRE (three concentric rings) configurations. The second numerical factor (C) was the CRE diameter presented at ten levels uniformly distributed in the range from 0.5 to 5 cm. One possible nuisance factor is the type of the FEM model used in this study which is known but uncontrollable [25]. Two response variables were the Relative Error and Maximum Error of Laplacian estimation computed using Eqs. (18) and (19) respectively for each of the $3 \times 2 \times 10 = 60$ combinations of levels for the three factors. Assumptions of ANOVA including normality, homogeneity of variance, and independence of observations were verified ensuring the validity of the analysis with no studentized residuals being outliers (falling outside of the $[-3, 3]$ range) [25]. Due to the deterministic nature of the FEM model randomizing the order of runs and adding replications were not feasible.

Results

Validating truncation term coefficient functions using ratios of truncation term coefficients for constant and linearly variable inter-ring distances TCRE and QCRE configurations

In [18] two special cases of variable inter-ring distances CREs: linearly increasing [Eq. (7)] and linearly decreasing [Eq. (8)] configurations were proposed and assessed. These two special cases were compared to constant inter-ring distances CREs. It was hypothesized that the ratios of constant inter-ring distances truncation term coefficients over the increasing inter-ring distances truncation term coefficients as well as the ratios of decreasing inter-ring distances truncation term coefficients over constant inter-ring distances truncation term coefficients calculated for TCRE and QCRE configurations will be comparable to the respective ratios of Relative and Maximum Errors of Laplacian estimation obtained using the FEM model. For constant inter-ring distances over increasing inter-ring distances, the truncation term coefficient ratios for the lowest nonzero truncation term for TCRE (sixth order) and QCRE (eighth order) configurations were calculated to be equal to 2.25 and 7.11 respectively which were comparable (difference of less than 5%) to the corresponding ratios of Relative and Maximum Errors obtained using the FEM model for TCRE (2.23 ± 0.02 and 2.22 ± 0.03 respectively) and QCRE (6.95 ± 0.14 and 6.91 ± 0.16) configurations [18]. For decreasing inter-ring distances over constant inter-ring distances, the coefficient truncation term coefficient ratios for the lowest nonzero truncation term for TCRE and QCRE configurations were calculated to be equal to 1.78 and 3.52 respectively which also were comparable (difference of less than 5%) to the corresponding ratios of Relative and Maximum Errors obtained using the FEM model for TCRE (1.75 ± 0.02 and 1.74 ± 0.03 respectively) and QCRE (3.41 ± 0.09 and 3.38 ± 0.11) configurations [18].

Without the truncation term coefficient functions from the general inter-ring distances optimization problem proposed in this study, in [18] all of the aforementioned analytic ratios had to be calculated independently from separate CRE setups while now they can be calculated using functions $c^{TCRE}(\alpha, k)$ and $c^{QCRE}(\alpha, \beta, k)$ from Eqs. (13) and (17) respectively. For constant inter-ring distances TCRE and QCRE configurations we have functions $c^{TCRE}\left(\frac{1}{2}, k\right)$ and $c^{QCRE}\left(\frac{1}{3}, \frac{2}{3}, k\right)$ respectively. For linearly increasing inter-ring distances TCRE and QCRE configurations we have functions $c^{TCRE}\left(\frac{1}{3}, k\right)$ and $c^{QCRE}\left(\frac{1}{6}, \frac{1}{2}, k\right)$ respectively. For linearly decreasing inter-ring distances TCRE and QCRE configurations we have functions $c^{TCRE}\left(\frac{2}{3}, k\right)$ and $c^{QCRE}\left(\frac{1}{2}, \frac{5}{6}, k\right)$ respectively.

To validate the proposed functions $c^{TCRE}(\alpha, k)$ and $c^{QCRE}(\alpha, \beta, k)$ from Eqs. (13) and (17) respectively, the aforementioned analytic ratios (2.25, 7.11, 1.78, and 3.52) of truncation term coefficients from [18] were recalculated for the lowest nonzero truncation term orders equal to 6 and 8 for TCREs and QCREs respectively and rounded to the nearest hundredth:

$$\frac{c^{TCRE}\left(\frac{1}{2}, 6\right)}{c^{TCRE}\left(\frac{1}{3}, 6\right)} = \frac{-1}{-\frac{4}{9}} = 2.25 \tag{20}$$

$$\frac{c^{QCRE}\left(\frac{1}{3}, \frac{2}{3}, 8\right)}{c^{QCRE}\left(\frac{1}{6}, \frac{1}{2}, 8\right)} = \frac{\frac{16}{81}}{\frac{1}{36}} = 7.11 \tag{21}$$

$$\frac{c^{TCRE}\left(\frac{2}{3}, 6\right)}{c^{TCRE}\left(\frac{1}{2}, 6\right)} = \frac{-\frac{16}{9}}{-1} = 1.78 \tag{22}$$

$$\frac{c^{QCRE}\left(\frac{1}{2}, \frac{5}{6}, 8\right)}{c^{QCRE}\left(\frac{1}{3}, \frac{2}{3}, 8\right)} = \frac{\frac{25}{36}}{\frac{16}{81}} = 3.52 \tag{23}$$

Solving inter-ring distances optimization problem for the TCRE configuration

Relationship between the absolute values of truncation term coefficients and middle ring radius coefficient α based on the function $c^{TCRE}(\alpha, k)$ for TCRE configuration and truncation term order k ranging from 6 to 12 is presented in Fig. 4. As described in "Methods" section, the 5th percentile (corresponding to the absolute value of the truncation term coefficient equal to 0.2) was used to determine the boundary value of α for the

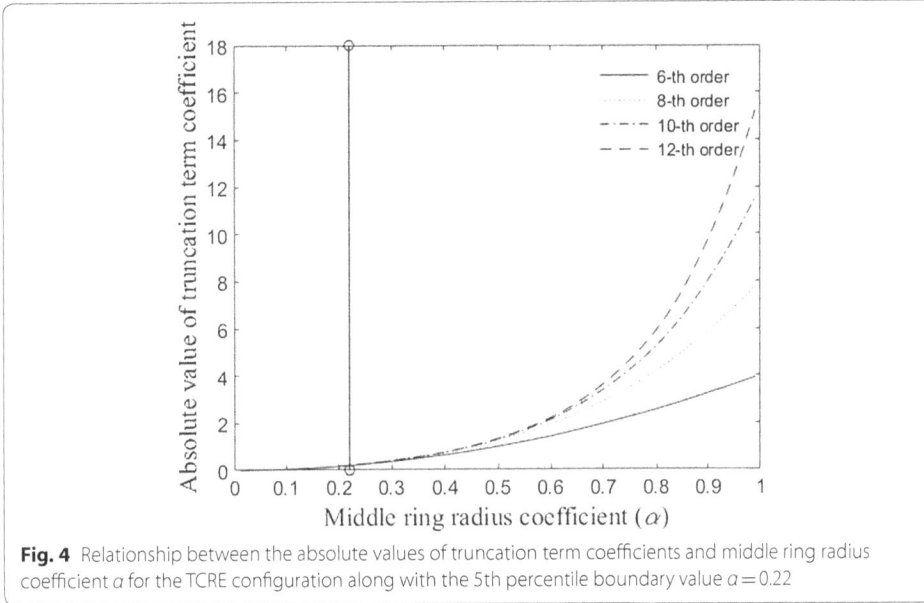

Fig. 4 Relationship between the absolute values of truncation term coefficients and middle ring radius coefficient a for the TCRE configuration along with the 5th percentile boundary value $a = 0.22$

lowest nonzero truncation term order equal to 6 and resulting in $\alpha = 0.22$. Therefore, the optimal range of distances between the central disc and the middle concentric ring of radius αr that keeps absolute values of the sixth order truncation term coefficients within the 5th percentile is determined by inequality $0 < \alpha \leq 0.22$.

Solving inter-ring distances optimization problem for the QCRE configuration

Absolute values of truncation term coefficients based on the function $c^{QCRE}(\alpha, \beta, k)$ for all the combinations of the first middle ring radius coefficient α and the second middle

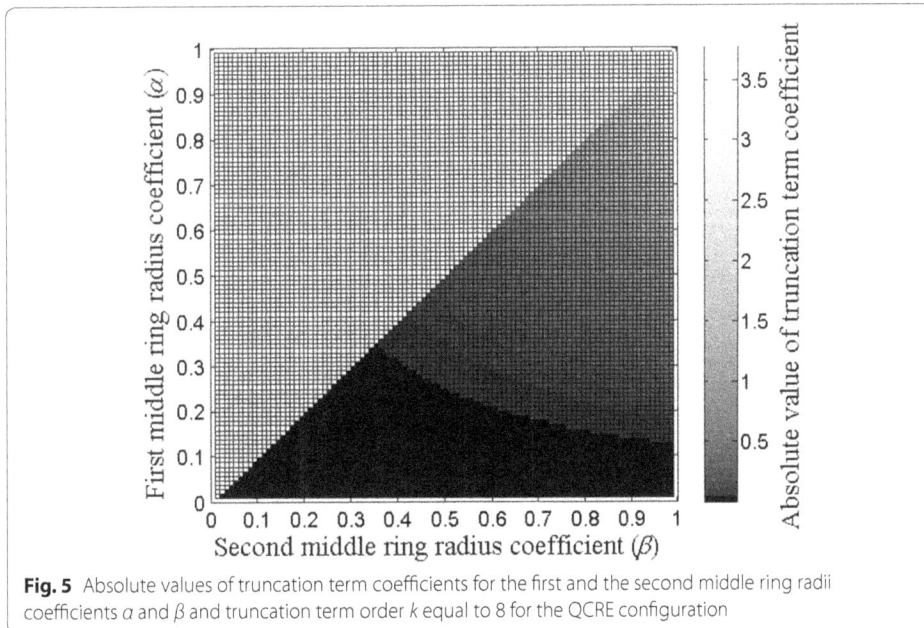

Fig. 5 Absolute values of truncation term coefficients for the first and the second middle ring radii coefficients a and β and truncation term order k equal to 8 for the QCRE configuration

ring radius coefficient β that satisfy $0 < \alpha < \beta < 1$ for QCRE configuration and the lowest nonzero truncation term order k equal to 8 are presented in Fig. 5.

As described in "Methods" section, the 5th percentile (corresponding to the absolute value of the truncation term coefficient equal to 0.19) was used to find the boundary values of α and β that determine the optimal range of distances between the central disc and both middle concentric rings with radii αr and βr respectively which keeps absolute values of the eighth order truncation term coefficients within the 5th percentile as presented in Fig. 6.

While the linear portion of the boundary in Fig. 6 is described by the inequality $\alpha < \beta$, the nonlinear portion had to be fitted with a curve first. Based on the shape of the nonlinear portion of the boundary, a rectangular hyperbola model had been chosen [26]. Even the simplest rectangular hyperbola model $\alpha = m/\beta$, where m is a real constant, provides a good fit to our data presented in Fig. 7 for $m = 0.21$. Goodness-of-fit metric R-squared indicates that the model fit explained 99.79% of the total variation in the data [25].

Therefore, the optimal range of distances between the central disc and the first and the second middle concentric rings with radii αr and βr that keeps absolute values of the eighth order truncation term coefficients within the 5th percentile is determined by two inequalities $0 < \alpha < \beta < 1$ and $\alpha \leq 0.21/\beta$ or, equivalently, $\alpha\beta \leq 0.21$.

FEM modeling

FEM modeling results for the two error measures computed for seven CRE configurations using Eqs. (18) and (19) are presented on a semi-log scale in Fig. 8 for CRE diameters ranging from 0.5 to 5 cm.

Figure 8 suggests that novel quadratically increasing inter-ring distances TCRE and QCRE configurations hold potential for an improvement in Laplacian estimation errors over previously proposed constant [17] and linearly increasing [18] inter-ring

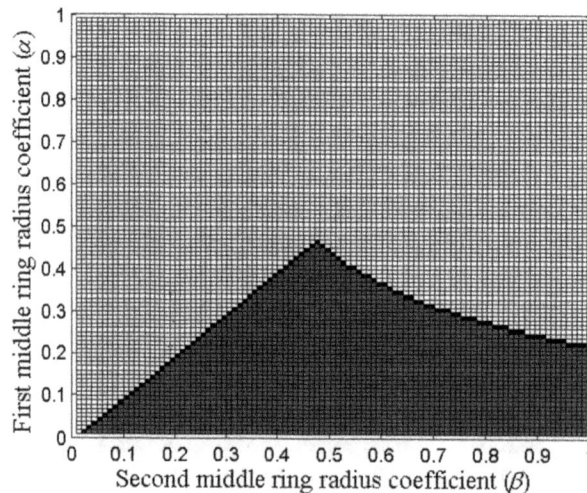

Fig. 6 Absolute values of truncation term coefficients within the 5th percentile (gray) along with the boundary (black) separating them from the values outside of the 5th percentile for the first and the second middle ring radii coefficients α and β

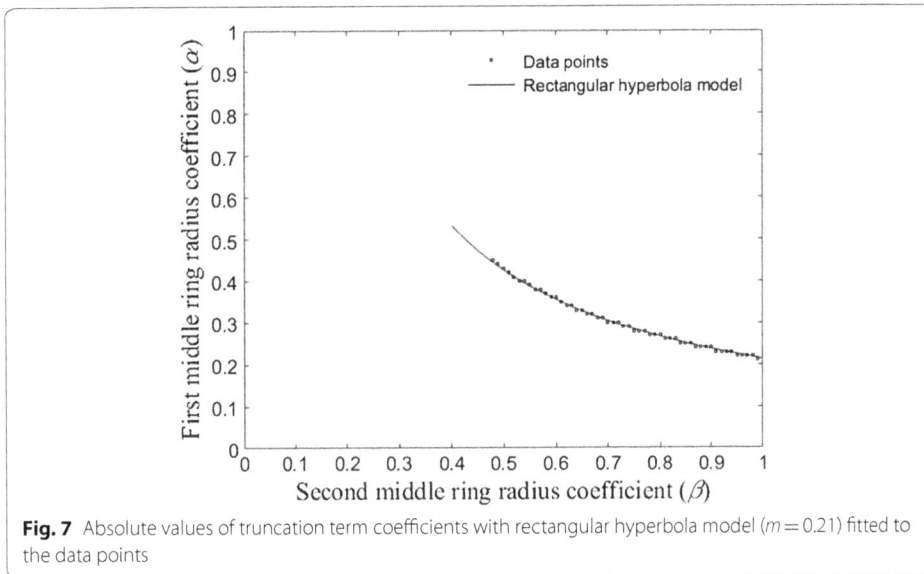

Fig. 7 Absolute values of truncation term coefficients with rectangular hyperbola model ($m = 0.21$) fitted to the data points

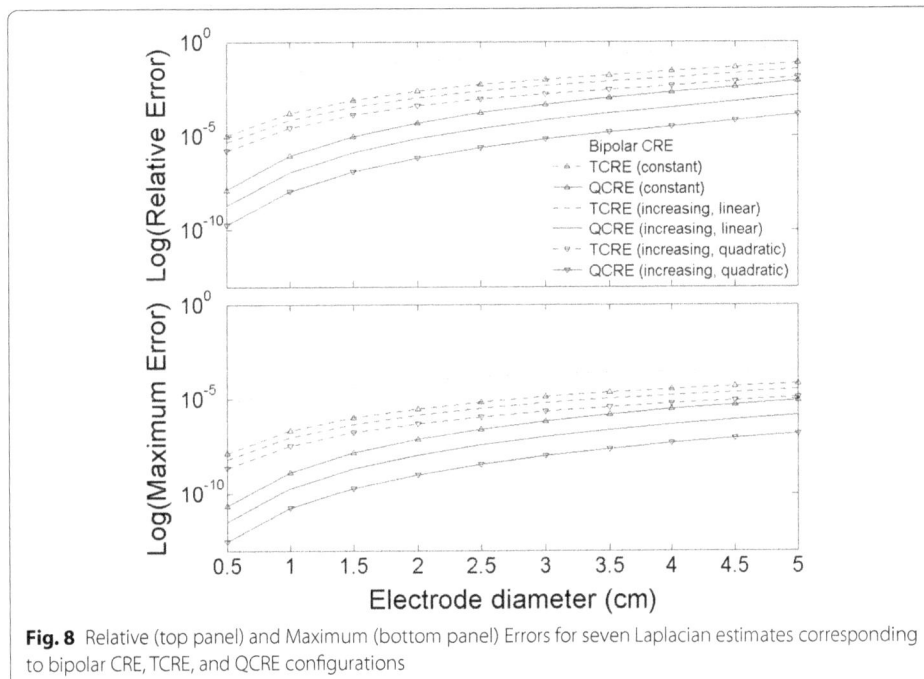

Fig. 8 Relative (top panel) and Maximum (bottom panel) Errors for seven Laplacian estimates corresponding to bipolar CRE, TCRE, and QCRE configurations

distances counterparts. Moreover, improvement appears to become more significant with the increase of the number of rings (i.e. there is more improvement for the QCRE configuration in comparison with the TCRE one). This stems from comparison of averages (mean ± standard deviation for 10 different sizes of each CRE configuration) of errors for linearly increasing inter-ring distances and quadratically increasing inter-ring distances CREs. Compared to their quadratically increasing inter-ring distances counterparts Relative and Maximum Errors are 2.73 ± 0.04 and 2.72 ± 0.05 times higher on average for linearly increasing inter-ring distances TCREa and

10.32 ± 0.3 and 10.23 ± 0.32 times higher on average for linearly increasing inter-ring distances QCREs respectively (Fig. 8).

These ratios of Relative and Maximum Errors involving the novel quadratically increasing inter-ring distances CREs were compared to analytic ratios of truncation term coefficients using Eqs. (13) and (17) respectively. For quadratically increasing inter-ring distances TCRE and QCRE configurations we have truncation term coefficient functions $c^{TCRE}\left(\frac{1}{5}, k\right)$ and $c^{QCRE}\left(\frac{1}{14}, \frac{5}{14}, k\right)$ respectively. The analytic ratios of trunca-

tion term coefficients for linearly increasing over quadratically increasing inter-ring distances TCRE and QCRE configurations calculated for the lowest nonzero truncation term orders equal to 6 and 8 respectively and rounded to the nearest hundredth are equal to:

$$\frac{c^{TCRE}\left(\frac{1}{3}, 6\right)}{c^{TCRE}\left(\frac{1}{5}, 6\right)} = \frac{-\frac{4}{9}}{-\frac{4}{25}} = 2.78 \tag{24}$$

$$\frac{c^{QCRE}\left(\frac{1}{6}, \frac{1}{2}, 8\right)}{c^{QCRE}\left(\frac{1}{14}, \frac{5}{14}, 8\right)} = \frac{\frac{1}{36}}{\frac{25}{9604}} = 10.67 \tag{25}$$

Consistent with the comparison between linearly decreasing, constant, and linearly increasing inter-ring distances CREs from [18], the FEM derived ratios of Relative and Maximum Errors involving the novel quadratically increasing inter-ring CREs are comparable (difference of less than 5%) to the respective analytic ratios of truncation term coefficients from Eqs. (24) and (25).

ANOVA results assessing the effect of factors A (inter-ring distances), B (CRE diameter), and C (number of rings) along with the effect of all possible two-factor interactions on Relative and Maximum Errors suggest that all three factors are statistically significant (Relative Error: $df=9$, $F=85.76$, $p<0.0001$; Maximum Error: $df=9$, $F=129.90$, $p<0.0001$) for the optimal transform being natural logarithmic function ($\lambda=0$ for both the Relative Error and the Maximum Error) as determined using the Box–Cox procedure [25]. Individual effects of the three factors are: A (Relative Error: $df=2$, $F=32.42$, $p<0.0001$; Maximum Error: $df=2$, $F=55.87$, $p<0.0001$), B (Relative Error: $df=1$, $F=251.24$, $p<0.0001$; Maximum Error: $df=1$, $F=311.89$, $p<0.0001$), and C (Relative Error: $df=1$, $F=427.55$, $p<0.0001$; Maximum Error: $df=1$, $F=422.95$, $p<0.0001$). Out of the three two-factor interactions assessed none had statistically significant effect for both response variables.

Discussion

This paper continues our work toward improving the accuracy of Laplacian estimation via multipolar CREs derived using the $(4n+1)$-point method proposed in [17] and modified for linearly variable inter-ring distances CREs in [18]. Prior to [18], inter-ring distances of a CRE were not considered to be a means of improving the accuracy of

Laplacian estimation with, to the best of the author's knowledge, all the previous CRE research having been based on assumption of constant inter-ring distances.

This research direction is important since ability to estimate the Laplacian at each electrode constitutes the primary biomedical significance of CREs. Further improvement of the accuracy of Laplacian estimation via optimized inter-ring distances CREs may contribute to the advancement of noninvasive electrophysiological electrode design with application areas not limited to EEG, ECG, EMG, etc. In particular, for the case of EEG, since "negative Laplacian is approximately proportional to cortical (or dura) surface potential" [27] and enhances the high spatial frequency components of the brain activity close to the electrode [28], Laplacian filtering has been proven to be a high-pass filter for cortical imaging [29, 30]. Ability to attenuate distant sources sharply is critical for location specific EEG applications such as brain–computer interface, seizure onset detection, and detection of high-frequency oscillations and seizure onset zones which is why superiority of tEEG via TCRE over EEG via conventional disc electrodes has been recently shown in these areas [4–9]. This superiority depends on the ability to estimate the surface Laplacian as accurately as possible which is why every application currently recording and utilizing surface Laplacian signals such as tEEG may benefit from more accurate Laplacian estimation. Therefore, this paper provides an innovative solution (ability to optimize the inter-ring distances of the CRE) to improve the accuracy of an acquired signal (surface Laplacian estimate) via improved design of the sensor (such as the novel quadratically increasing inter-ring distances design) selected from the class of all the optimized inter-ring distances designs defined by the solutions of the proposed general inter-ring distances optimization problem. This work may provide insight for future sensor design in noninvasive electrophysiological measurement systems that use CREs to acquire electrical signals such as from the brain, intestines, heart or uterus for diagnostic purposes [4–16].

The contribution of this paper is threefold. First, analytic ratios of truncation term coefficients for linearly increasing, linearly decreasing, and constant inter-ring distances TCRE and QCRE configurations from [18] were recalculated using truncation term coefficient functions derived for the proposed general inter-ring distances optimization problem in order to validate those functions. In [18] it has been shown that these analytic ratios are comparable (difference of less than 5%) to the respective ratios of Relative and Maximum Errors of Laplacian estimation computed using the FEM model. Therefore, it was important to integrate this relationship between analytic and FEM results established in [18] into the framework of the proposed general inter-ring distances optimization problem for the $(4n + 1)$-point method of Laplacian estimation since it allows quantifying the expected improvement in FEM Laplacian estimation accuracy analytically. Furthermore, an identical result was obtained for ratios involving the novel quadratically increasing inter-ring distances TCRE and QCRE configurations proposed in this study.

Second, the general inter-ring distances optimization problem has been solved for TCRE and QCRE configurations. The same approach can be applied to solve corresponding problems for higher numbers of concentric rings in pentapolar, sextapolar, etc. CRE configurations even though the number of decision variables will increase by one for each additional concentric ring. This is a fundamental improvement over preliminary work such as [17] where just constant inter-ring distances have been considered and [18] where only two specific cases of linearly variable inter-ring distances were proposed and assessed in that it allows to further improve the surface Laplacian estimation accuracy via optimized inter-ring distances CREs. As was hypothesized in [18], solutions of the general inter-ring distances optimization problem correspond to nonlinear relationships between inter-ring distances as opposed to the linear relationship considered in [18].

For the TCRE configuration, the optimal range of distances between the central disc and the middle concentric ring of radius αr that keeps absolute values of the sixth order truncation term coefficients within the 5th percentile was determined by inequality $0 < \alpha \leq 0.22$. Currently used constant inter-ring distances TCREs [1–9] correspond to $\alpha = 0.5$ while linearly increasing and linearly decreasing inter-ring distances TCREs from [18] correspond to $\alpha = 0.33$ and $\alpha = 0.67$ respectively rounded to the nearest hundredth. Therefore, all three previously considered TCRE configurations fall outside the 5th percentile range corresponding to optimized inter-ring distances. For the QCREs configuration, the optimal range of distances between the central disc and the first and the second middle concentric rings with radii αr and βr respectively that keeps absolute values of the eighth order truncation term coefficients within the 5th percentile is determined by two inequalities $0 < \alpha < \beta < 1$ and $\alpha\beta \leq 0.21$. Constant inter-ring distances QCREs correspond to $\alpha = 0.33$ and $\beta = 0.67$ while linearly increasing and decreasing inter-ring distances QCREs from [18] correspond to $\alpha = 0.17$ and $\beta = 0.5$ and $\alpha = 0.5$ and $\beta = 0.83$ respectively rounded to the nearest hundredth. Therefore, out of three previously considered QCRE configurations only linearly increasing inter-ring distances configuration falls within the 5th percentile range corresponding to optimized inter-ring distances. For the novel quadratically increasing inter-ring distances CREs proposed in this paper both TCRE ($\alpha = 0.2$) and QCRE ($\alpha = 0.07$ and $\beta = 0.36$) configurations fall within the 5th percentile range corresponding to optimized inter-ring distances.

Finally, full factorial ANOVA was used to confirm the statistical significance of FEM results obtained for CRE configurations including the optimized quadratically increasing inter-ring distances CREs. The ANOVA results for comparison of surface Laplacian estimates corresponding to different CRE configurations showed statistical significance of all three factors included in the study. It was important to confirm that the accuracy of Laplacian estimation increases (Relative and Maximum Errors decrease) with an increase in the number of rings n (factor B) and decreases (Relative and Maximum Errors increase) with an increase of the CRE diameter (factor C), which is consistent with the ANOVA results obtained in [17, 20]. However, the most important ANOVA result obtained was that, for the case of inter-ring distances (factor A), the Laplacian estimates for novel quadratically increasing inter-ring distances CREs are significantly more accurate than the ones for their constant and linearly increasing inter-ring distances counterparts ($p < 0.0001$). In particular, more than two- and tenfold decreases in estimation error are expected for optimized quadratically increasing inter-ring distances TCREs and QCREs respectively compared to corresponding linearly increasing inter-ring distances CRE configurations from [18]. This result further suggests the potential of using the distances between the rings as a means of improving the accuracy of surface Laplacian estimation via CREs.

Directions of future work are twofold. The first one is based on the limitation of the $(4n + 1)$-point method. At this point of time the widths of concentric rings and the radius of the central disc are not taken into account and therefore cannot be optimized. Moreover, assuming these parameters to be negligible is inconsistent with the design of currently used TCREs (Fig. 1b). In order to pursue the ultimate goal of optimizing all of the CRE parameters simultaneously, the first direction is to include these parameters into future modifications of the $(4n+1)$-point method along with the currently included number of rings and inter-ring distances. The first step in this direction has been taken in [31] by deriving a Laplacian estimate for a proof of concept TCRE with incorporated radius of the central disc and the widths of the concentric rings. However, it remains unclear how this proof of concept could be practically incorporated into a modification of the $(4n+1)$-point method and/or used for design optimization purposes due to associated increases in complexity of the linear algebra involved and in the number of decision variables in the optimization problem.

The second direction is to build prototypes of optimized inter-ring distances CREs and assess them on real life data: phantom, animal model, and human. These prototypes will allow quantifying the translation of truncation error of Laplacian estimation assessed in this paper into improvement of spatial selectivity, signal-to-noise ratio, source mutual information, etc. the same way it has been quantified for tEEG via TCREs compared to EEG with conventional disc electrodes in [3]. The first step in this direction has been taken in [19] by assessing stencil printed TCRE prototypes closely resembling the linearly increasing inter-ring distances design proposed in [18] on human EEG, ECG, and EMG data with obtained results suggesting enhanced spatial resolution and localization of signal sources. To the best of the author's knowledge these are the first physical prototypes of variable inter-ring distances CREs and they stemmed from the analytical and modeling results in [18]. Next, prototypes of optimized inter-ring distances CRE designs such as the quadratically increasing inter-ring distances TCREs and QCREs proposed in this paper are needed. These prototypes need to be compared directly to their constant and linearly increasing inter-ring distances counterparts in addition to comparison against the conventional disc electrodes drawn in [19]. Moreover, the question of how small can the distances between concentric rings become without partial shorting due to salt bridges becoming a significant factor affecting the Laplacian estimation can be answered using physical CRE prototypes as well. If prototype assessment results would suggest that physical considerations render the inter-ring distances within the 5th percentile region impractical, then inter-ring distances within the higher percentile region will be considered such as, for example, the 10th percentile region resulting in $0 < \alpha \leq 0.31$ for the TCRE configuration and $0 < \alpha < \beta < 1$ and $\alpha\beta \leq 0.3$ for the QCRE configuration.

Conclusions

As noninvasive tripolar concentric ring electrodes are gaining increased recognition in a range of applications related to electrophysiological measurement due to their unique capabilities this paper establishes a theoretical basis for optimization of variable inter-ring distances in concentric ring electrode design. Previous findings for constant and linearly variable inter-ring distances electrode configurations are integrated into the framework of the general inter-ring distances optimization problem. The problem is solved for tripolar and quadripolar concentric ring electrode configurations and solutions, in the form of optimal ranges for inter-ring distances, may offer more accurate surface Laplacian estimates for electrophysiological measurement systems based on optimized inter-ring distances concentric ring electrodes. Full factorial analysis of variance is used to assess finite element method modeling results obtained for concentric ring electrode configurations including the optimized inter-ring distances ones. It showed statistical significance of the effect of three factors included in this study on the estimation accuracy of surface Laplacian including the inter-ring distances suggesting the potential of using optimization of inter-ring distances to improve the concentric ring electrode design.

Abbreviations

CRE: concentric ring electrode; EEG: electroencephalography; TCRE: tripolar concentric ring electrode; FPM: five-point method; tEEG: Laplacian electroencephalography via tripolar concentric ring electrode; ECG: electrocardiography; FEM: finite element method; QCRE: quadripolar concentric ring electrode; EMG: electromyography; ANOVA: analysis of variance.

Authors' contributions

OM conceived and designed the experiments, performed the experiments and analyzed the data, and wrote the paper. The author read and approved the final manuscript.

Authors' information

Oleksandr Makeyev received the B.S. degree in mathematics and the M.S. degree in statistics from Taras Shevchenko National University of Kyiv, Kyiv, Ukraine, in 2003 and 2005, respectively, and the Ph.D. degree in engineering science from Clarkson University, Potsdam, NY, USA, in 2010. After completing a postdoctoral fellowship at the Department of Electrical, Computer, and Biomedical Engineering, University of Rhode Island, Kingston, RI, USA in 2014 he joined the Department of Mathematics, Diné College, Tsaile, AZ, USA. His National Science Foundation funded Mathematics for Engineering Applications laboratory performs research related to the development and application of computational intelligence and statistics-based signal processing and pattern recognition methods to engineering problems with an emphasis on biomedical and neural engineering.

Acknowledgements

The author thanks Dr. Walter G. Besio from the University of Rhode Island, Dr. Chengde Wang from Diné College, and Dr. Ernst Kussul from the National Autonomous University of Mexico, Mexico City, Mexico for the constructive discussions and helpful comments on this research.

Competing interests

The author declares no competing interests.

Funding

This work was supported by the National Science Foundation (NSF) Division of Human Resource Development (HRD) Tribal Colleges and Universities Program (TCUP) via Award Number 1622481 to Oleksandr Makeyev.

References

1. Besio WG, Koka K, Aakula R, Dai W. Tri-polar concentric ring electrode development for Laplacian electroencephalography. IEEE Trans Biomed Eng. 2006;53:926–33.
2. Besio WG, Aakula R, Koka K, Dai W. Development of a tri-polar concentric ring electrode for acquiring accurate laplacian body surface potentials. Ann Biomed Eng. 2006;34:426–35.
3. Koka K, Besio WG. Improvement of spatial selectivity and decrease of mutual information of tri-polar concentric ring electrodes. J Neurosci Methods. 2007;165:216–22.
4. Besio WG, Cao H, Zhou P. Application of tripolar concentric electrodes and prefeature selection algorithm for brain–computer interface. IEEE Trans Neural Syst Rehabil Eng. 2008;16:191–4.
5. Boudria Y, Feltane A, Besio W. Significant improvement in one-dimensional cursor control using Laplacian electroencephalography over electroencephalography. J Neural Eng. 2014;11:035014.
6. Makeyev O, Liu X, Luna-Munguia H, Rogel-Salazar G, Mucio-Ramirez S, Liu Y, et al. Toward a noninvasive automatic seizure control system in rats with transcranial focal stimulations via tripolar concentric ring electrodes. IEEE Trans Neural Syst Rehabil Eng. 2012;20:422–31.

7. Feltane A, Boudreaux-Bartels GF, Besio WG. Automatic seizure detection in rats using laplacian eeg and verification with human seizure signals. Ann Biomed Eng. 2012;41:645–54.

8. Besio WG, Martinez-Juarez IE, Makeyev O, Gaitanis JN, Blum AS, Fisher RS, et al. High-frequency oscillations recorded on the scalp of patients with epilepsy using tripolar concentric ring electrodes. IEEE J Transl Eng Health Med. 2014;2:1–11.

9. Makeyev O, Musngi M, Lee F, Tamayo M. Recent Advances in high-frequency oscillations and seizure onset detection using laplacian electroencephalography via tripolar concentric ring electrodes. Proceedings. 2017;2:117.

10. Prats-Boluda G, Garcia-Casado J, Martinez-de-Juan JL, Ye-Lin Y. Active concentric ring electrode for non-invasive detection of intestinal myoelectric signals. Med Eng Phys. 2011;33:446–55.

11. Garcia-Casado J, Zena-Gimenez V, Prats-Boluda G, Ye-Lin Y. Enhancement of non-invasive recording of electroen-terogram by means of a flexible array of concentric ring electrodes. Ann Biomed Eng. 2013;42:651–60.

12. Besio WG, Chen T. Tripolar Laplacian electrocardiogram and moment of activation isochronal mapping. Physiol Meas. 2007;28:515.

13. Prats-Boluda G, Ye-Lin Y, Garcia-Breijo E, Ibañez J, Garcia-Casado J. Active flexible concentric ring electrode for non-invasive surface bioelectrical recordings. Meas Sci Technol. 2012;23:125703.

14. Prats-Boluda G, Ye-Lin Y, Bueno-Barrachina JM, de Sanabria RR, Garcia-Casado J. Towards the clinical use of concentric electrodes in ECG recordings: influence of ring dimensions and electrode position. Meas Sci Technol. 2016;27:025705.

15. Ye-Lin Y, Bueno-Barrachina JM, Prats-boluda G, Rodriguez de Sanabria R, Garcia-Casado J. Wireless sensor node for non-invasive high precision electrocardiographic signal acquisition based on a multi-ring electrode. Measurement. 2017;97:195–202.

16. Ye-Lin Y, Alberola-Rubio J, Prats-boluda G, Perales A, Desantes D, Garcia-Casado J. Feasibility and analysis of bipolar concentric recording of electrohysterogram with flexible active electrode. Ann Biomed Eng. 2014;43:968–76.

17. Makeyev O, Ding Q, Besio WG. Improving the accuracy of Laplacian estimation with novel multipolar concentric ring electrodes. Measurement. 2016;80:44–52.

18. Makeyev O, Besio WG. Improving the accuracy of Laplacian estimation with novel variable inter-ring distances concentric ring electrodes. Sensors. 2016;16:858.

19. Wang K, Parekh U, Pailla T, Garudadri H, Gilja V, Ng TN. Stretchable dry electrodes with concentric ring geometry for enhancing spatial resolution in electrophysiology. Adv Healthc Mater. 2017;6:1700552.

20. Makeyev O, Joe C, Lee C, Besio WG. Analysis of variance to assess statistical significance of Laplacian estimation accuracy improvement due to novel variable inter-ring distances concentric ring electrodes. In: 39th Annu Int Conf IEEE Eng Med Biol Soc. 2017. p. 4110–3.

21. Huiskamp G. Difference formulas for the surface Laplacian on a triangulated surface. J Comput Phys. 1991;95:477–96.

22. Weisstein EW. Triangular number. http://mathworld.wolfram.com/TriangularNumber.html. Accessed 11 Feb 2016.

23. King MR, Mody NA. Numerical and statistical methods for bioengineering: applications in MATLAB. Cambridge: Cambridge University Press; 2010.

24. Besio WG, Fasiuddin M. Quantizing the depth of bioelectrical sources for non-invasive 3D imaging. J Bioelectro-magn. 2005;7:90–3.

25. Montgomery DC. Design and analysis of experiments. New York: Wiley; 2008.

26. Wohlfart B, Edman KAP. Rectangular hyperbola fitted to muscle force-velocity data using three-dimensional regression analysis. Exp Physiol. 1994;79:235.

27. Tong S, Thakor NV. Quantitative EEG analysis methods and clinical applications. Norwood: Artech House; 2009.

28. Babiloni F, Babiloni C, Fattorini L, Carducci F, Onorati P, Urbano A. Performances of surface Laplacian estimators: a study of simulated and real scalp potential distributions. Brain Topogr. 1995;8:35–45.

29. Srinivasan R. Methods to improve the spatial resolution of EEG. Int J Bioelectromagn. 1999;1:102–11.

30. Kramer MA, Szeri AJ. Quantitative approximation of the cortical surface potential from EEG and ECoG measurements. IEEE Trans Biomed Eng. 2004;51:1358–65.

31. Makeyev O, Lee C, Besio WG. Proof of concept Laplacian estimate derived for noninvasive tripolar concentric ring electrode with incorporated radius of the central disc and the widths of the concentric rings. In: 39th Annu Int Conf IEEE Eng Med Biol Soc. 2017. p. 841–4.

7

Modeling and classification of gait patterns between anterior cruciate ligament deficient and intact knees based on phase space reconstruction, Euclidean distance and neural networks

Wenbao Wu[1†], Wei Zeng[2*†] ⓘ, Limin Ma[3], Chengzhi Yuan[4] and Yu Zhang[5†]

*Correspondence:
zengwei@lyun.edu.cn
†Wenbao Wu, Wei Zeng
and Yu Zhang contributed
equally to this work
[2] School of Physics
and Mechanical & Electrical
Engineering, Longyan
University, Longyan 364012,
China
Full list of author information
is available at the end of the
article

Abstract

Background: The anterior cruciate ligament (ACL) plays an important role in stabilizing translation and rotation of the tibia relative to the femur. ACL injury alters knee kinematics and usually links to the alternation of gait patterns. The aim of this study is to develop a new method to distinguish between gait patterns of patients with anterior cruciate ligament deficient (ACL-D) knees and healthy controls with ACL-intact (ACL-I) knees based on nonlinear features and neural networks. Therefore ACL injury will be automatically and objectively detected.

Methods: First knee rotation and translation parameters are extracted and phase space reconstruction (PSR) is employed. The properties associated with the gait system dynamics are preserved in the reconstructed phase space. For the purpose of classification of ACL-D and ACL-I knee gait patterns, three-dimensional (3D) PSR together with Euclidean distance computation has been used. These measured parameters show significant difference in gait dynamics between the two groups and have been utilized to form a feature set. Neural networks are then constructed to identify gait dynamics and are utilized as the classifier to distinguish between ACL-D and ACL-I knee gait patterns based on the difference of gait dynamics between the two groups.

Results: Experiments are carried out on a database containing 18 patients with ACL injury and 28 healthy controls to assess the effectiveness of the proposed method. By using the twofold and leave-one-subject-out cross-validation styles, the correct classification rates for ACL-D and ACL-I knees are reported to be 91.3% and 95.65%, respectively.

Conclusion: Compared with other state-of-the-art methods, the results demonstrate that gait alterations in the presence of ACL deficiency can be detected with superior performance. The proposed method is a potential candidate for the automatic and non-invasive classification between patients with ACL deficiency and healthy subjects.

Keywords: Gait analysis, Anterior cruciate ligament, Movement disorders, Phase space reconstruction (PSR), Euclidean distance (ED), Neural networks

Background

Knowledge of spatiotemporal knee motion is important for understanding normal functions as well as addressing clinical problems, including instability after anterior cruciate ligament (ACL) injury. ACL plays an important role in controlling knee joint stability, not only by limiting tibia anterior translation, but also by controlling knee axial rotation and varus movement [1]. Numerous studies have been carried to provide information on biomechanical changes in the ACL-deficient (ACL-D) knees [2–7], which revealed that ACL-D knees would exhibit altered joint kinematics. Currently, the most widely accepted method for assessing joint movement patterns is gait analysis, which offers a unique means of providing insight into mechanisms of ACL-D progression by measuring the kinematic and kinetic parameters [8]. Gait analysis also provides important information concerning motion variability in ACL-D and ACL-intact (ACL-I) knees [9].

Many studies have addressed gait pattern classification and there are several reviews on this subject [10–14]. However, the research work dealing specifically with ACL-D knees is not sufficient [15–18]. Biomechanics plays an important role in the progression of ACL-D knees and many studies have been carried out in gait laboratories to ascertain which parameters are affected by ACL-D knees compared to healthy controls with bilateral ACL-I knees [19–31]. These gait parameters may be adopted as gait features for the classification of gait patterns between ACL-D and ACL-I knees. In the study by Gao et al. [1], spatiotemporal gait and knee joint kinematic variables were calculated and further analyzed. The ACL-D knees exhibited a significant extension deficit compared to the ACL-I knees. A more varus and internally rotated tibial position was also identified in the ACL-D knees during both stair ascent and descent. Knoll et al. [19] revealed a quadriceps-avoidance gait pattern in acute ACL-D patients. Chronic ACL-D individuals demonstrated a significantly different gait pattern. Robinson et al. [32] investigated whether using a direct kinematic or inverse kinematic modeling approach could influence the estimation of knee joint kinematics and kinetics. The similarity between kinematic and kinetic waveforms was evaluated using the root mean square difference and the one-dimensional statistical parametric mapping. Atarod et al. [33] investigated the interactions between different kinematic degree of freedom during normal gait and determined how these interactions would change over time following ACL transection in vivo. They claimed that ACL deficiency would significantly alter the kinematic and kinetic interactions during in vivo gait. Clinical imaging studies of ACL-D individuals versus healthy controls have found greater medial–lateral posterior tibial slope in injured population, with stronger evidence on the lateral plateau slope. To quantify these effects, Marouane et al. [34] used a lower extremity musculoskeletal model which included a detailed finite element model of the knee joint. It was used to compute the role of changes in medial and/or lateral posterior tibial slope on knee joint biomechanics.

The current study has two aims. First, to provide further evidence to support the claim that ACL-D knees demonstrate altered gait patterns compared to ACL-I knees. Second, to provide an automatic and objective method to distinguish between ACL-D and ACL-I knees. Based on the nonlinear and non-stationary nature of knee kinematic signals [35], a popular nonlinear method named phase space reconstruction (PSR), is a valuable tool for the studies of this kind of signals [36–41]. The principle of PSR is to transform the properties of a time series into topological properties of a geometrical object which is embedded in a space, wherein all possible states of the system are represented, each state corresponds to a unique point, and this reconstructed space sharing the same topological properties as the original space. The dynamics in the reconstructed state space is equivalent to the original dynamics. Hence reconstructed phase space is a very useful tool to extract nonlinear dynamics of the signal [36–41]. It is hypothesized that gait dynamics between ACL-D and ACL-I knee gait patterns is significantly different, which implies that PSR offers the potential to compute the difference and classify the two groups.

In this paper, we present a new method using gait analysis to distinguish between ACL-D and ACL-I knees. First knee rotation and translation parameters are extracted and phase space is reconstructed. The properties associated with the gait system dynamics are preserved in the reconstructed phase space. For the purpose of classification of ACL-D and ACL-I knee gait patterns, three-dimensional (3D) PSR together with Euclidean distance (ED) computation has been used. These measured parameters show significant difference in gait dynamics between the two groups and have been utilized to form a feature set. Neural networks are then constructed to identify gait dynamics and are utilized as the classifier, in which the feature set is embedded, to distinguish between ACL-D and ACL-I knee gait patterns based on the difference of gait dynamics between the two groups.

Methods

In this section, we propose a method for the classification of ACL-D knees using the information obtained from gait dynamics. Two groups of subjects (patients with ACL-D knees and healthy controls with ACL-I knees) are recruited and tested in this study. The method is divided into the training stage and the classification stage and follows the following steps. In the first step, knee kinematic signals are extracted by using a motion capture system. In the second step, PSR is applied to extract nonlinear dynamics of lower extremities signals. Euclidean distances are computed to extract gait features. Finally, feature vectors are fed into the neural networks for the modeling and identification of gait dynamics. The difference of gait dynamics will be derived from a set of estimators constructed by neural networks and be applied to distinguish between ACL-D and ACL-I knees. The outline of the proposed method is illustrated in Fig. 1.

Data measurement

Our database consists of 46 participants: 28 healthy controls with ACL-I knees and 18 patients with ACL-D knees. The mean value and the standard deviation (SD) of the age, height, weight and sex for the participants are depicted in Table 1. As the control group, healthy subjects who had bilateral ACL-I knees and no history of musculoskeletal diseases on the lower extremities were included. The ACL-D subjects documented via MRI and a clinical examination had no accompanying damage to the posterior cruciate and collateral ligaments, no more than 30% the meniscus removed, no injuries on the contralateral limb, and no difficulty or pain in performing activities of daily living including walking. A single experienced orthopaedic surgeon performed the physical examination and made the MRI diagnosis.

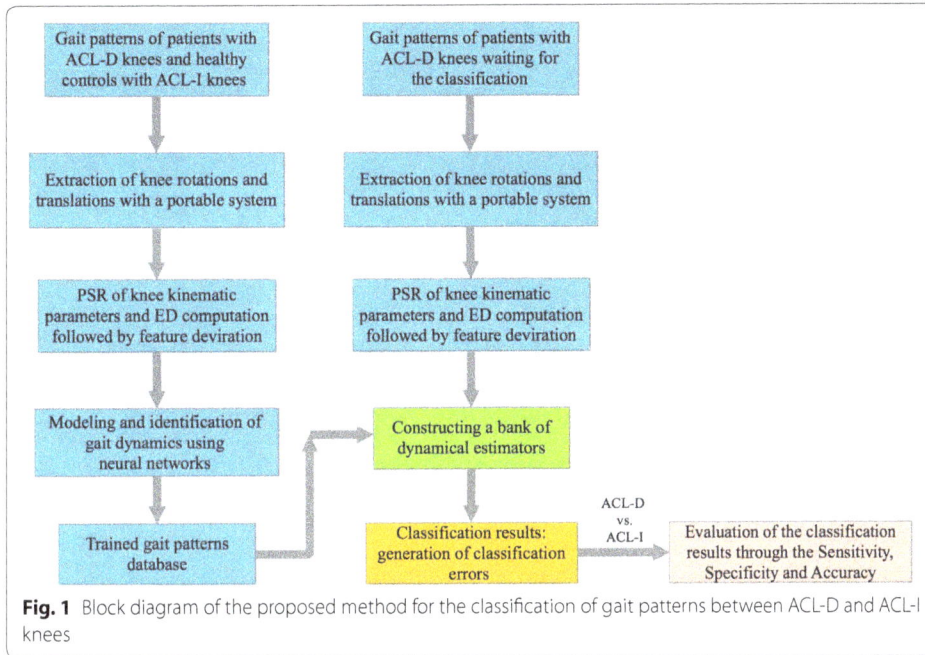

Fig. 1 Block diagram of the proposed method for the classification of gait patterns between ACL-D and ACL-I knees

Table 1 Descriptive characteristics of the ACL-D and ACL-I subjects

	Healthy controls with ACL-I knees	Patients with ACL-D knees	p value
Age (years), mean (SD)	38.6 (5.9)	40.3 (6.1)	0.352
Height (cm), mean (SD)	165.4 (9.6)	164.1 (7.6)	0.630
Weight (kg), mean (SD)	65.7 (10.5)	63.5 (9.4)	0.474
Male/female	14/14	11/7	—

The kinematic data of the knees in six-degree-of-freedom (6DOF) were captured using a portable marker-based motion analysis system (Opti_ Knee®, Innomotion Inc., Shanghai, China), which has been utilized and validated before [42–45], as illustrated in Fig. 2. These tibiofemoral kinematics include varus–valgus (VV), internal–external (IE) rotation and flexion–extension (FE); anterior–posterior (AP), proximal–distal (PD) and medial–lateral (ML) translations.

Each subject was required to undergo a 3-min treadmill gait training. Then data were collected with the sampling frequency of 60 Hz for 15 s and all the participants were guided to walk at the speed of 3 km/h. The detailed procedure about data extraction can be seen in the study by Zhang et al. [42]. The study was approved by the ethical review board and a written informed consent was obtained from each participant before data collection began.

Data description

Here in Table 2 we give the measures of the range of motion (ROM) of knee rotations and translations in patients with ACL-D knees and healthy controls with ACL-I knees.

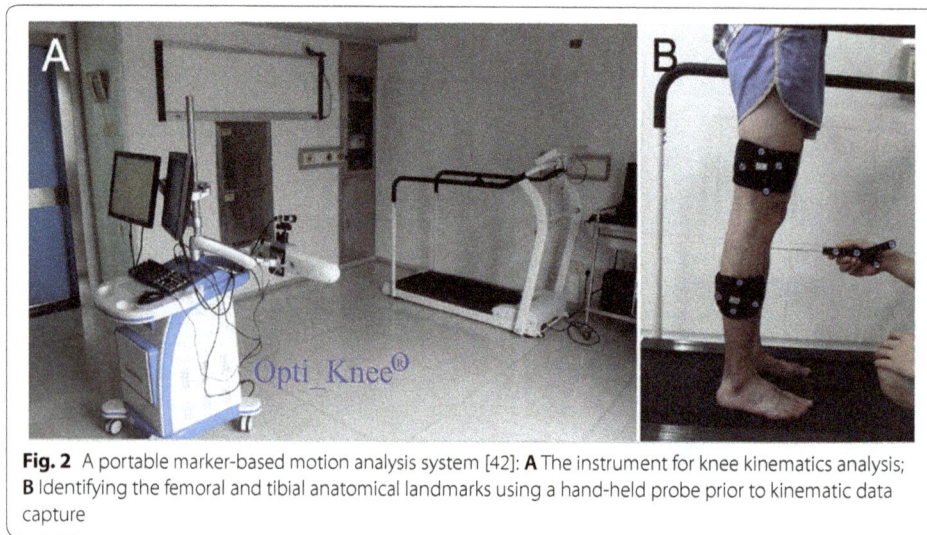

Fig. 2 A portable marker-based motion analysis system [42]: **A** The instrument for knee kinematics analysis; **B** Identifying the femoral and tibial anatomical landmarks using a hand-held probe prior to kinematic data capture

Table 2 Mean, SD, significant statistical difference *p* and effect sizes of the range of motion (ROM) of tibiofemoral rotations and translations for 28 healthy controls with ACL-I knees and 18 patients with ACL-D knees

Parameters	Groups		Difference between groups	Effect size
	ACL-D knees	ACL-I knees	*p*-value	Cohen's *d*
ROM of VV (degree)	13.01 (5.45)	15.40 (4.17)	0.1	0.51
ROM of IE rotation (degree)	18.87 (5.77)	22.45 (4.69)	0.03	0.70
ROM of FE (degree)	59.18 (8.49)	71.76 (6.93)	< 0.001	1.66
ROM of AP translation (cm)	2.41 (0.81)	1.95 (0.52)	0.02	− 0.71
ROM of PD translation (cm)	1.94 (0.74)	2.38 (0.44)	0.01	0.77
ROM of ML translation (cm)	1.84 (0.49)	1.86 (0.37)	0.88	0.05

Kinematic variations during walking were observed in 3-D rotations and translations between ACL-D and ACL-I knees, as shown in Fig. 3. For each of the rotational or translational kinematic component, 101 discrete points corresponding to 0–100% gait cycle at 1% interval were extracted using one-dimensional interpolation for statistical analysis. Measures of each spatiotemporal variable as well as each discrete kinematic point were compared between ACL-D and ACL-I knees using an independent *t*-test analysis of variance (SPSS Inc., IL, USA). A *p* value of < 0.05 was considered to indicate statistical significance.

It is observed from Table 2 that: (1) In the sagittal plane patients with ACL-D knees showed less range of flexion–extension than healthy controls with ACL-I knees (59.18 (8.49) and 71.76 (6.93), respectively, $p < 0.001$). (2) In the frontal plane, patients with ACL-D knees showed less range of internal–external rotation than healthy controls with ACL-I knees (18.87 (5.77) and 22.45 (4.69), respectively, $p = 0.03$). (3) The range of PD translation was lower in the ACL-D knees group compared to ACL-I knees group while the range of AP translation was higher in the ACL-D knees group compared to ACL-I knees group (Table 2). (4) Whereas statistical tests of significance tell us the likelihood that experimental results differ from chance expectations, effect-size measurements

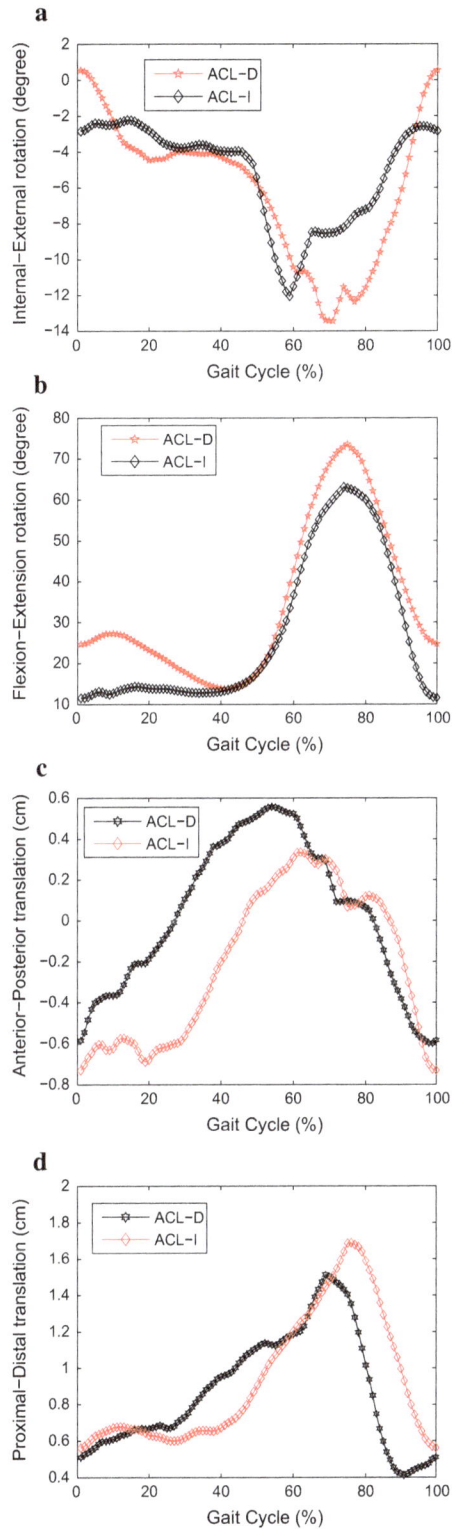

Fig. 3 The 3-D joint rotations and translations during walking of ACL-D and ACL-I knees. Ensemble curves of each subject group were normalized from heel strike to heel strike in a gait cycle. **a** IE rotation; **b** FE; **c** AP translation; **d** PD translation

tell us the relative magnitude of the experimental treatment. In essence, an effect size is the difference between two means divided by the standard deviation of the two conditions [46]. Cohen's d from t-test [47] was used to describe the effect sizes of the ROM of knee kinematic data, which have been shown in Table 2. The effect sizes were traditionally considered small ($d = 0.2$), medium ($d = 0.5$), and large ($d = 0.8$) [48, 49]. It is seen from Table 2 that IE, FE, AP and PD are with nearly large effect sizes compared to VV and ML, which also means there exist significant differences in IE, FE, AP and PD between ACL-deficient patients and healthy controls. The results are in accordance with the p-value analysis.

It is seen from the statistical analysis in Table 2 that IE rotation, FE, AP and PD translations between ACL-D and ACL-I knees are significantly different, which means gait dynamics of the two groups represented by the knee motion are significantly different. Hence these four signals are utilized as reference variables to carry out the following phase space reconstruction.

Phase space reconstruction (PSR)

It is sometimes necessary to search for patterns in a time series and in a higher dimensional transformation of the time series [50]. Phase space reconstruction (PSR) is a method used to reconstruct the so-called phase space. The concept of phase space is a useful tool for characterizing any low-dimensional or high-dimensional dynamic system. A dynamic system can be described using a phase space diagram, which essentially provides a coordinate system where the coordinates are all the variables comprising mathematical formulation of the system. Mathematically, the states of an d-dimensional dynamic system can only be characterized by d independent quantities. Such a set of d independent quantities represents the coordinates of the phase space. One of the most used methods of PSR is the time-delay embedding. Since this method does not require that the treated system could be mathematically defined, explicitly, it fits well with 1-dimensional time series. A point in the phase space represents the state of the system at any given time [50, 51]. Knee kinematic signals can be written as the time series vector $V = \{v_1, v_2, v_3, ..., v_K\}$, where K is the total number of data points. A new sequence of phase space vectors based on delay-coordinate embedding method is expressed as follows [50]:

$$Y_j = (V_j, V_{j+\tau}, V_{j+2\tau}, ..., V_{j+(d-1)\tau}) \tag{1}$$

where $j = 1, 2, ..., K - (d-1)\tau$, d is the embedding dimension of the phase space and τ is a time lag. Y_j means the jth reconstructed vector with embedding dimension d. Finally, we obtain a reconstructed phase space Y containing totally $K - (d-1)\tau$ vector points as the following trajectory matrix:

$$Y = \begin{bmatrix} Y_1 \\ Y_2 \\ \cdots \\ Y_M \end{bmatrix} = \begin{bmatrix} V_1 & V_{1+\tau} & \cdots & V_{1+(d-1)\tau} \\ V_2 & V_{2+\tau} & \cdots & V_{2+(d-1)\tau} \\ \cdots & \cdots & \cdots & \cdots \\ V_M & V_{M+\tau} & \cdots & V_{M+(d-1)\tau} \end{bmatrix} \tag{2}$$

where $M = K - (d-1)\tau$. It is worthwhile to mention that the properties associated with the gait system's dynamics are preserved in the reconstructed phase space. The

d-dimensional space of delay coordinates serves as a pseudo state-space which provides a natural setting to approximate the quantitative aspects of the dynamics.

The behavior of the signal over time can be visualized using PSR (especially when $d = 2$ or 3). In this work, we have confined our discussion to the value of embedding dimension $d = 3$, because of their visualization simplicity. For τ setting, we either utilized the first-zero crossing of the autocorrelation function for each time series or the average τ value obtained from all the time series in the training dataset by using the method depicted in [52]. In the present study we set the values of time lag $\tau = 1$ to test the classification performance. PSR for $d = 3$ has been referred as 3D PSR.

3D PSR is the plot of three delayed vectors V_j, V_{j+1} and V_{j+2} to visualize the dynamics of human gait system. Euclidian distance (ED) of a point (V_j, V_{j+1}, V_{j+2}), which is the distance of the point from origin in 3D PSR and can be defined as [50]

$$ED_j = \sqrt{V_j^2 + V_{j+1}^2 + V_{j+2}^2} \tag{3}$$

ED measures can be used in features extraction and have been studied and applied in many fields, such as clustering algorithms and induced aggregation operators [53].

Feature extraction and selection

Reconstructed phase spaces have been proven to be topologically equivalent to the original system and therefore are capable of recovering the nonlinear dynamics of the generating system [36, 37]. This implies that the full dynamics of the gait system are accessible in this space, and for this reason, the features extracted from it can potentially contain more and/or different information than the common features extraction method [8]. In order to get a more efficient features set, this paper proposes the following features extraction scheme using ED computation.

(1) Reconstruct the phase space for the above mentioned reference variables including knee IE rotation, FE, AP and PD translations with selected values of d and τ for each gait trial;

(2) Compute ED of 3D PSR of knee IE rotation, FE, AP and PD translations as gait features. Concatenate these features to form a feature vector $[ED_j^{IE}, ED_j^{FE}, ED_j^{AP}, ED_j^{PD}]^T$ and the dimension of feature space would be four.

For our dataset, IE rotation, FE, AP and PD translations of two groups (ACL-D and ACLI knees) are analyzed and signal dynamics are extracted by using 3D PSR. Samples of the 3D PSR of knee IE rotation, FE, AP and PD translations are exhibited in Fig. 4. After 3D PSR, features of $[ED_j^{IE}, ED_j^{FE}, ED_j^{AP}, ED_j^{PD}]^T$ for ACL-D and ACL-I knee gait patterns are derived through ED computation, as shown in Fig. 5. As we have analyzed before, significant difference in knee gait dynamics have been

(See figure on next page.)
Fig. 4 Samples of 3D PSR of the knee kinematic signals from ACL-D and ACL-I gait patterns: **a** 3D PSR of the IE rotation; **b** 3D PSR of the FE; **c** 3D PSR of the AP translation; **d** 3D PSR of the PD translation

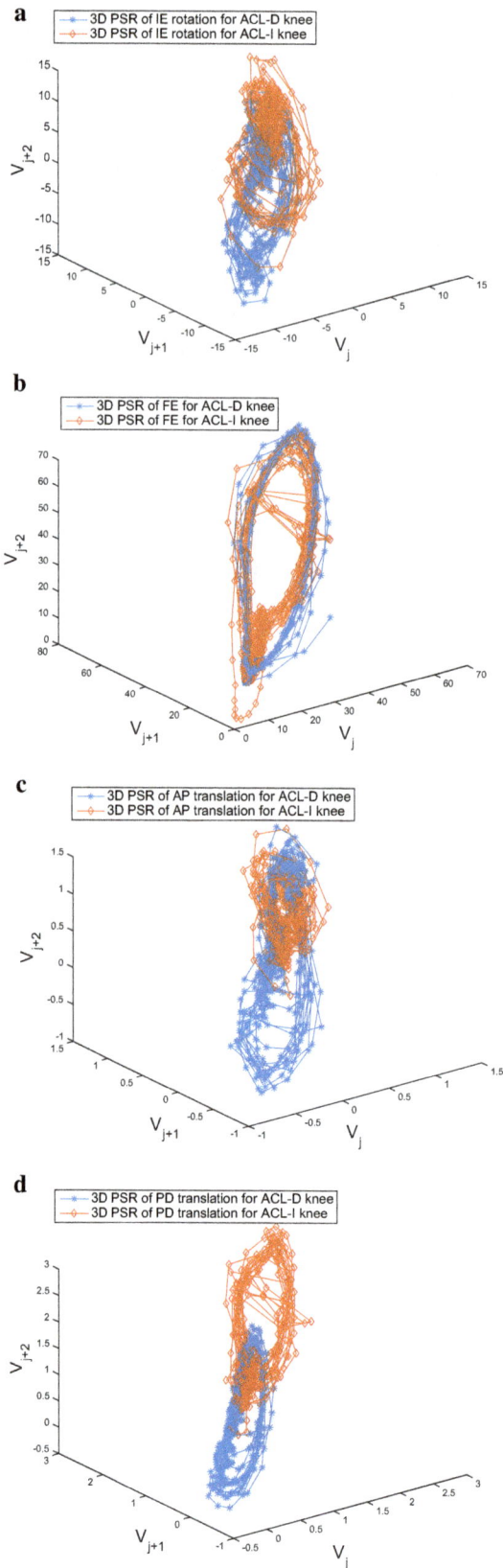

a 3D PSR of IE rotation for ACL-D knee
   3D PSR of IE rotation for ACL-I knee

b 3D PSR of FE for ACL-D knee
   3D PSR of FE for ACL-I knee

c 3D PSR of AP translation for ACL-D knee
   3D PSR of AP translation for ACL-I knee

d 3D PSR of PD translation for ACL-D knee
   3D PSR of PD translation for ACL-I knee

reported between ACL-D and ACL-I knees, which can also be seen obviously from Fig. 4.

Training and modeling mechanism based on selected features

In this section, we present a scheme for modeling and identification of gait dynamics of ACL-I and ACL-D knees based on the above mentioned features.

Consider a general nonlinear human gait system dynamics in the following form:

$$\dot{x} = F(x; p) + v(x; p) \tag{4}$$

where $x = [x_1, \ldots, x_n]^T \in R^n$ are the system states which represent the features $[ED_j^{IE}, ED_j^{FE}, ED_j^{AP}, ED_j^{PD}]^T$, p is a constant vector of system parameters. $F(x; p) = [f_1(x; p), \ldots, f_n(x; p)]^T$ is a smooth but unknown nonlinear vector representing the gait system dynamics, $v(x; p)$ is the modeling uncertainty. Since the modeling uncertainty $v(x; p)$ and the gait system dynamics $F(x; p)$ cannot be decoupled from each other, we consider the two terms together as an undivided term, and define $\phi(x; p) := F(x; p) + v(x; p)$ as the general gait system dynamics. Then, the following steps are taken to model and derive the gait system dynamics via deterministic learning theory [54–56].

In the first step, standard RBF neural networks are constructed in the following form

$$f_{nn}(Z) = \sum_{i=1}^{N} w_i s_i(Z) = W^T S(Z), \tag{5}$$

where Z is the input vector, $W = [w_1, \ldots, w_N]^T \in R^N$ is the weight vector, N is the node number of the neural networks, and $S(Z) = [s_1(\| Z - \mu_1 \|), \ldots, s_N(\| Z - \mu_N \|)]^T$, with $s_i(\| Z - \mu_i \|) = \exp[\frac{-(Z-\mu_i)^T(Z-\mu_i)}{\eta_i^2}]$ being a Gaussian function, $\mu_i(i = 1, \ldots, N)$ being distinct points in state space, and η_i being the width of the receptive field.

In the second step, the following dynamical RBF neural networks are employed to model and derive the gait system dynamics $\phi(x; p)$:

$$\dot{\hat{x}} = -A(\hat{x} - x) + \hat{W}^T S(x) \tag{6}$$

where $\hat{x} = [\hat{x}_1, \ldots, \hat{x}_n]$ is the state vector of the dynamical RBF neural networks, $A = diag[a_1, \ldots, a_n]$ is a diagonal matrix, with $a_i > 0$ being design constants, localized RBF neural networks $\hat{W}^T S(x) = [\hat{W}_1^T S_1(x), \ldots, \hat{W}_n^T S_n(x)]^T$ are used to approximate the unknown $\phi(x; p)$.

The following law is used to update the neural weights

$$\dot{\hat{W}}_i = \dot{\tilde{W}}_i = -\Gamma_i S(x)\tilde{x}_i - \sigma_i \Gamma_i \hat{W}_i \tag{7}$$

(See figure on next page.)
Fig. 5 Samples of Euclidian distance of 3D PSR of the knee kinematic signals from ACL-D and ACL-I gait patterns: **a** Euclidian distance of 3D PSR of the IE rotation; **b** Euclidian distance of 3D PSR of the FE; **c** Euclidian distance of 3D PSR of the AP translation; **d** Euclidian distance of 3D PSR of the PD translation

a

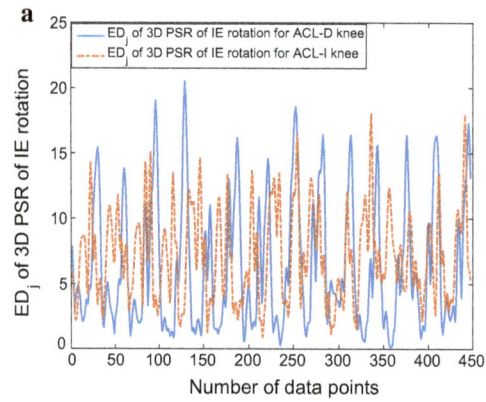

b

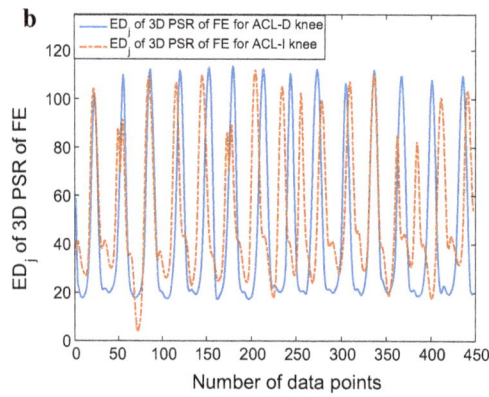

c

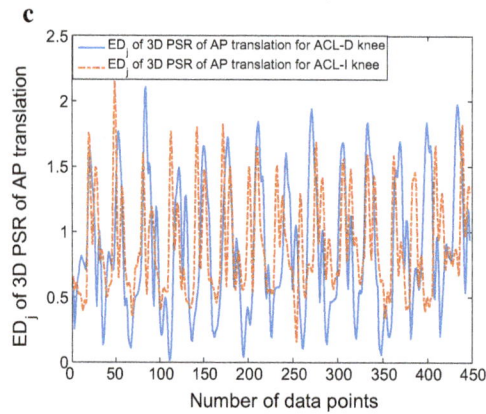

d

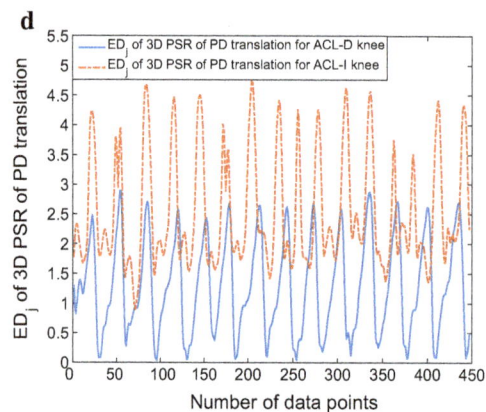

where $\tilde{x}_i = \hat{x}_i - x_i$, $\tilde{W}_i = \hat{W}_i - W_i^*$, W_i^* is the ideal constant weight vector such that $\phi_i(x; p) = W_i^{*T} S(x) + \epsilon_i(x)$, $\epsilon_i(x) < \epsilon^*$ represents the neural network modeling error, $\Gamma_i = \Gamma_i^T > 0$, and $\sigma_i > 0$ is a small value.

With Eqs. (4–6), the derivative of the state estimation error $\tilde{x}_i$ satisfies

$$\dot{\tilde{x}}_i = -a_i\tilde{x}_i + \hat{W}_i^T S(x) - \phi_i(x; p) = -a_i\tilde{x}_i + \tilde{W}_i^T S(x) - \epsilon_i \tag{8}$$

In the third step, by using the local approximation property of RBF neural networks, the overall system consisting of dynamical model (8) and the neural weight updating law (7) can be summarized into the following form in the region Ω_ζ

$$\begin{bmatrix} \dot{\tilde{x}}_i \\ \dot{\tilde{W}}_{\zeta i} \end{bmatrix} = \begin{bmatrix} -a_i & S_{\zeta i}(x)^T \\ -\Gamma_{\zeta i}S_{\zeta i}(x) & 0 \end{bmatrix} \begin{bmatrix} \tilde{x}_i \\ \tilde{W}_{\zeta i} \end{bmatrix} + \begin{bmatrix} -\epsilon_{\zeta i} \\ -\sigma_i\Gamma_{\zeta i}\hat{W}_{\zeta i} \end{bmatrix} \tag{9}$$

and

$$\dot{\hat{W}}_{\bar{\zeta} i} = \dot{\tilde{W}}_{\bar{\zeta} i} = -\Gamma_{\bar{\zeta} i}S_{\bar{\zeta} i}(x)\tilde{x}_i - \sigma_i\Gamma_{\bar{\zeta} i}\hat{W}_{\bar{\zeta} i} \tag{10}$$

where $\epsilon_{\zeta i} = \epsilon_i - \tilde{W}_{\bar{\zeta} i}^T S_{\bar{\zeta}}(x)$. The subscripts $(\cdot)_\zeta$ and $(\cdot)_{\bar{\zeta}}$ are used to stand for terms related to the regions close to and far away from the trajectory $\varphi_\zeta(x_0)$. The region close to the trajectory is defined as $\Omega_\zeta := Z|\text{dist}(Z, \varphi_\zeta) \leq d_\iota$, where $Z = x, d_\iota > 0$ is a constant satisfying $s(d_\iota) > \iota$, $s(\cdot)$ is the RBF used in the network, ι is a small positive constant. The related subvectors are given as: $S_\zeta(x) = [s_{j1}(x), \ldots, s_{j\zeta}(x)]^T \in R^{N_\zeta}$, with the neurons centered in the local region Ω_ζ, and $W_\zeta^* = [w_{j1}^*, \ldots, w_{j\zeta}^*]^T \in R^{N_\zeta}$ is the corresponding weight subvector, with $N_\zeta < N$. For localized RBF neural networks, $|\tilde{W}_{\bar{\zeta} i}^T S_{\bar{\zeta}}(x)|$ is small, so $\epsilon_{\zeta i} = O(\epsilon_i)$.

Finally, according to Theorem 1 in [57], the regression subvector $S_{\zeta i}(x)$ satisfies the persistent excitation condition almost always. This will lead to exponential stability of $(\tilde{x}_i, \tilde{W}_{\zeta i}) = 0$ of the nominal part of system (9) [58]. Based on the analysis results given in [57], the neural network weight estimate error $\tilde{W}_{\zeta i}$ converges to small neighborhoods of zero, with the sizes of the neighborhoods being determined by $\epsilon_{\zeta i}$ and $\|\sigma_i\Gamma_{\zeta i}W_{\zeta i}^*\|$, both of which are small values. This means that the entire RBF network $\hat{W}_i^T S(x)$ can approximate the unknown $\phi_i(x; p)$ along the trajectory φ_ζ, and

$$\phi_i(x; p) = \hat{W}_i^T S(x) + \epsilon_{i1} \tag{11}$$

where $\epsilon_{i1} = O(\epsilon_{\zeta i})$.

By the convergence result, we can obtain a constant vector of neural weights according to

$$\bar{W}_i = mean_{t \in [t_a, t_b]}\hat{W}_i(t) \tag{12}$$

where $t_b > t_a > 0$ represent a time segment after the transient process. Therefore, we conclude that accurate identification of the function $\phi_i(x; p)$ is obtained along the trajectory $\varphi_\zeta(x_0)$ by using $\bar{W}_i^T S_i(x)$, i.e.,

$$\phi_i(x; p) = \bar{W}_i^T S(x) + \epsilon_{i2} \tag{13}$$

where $\epsilon_{i2} = O(\epsilon_{i1})$ and subsequently $\epsilon_{i2} = O(\epsilon^*)$.

Classification mechanism

In this section, we present a scheme to distinguish between ACL-I and ACL-D knees.

Consider a training dataset consisting of gait patterns φ_ζ^k, $k = 1, \ldots, M$, with the *kth* training pattern φ_ζ^k generated from

$$\dot{x} = F^k(x; p^k) + v^k(x; p^k), \quad x(t_0) = x_{\zeta 0} \tag{14}$$

where $F^k(x; p^k)$ denotes the gait system dynamics, $v^k(x; p^k)$ denotes the modeling uncertainty, p^k is the system parameter vector.

As shown in the above subsection, the general gait system dynamics $\phi^k(x; p^k) := F^k(x; p^k) + v^k(x; p^k)$ can be accurately derived and preserved in constant RBF neural networks $\bar{W}^{k^T} S(x)$. By utilizing the learned knowledge obtained in the training stage, a bank of M estimators is constructed for the training gait patterns as follows:

$$\dot{\bar{\chi}}^k = -B(\bar{\chi}^k - x) + \bar{W}^{k^T} S(x) \tag{15}$$

where $k = 1, \ldots, M$ is used to stand for the *kth* estimator, $\bar{\chi}^k = [\bar{\chi}_1^k, \ldots, \bar{\chi}_n^k]^T$ is the state of the estimator, $B = diag[b_1, \ldots, b_n]$ is a diagonal matrix which is kept the same for all estimators, x is the state of an input test gait pattern generated from Eq. (4).

In the classification phase, by comparing the test gait pattern (standing for an ACL-D or an ACL-I gait pattern) generated from gait system (4) with the set of M estimators (15), we obtain the following test error systems:

$$\dot{\tilde{\chi}}_i^k = -b_i \tilde{\chi}_i^k + \bar{W}_i^{k^T} S_i(x) - \phi_i(x; p), \\ i = 1, \ldots, n, \quad k = 1, \ldots, M \tag{16}$$

where $\tilde{\chi}_i^k = \bar{\chi}_i^k - x_i$ is the state estimation (or synchronization) error. We compute the average L_1 norm of the error $\tilde{\chi}_i^k(t)$

$$\|\tilde{\chi}_i^k(t)\|_1 = \frac{1}{T_c} \int_{t-T_c}^t |\tilde{\chi}_i^k(\tau)| d\tau, \quad t \geq T_c \tag{17}$$

where T_c is the cycle of human gait.

The fundamental idea of the classification between ACL-D and ACL-I knees is that if a test gait pattern generated from a certain ACL-D or ACL-I knee is similar to the trained gait pattern s ($s \in \{1, \ldots, k\}$), the constant RBF network $\bar{W}_i^{s^T} S_i(x)$ embedded in the matched estimator s will quickly recall the learned knowledge by providing accurate approximation to gait system dynamics. Thus, the corresponding error $\|\tilde{\chi}_i^s(t)\|_1$ will become the smallest among all the errors $\|\tilde{\chi}_i^k(t)\|_1$. Based on the smallest error principle, the appearing test gait pattern can be classified. We have the following classification scheme.

Classification scheme: If there exists some finite time t^s, $s \in \{1, \ldots, k\}$ and some $i \in \{1, \ldots, n\}$ such that $\|\tilde{\chi}_i^s(t)\|_1 < \|\tilde{\chi}_i^k(t)\|_1$ for all $t > t^s$, then the appearing gait pattern can be classified.

Experimental results

The classification performance of ACL-D knees against ACL-I knees is evaluated on several experiments. Three measurements, including the Sensitivity, the Specificity and the Accuracy, are employed for the evaluation, which are defined as follows:

$$\text{Sensitivity} = \frac{\text{TP}}{\text{TP} + \text{FN}} \tag{18}$$

$$\text{Specificity} = \frac{\text{TN}}{\text{TN} + \text{FP}} \tag{19}$$

$$\text{Accuracy} = \frac{\text{TP} + \text{TN}}{\text{TP} + \text{TN} + \text{FN} + \text{FP}} \tag{20}$$

where TP is the number of true positives, FN is the number of false negatives, TN is the number of true negatives and FP is the number of false positives.

The classification results of ACL-D knees will be evaluated in the twofold cross-validation and leave-one-subject-out cross-validation styles, respectively. In the experiment of twofold cross-validation style, we randomly select half of the group of patients with ACL-D knees and half of the group of the healthy controls with ACL-I knees to constitute the training dataset, the rest of the subjects in the two groups are selected as the test dataset. That means there are 9 patients with ACL-D knees and 14 healthy controls with ACL-I knees in the training dataset. In the experiment of leave-one-subject-out cross-validation, each time we select one subject for classification, the rest of the 45 subjects for training. This process is repeated 46 times and the leave-one-subject-out classification accuracy is calculated as the average of the classification accuracy of all of the individually left-out subjects.

In the training phase, the RBF network $\hat{W}_i^T S_i(x)$ is constructed in a regular lattice, with nodes $N = 83521$, the centers μ_i evenly spaced on $[-1.2, 1.2] \times [-1.2, 1.2] \times [-1.2, 1.2] \times [-1.2, 1.2]$ so as to cover all the trajectories of the input vectors, and the widths $\eta = 0.15$. The weights of the RBF neural networks are updated according to Eq. (7). The initial weights $\hat{W}_i(0) = 0$. The design parameters for (6) and (7) are $a_i = 0.5$, $\Gamma = diag\{1.5, 1.5, 1.5, 1.5\}$, $\sigma_i = 10$, $(i = 1, \ldots, 4)$.

In the classification phase, by using the constant networks $\bar{W}_i^{k^T} S_i(x)$, RBF network estimators are constructed based on Eq. (15). The parameters in Eqs. (15) and (17) are $b_i = -30$ $(i = 1, \ldots, 4)$, $T_c = 1.08s$. Experimental results are illustrated in Tables 3 and 4, and Fig. 6. Tables 3 and 4 shows the confusion matrix of gait pattern classification between ACL-D and ACL-I knees by using twofold and leave-one-subject-out cross-validation styles. Figure 6 shows the classification results. By using the twofold cross-validation and leave-one-subject-out cross-validation styles, the correct classification rates for ACL-D knees are reported to be 91.3% and 95.65%, respectively.

Table 3 Confusion matrix of gait pattern classification between ACL-D and ACL-I knees by using twofold cross-validation method

	ACL-D knees	ACL-I knees
ACL-D knees	8	1
ACL-I knees	1	13

Table 4 Confusion matrix of gait pattern classification between ACL-D and ACL-I knees by using leave-one-subject-out cross-validation method

	ACL-D knees	ACL-I knees
ACL-D knees	17	1
ACL-I knees	1	27

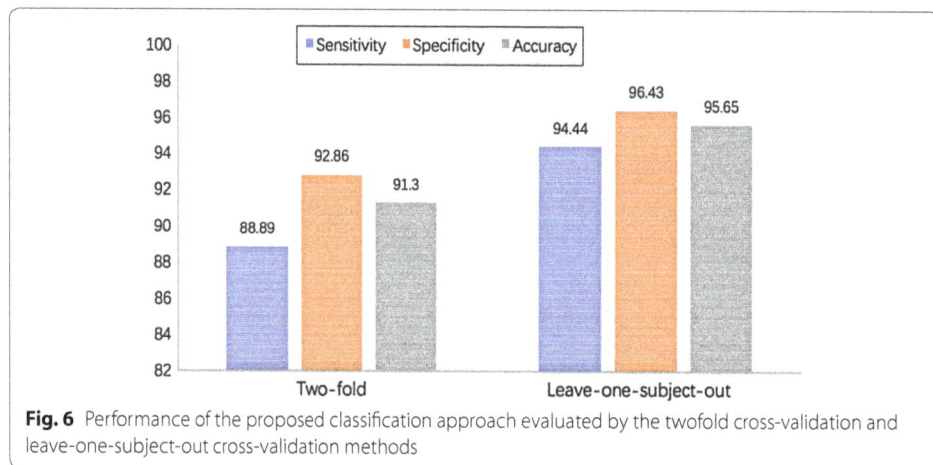

Fig. 6 Performance of the proposed classification approach evaluated by the twofold cross-validation and leave-one-subject-out cross-validation methods

Discussion

The methodology described in this study is expected to provide the clinicians with an efficient tool for assisted diagnosis of ACL-D knees. In comparison to other methods reported in [15, 16, 20, 24, 59–61], the proposed method focuses not only on providing evidence to support the claim that ACL-D knees demonstrate altered gait patterns compared to ACL-I knees, but also on providing an automatic and objective method to distinguish between patients with ACL-D knees and healthy controls with ACL-I knees. Almosnino et al. [15] aimed to identify, using Principal Component Analysis, strength curve features that explain the majority of variation between the injured and uninjured knee, and to assess the capabilities of these features to detect the presence of injury. 43 unilateral ACL deficient patients were included in the experiments to discern between the ACL-D and contra lateral, healthy knees. The specificity, sensitivity and accuracy are reported to be 60.5%, 60.5% and 62%, respectively. Christian et al. [16] showed the potential of a pattern recognition system for the diagnoses of kinematic gait patterns in patients due to a recently ruptured ACL. Principal component analysis and recursive feature elimination were used to extract features from 3D marker trajectories. Seven patients with acute ACL rupture were included in the

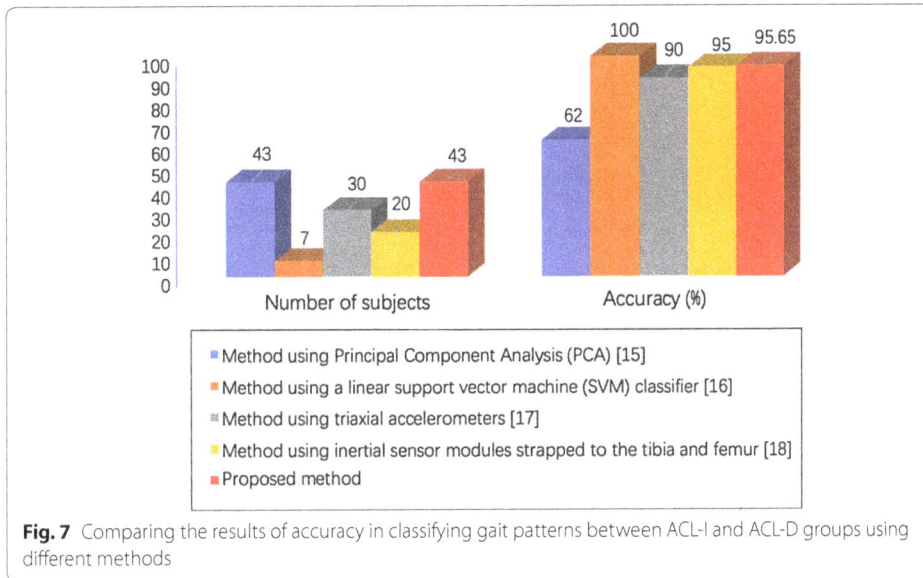

Fig. 7 Comparing the results of accuracy in classifying gait patterns between ACL-I and ACL-D groups using different methods

experiment and cross validation yielded 100% accuracy. However, the database used is too small which may weaken the persuasion of the classification performance. Berruto et al. [17] used tibial accelerometers to quantify pivot-shift differences between knees for subjects with unilateral ACL injuries. They considered only acceleration-based metrics, and discrimination of the side of ACL deficiency was accomplished by comparing the magnitudes of accelerations measured for the two legs. Accuracy of correctly identifying the injured knee was roughly 90%. Kopf et al. [18] performed a study similar to [17], in which 20 subjects with unilateral ACL deficiency were graded with inertial sensor modules strapped to the tibia and femur. All 3 metrics based on accelerometer measurements were found to be significantly different between injured and uninjured knees of subjects with unilateral ACL deficiency. They did not explicitly state accuracies in determining the side of ACL injury, but examination of their results suggested an accuracy of 95% (19 of 20) based on acceleration difference. Comparison of the classification performance to other state-of-the-art methods between ACL-I and ACL-D groups is shown in Fig. 7.

Different from the methods in the above-mentioned literature, our method focused on modeling the human gait and extracting the disparity of gait system dynamics between ACL-D and ACL-I knees for the discrimination task. It abandoned the traditional and direct comparison of lower extremity motion parameters between ACL-D and ACL-I knees and adopted instead the modeling, identification and classification of gait dynamics based on motion parameters. This may better explain and reveal the motion principle of pathological and healthy gaits hidden underneath the parameters extracted through PSR and ED. The proposed method serves not only as a measure of kinematic variability and discrimination between two groups of patients with ACL deficiency and healthy controls, but also as a non-invasive, objective and assistant technical means to other diagnostic approaches such as X-rays, MRI, arthroscopy, etc.

However, there are some limitations in the present study which need further improvement. Experiments were carried out on a small database and more participants need to be recruited to verify the effectiveness. At current stage, the proposed method is more suitable to be a tool applicable for gait reeducation on previously diagnosed patients. It is not easy for the clinicians to distinguish a deficiency of the ACL from a possible injury of another structure of the knee, such as posterior cruciate ligament or collateral ligaments, since these injuries may also lead to the same gait patterns. Only when the patients were highly suspected to have the ACL injury, can the proposed method be used to diagnose it as an assistant tool. In future work, injury of other structures of the knee, including posterior cruciate ligament or collateral ligaments injury and their related gait patterns, may also be included in our study to assist in diagnosing the knee injury more accurately. In addition, other parameters regarding different knee lessons can be adopted to improve the classification accuracy.

Conclusions

The results of this study indicate that the pattern classification of knee kinematic data can offer an objective and invasive method to assess the gait disparity between ACL-D and ACL-I knees. These results demonstrate the potential of the proposed technique for detecting pathological gait patterns caused by ACL deficiency by analysing and measuring the disparity of gait system dynamics using PSR, ED and neural networks. PSR is one of the most used methods which is the time-delay embedding and fits well with 1-dimensional time series. The d-dimensional space of delay coordinates serves as a pseudo state-space which provides a natural setting to approximate the quantitative aspects of the gait system dynamics. PSR plots gait system dynamics along the gait signal trajectory in a 3D phase space diagram and visualizes the gait system dynamics. ED measures and derives gait features, which are fed into RBF neural networks for the modeling, identification and classification of gait system dynamics between ACL-D and ACL-I knees. However, some limitations such as the small size of the database, the regulation principle of the embedding dimension and time lag, still need to be improved and overcome. Future work will include a clinical validation of the proposed technique with a larger number of patients with ACL deficiency and age-matched healthy controls. In the present study, PSR parameters such as the time lag and embedding dimension are with fixed values. Assessments of the relationship between the embedding dimension, time lag and the classification accuracy can also be considered in future investigations.

Authors' contributions
Study concept and design (WZ); drafting of the manuscript (WZ and WW); critical revision of the manuscript for important intellectual content (WZ, WW and CY); obtained funding (WW, WZ and YZ); administrative, technical, and material support (LM); study supervision (WZ and YZ). WW, WZ and YZ contributed equally. All authors read and approved the final manuscript.

Author details
[1] Department of Acupuncture, Longyan First Hospital, Longyan 364000, China. [2] School of Physics and Mechanical & Electrical Engineering, Longyan University, Longyan 364012, China. [3] Department of Orthopaedic Surgery, Guangzhou General Hospital of Guangzhou Military Command, Guangzhou 510010, China. [4] Department of Mechanical, Industrial and Systems Engineering, University of Rhode Island, Kingston, RI 02881, USA. [5] Department of Orthopedics, Guangdong General Hospital, Guangdong Academy of Medical Sciences, Guangzhou 510080, China.

Acknowledgements
This work was supported by the National Natural Science Foundation of China (Grant Nos. 61773194, 61304084, 31700880), by the Natural Science Foundation of Fujian Province (Grant No. 2018J01542), by the Program for New Century Excellent Talents in Fujian Province University, by the Science and Technology Planning Project of Guangzhou city (Grant No. 201803010106) and by the Science and Technology Project of Longyan City (Grant No. 2017LY85).

Competing interests
The authors declare that they have no competing interests.

References

1. Gao B, Cordova ML, Zheng NN. Three-dimensional joint kinematics of ACL-deficient and ACL-reconstructed knees during stair ascent and descent. Hum Mov Sci. 2012;31(1):222–35.
2. Ren S, Yu Y, Shi H, Miao X, Jiang Y, Liang Z, Ao Y. Three dimensional knee kinematics and kinetics in ACL-deficient patients with and without medial meniscus posterior horn tear during level walking. Gait Posture. 2018;66:26–31.
3. Musahl V, Getgood A, Neyret P, Claes S, Burnham JM, Batailler C, Karlsson J. Contributions of the anterolateral complex and the anterolateral ligament to rotatory knee stability in the setting of ACL injury: a roundtable discussion. Knee Surg Sports Traumatol Arthrosc. 2017;25(4):997–1008.
4. Adouni M, Shirazi-Adl A, Marouane H. Role of gastrocnemius activation in knee joint biomechanics: gastrocnemius acts as an ACL antagonist. Comput Methods Biomech Biomed Eng. 2016;19(4):376–85.
5. Wellsandt E, Arundale A, Manal K, Buchanan TS, Snyder-Mackler L. Lower hop scores related to gait asymmetries after ACL injury: identifying associations related to the development of early onset knee OA. Osteoarthritis Cartilage. 2015;23:A279.
6. Kiapour AM, Kiapour A, Goel VK, Quatman CE, Wordeman SC, Hewett TE, Demetropoulos CK. Uni-directional coupling between tibiofemoral frontal and axial plane rotation supports valgus collapse mechanism of ACL injury. J Biomech. 2015;48(10):1745–51.
7. Weiss K, Whatman C. Biomechanics associated with patellofemoral pain and ACL injuries in sports. Sports Med. 2015;45(9):1325–37.
8. Chen HC, Wu CH, Wang CK, Lin CJ, Sun YN. A Joint-constraint model-based system for reconstructing total knee motion. IEEE Trans Biomed Eng. 2014;61(1):171–81.
9. Lam MH, Fong DTP, Yung PSH, Ho EP, Chan WY, Chan KM. Knee stability assessment on anterior cruciate ligament injury: clinical and biomechanical approaches. Sports Med Arthrosc Rehab Therapy Technol. 2009;1(1):1.
10. Altilio R, Paoloni M, Panella M. Selection of clinical features for pattern recognition applied to gait analysis. Med Biol Eng Comput. 2017;55(4):685–95.
11. El Habachi A, Moissenet F, Duprey S, Cheze L, Dumas R. Global sensitivity analysis of the joint kinematics during gait to the parameters of a lower limb multi-body model. Med Biol Eng Comput. 2015;53(7):655–67.
12. Wolf A, Degani A. Recognizing knee pathologies by classifying instantaneous screws of the six degrees-of-freedom knee motion. Med Biol Eng Comput. 2007;45(5):475–82.
13. Padole C, Proenca H. An aperiodic feature representation for gait recognition in cross-view scenarios for unconstrained biometrics. Patt Anal Appl. 2017;20(1):73–86.
14. Jensen U, Kugler P, Ring M, Eskofier BM. Approaching the accuracy-cost conflict in embedded classification system design. Patt Anal Appl. 2016;19(3):839–55.
15. Almosnino S, Brandon SC, Day AG, Stevenson JM, Dvir Z, Bardana DD. Principal component modeling of isokinetic moment curves for discriminating between the injured and healthy knees of unilateral ACL deficient patients. J Electromyog Kinesiol. 2014;24(1):134–43.
16. Christian J, Kroll J, Strutzenberger G, Alexander N, Ofner M, Schwameder H. Computer aided analysis of gait patterns in patients with acute anterior cruciate ligament injury. Clin Biomech. 2016;33:55–60.
17. Berruto M, Uboldi F, Gala L, Marelli B, Albisetti W. Is triaxial accelerometer reliable in the evaluation and grading of knee pivot shift phenomenon? Knee Surg Sports Traumatol Arthrosc. 2013;21(4):981–5.
18. Kopf S, Kauert R, Halfpaap J, Jung T, Becker R. A new quantitative method for pivot shift grading. Knee Surg Sports Traumatol Arthrosc. 2012;20(4):718–23.
19. Knoll Z, Kocsis L, Kiss RM. Gait patterns before and after anterior cruciate ligament reconstruction. Knee Surg Sports Traumatol Arthrosc. 2004;12(1):7–14.
20. Houck JR, Duncan A, Haven KED. Knee and hip angle and moment adaptations during cutting tasks in subjects with anterior cruciate ligament deficiency classified as noncopers. J Orthop Sports Phys Therapy. 2005;35(8):531–40.
21. Houck JR, De Haven KE, Maloney M. Influence of anticipation on movement patterns in subjects with ACL deficiency classified as noncopers. J Orthop Sports Phys Therapy. 2007;37(2):56–64.
22. Takeda K, Hasegawa T, Kiriyama Y, Matsumoto H, Otani T, Toyama Y, Nagura T. Kinematic motion of the anterior cruciate ligament deficient knee during functionally high and low demanding tasks. J Biomech. 2014;47(10):2526–30.
23. Ismail SA, Button K, Simic M, Van Deursen R, Pappas E. Three-dimensional kinematic and kinetic gait deviations in individuals with chronic anterior cruciate ligament deficient knee: A systematic review and meta-analysis. Clin Biomech. 2016;35:68–80.
24. Andriacchi TP, Dyrby CO. Interactions between kinematics and loading during walking for the normal and ACL deficient knee. J Biomech. 2005;38(2):293–8.
25. Kvist J, Good L, Tagesson S. Changes in knee motion pattern after anterior cruciate ligament injury—a case report. Clin Biomech. 2007;22(5):551–6.
26. Lindström M, Felländer-Tsai L, Wredmark T, Henriksson M. Adaptations of gait and muscle activation in chronic ACL deficiency. Knee Surg Sports Traumatol Arthrosc. 2010;18(1):106–14.
27. Fuentes A, Hagemeister N, Ranger P, Heron T, de Guise JA. Gait adaptation in chronic anterior cruciate ligament-deficient patients: pivot-shift avoidance gait. Clin Biomech. 2011;26(2):181–7.
28. Chen CH, Li JS, Hosseini A, Gadikota HR, Gill TJ, Li G. Anteroposterior stability of the knee during the stance phase of gait after anterior cruciate ligament deficiency. Gait Posture. 2012;35(3):467–71.
29. Gardinier ES, Manal K, Buchanan TS, Snyder-Mackler L. Clinically-relevant measures associated with altered contact forces in patients with anterior cruciate ligament deficiency. Clin Biomech. 2014;29(5):531–6.
30. Shabani B, Bytyqi D, Lustig S, Cheze L, Bytyqi C, Neyret P. Gait changes of the ACL-deficient knee 3D kinematic assessment. Knee Surg Sports Traumatol Arthrosc. 2015;23(11):3259–65.
31. Vairis A, Stefanoudakis G, Petousis M, Vidakis N, Tsainis AM, Kandyla B. Evaluation of an intact, an ACL-deficient, and a reconstructed human knee joint finite element model. Comput Methods Biomech Biomed Eng. 2016;19(3):263–70.
32. Robinson MA, Donnelly CJ, Tsao J. Impact of knee modeling approach on indicators and classification of anterior cruciate ligament injury risk. Med Sci Sports Exerc. 2014;46(7):1269–76.
33. Atarod M, Frank CB, Shrive NG. Kinematic and kinetic interactions during normal and ACL-Deficient gait: a longitudinal in vivo study. Ann Biomed Eng. 2014;42(3):566–78.

34. Marouane H, Shirazi-Adl A, Hashemi J. Quantification of the role of tibial posterior slope in knee joint mechanics and ACL force in simulated gait. J Biomech. 2015;48(10):1899–905.
35. McCarthy I, Hodgins D, Mor A, Elbaz A, Segal G. Analysis of knee flexion characteristics and how they alter with the onset of knee osteoarthritis: a case control study. BMC Musculoskelet Disord. 2013;14(1):169.
36. Takens F. Detecting strange attractors in turbulence, in: Dynamical Systems and Turbulence, Warwick 1980, Berlin: Springer; 1981. p. 366–81.
37. Xu B, Jacquir S, Laurent G, Bilbault JM, Binczak S. Phase space reconstruction of an experimental model of cardiac field potential in normal and arrhythmic conditions. In: 35th Annual international conference of the IEEE engineering in medicine and biology society. 2013. p. 3274–7.
38. Sharma R, Pachori RB. Classification of epileptic seizures in EEG signals based on phase space representation of intrinsic mode functions. Expert Syst Appl. 2015;42(3):1106–17.
39. Alkjaer T, Raffalt PC, Dalsgaard H, Simonsen EB, Petersen NC, Bliddal H, Henriksen M. Gait variability and motor control in people with knee osteoarthritis. Gait Posture. 2015;42(4):479–84.
40. Piórek M, Josiński H, Michalczuk A, Świtoński A, Szczesna A. Quaternions and joint angles in an analysis of local stability of gait for different variants of walking speed and treadmill slope. Inf Sci. 2017;384:263–80.
41. Georgescu M, Petcu A, Tarnita D. Influences of speed and treadmill inclination on the local dynamic stability of human knee joint. In: Applied mechanics and materials (Vol. 880). Switzerland: Trans Tech Publications; 2018. p. 130–5.
42. Zhang Y, Yao Z, Wang S, Huang W, Ma L, Huang H, Xia H. Motion analysis of Chinese normal knees during gait based on a novel portable system. Gait Posture. 2015;41(3):763–8.
43. Fung P, Mok K, Leow R, Fu S, Yung P, Chan K. Knee kinematics of ACL-deficient patients: a development of a portable motion analysis system. J Hum Sport Exerc. 2018. https://doi.org/10.14198/jhse.2018.134.11.
44. Zeng X, Ma L, Lin Z, Huang W, Huang Z, Zhang Y, Mao C. Relationship between Kellgren-Lawrence score and 3D kinematic gait analysis of patients with medial knee osteoarthritis using a new gait system. Sci Rep. 2017;7(1):4080.
45. Yeung MY, Fu SC, Chua EN, Mok KM, Yung PSH, Chan KM. Use of portable motion analysis system for knee stability assessment in ACL deficiency during single-leg-hop. Asia Pac J Sports Med Arthrosc Rehabil Technol. 2016;6:24.
46. Thalheimer W, Cook S. How to calculate effect sizes from published research: a simplified methodology. Somerville: Work-Learning Research Inc; 2002.
47. Lakens D. Calculating and reporting effect sizes to facilitate cumulative science: a practical primer for t-tests and ANOVAs. Front Psychol. 2013;4:863.
48. Cohen J. The statistical power of abnormal-social psychological research: a review. J Abnorm Soc Psychol. 1962;65:145–53.
49. Sedlmeier P, Gigerenzer G. Do studies of statistical power have an effect on the power of the studies? Psychol Bull. 1989;105:309–16.
50. Lee SH, Lim JS, Kim JK, Yang J, Lee Y. Classification of normal and epileptic seizure EEG signals using wavelet transform, phase-space reconstruction, and Euclidean distance. Comput Methods Programs Biomed. 2014;116(1):10–25.
51. Sivakumar B. A phase-space reconstruction approach to prediction of suspended sediment concentration in rivers. J Hydrol. 2002;258(1–4):149–62.
52. Michael S. Applied nonlinear time series analysis: applications in physics, physiology and finance (Vol. 52). World Scientific; 2005.
53. Merigó JM, Casanovas M. Induced aggregation operators in the Euclidean distance and its application in financial decision making. Expert Syst Appl. 2011;38:7603–8.
54. Wang C, Hill DJ. Learning from neural control. IEEE Trans Neural Netw. 2006;17(1):130–46.
55. Wang C, Hill DJ. Deterministic learning and rapid dynamical pattern recognition. IEEE Trans Neural Netw. 2007;18(3):617–30.
56. Wang C, Hill DJ. Deterministic learning theory for identification, recognition and control. Boca Raton: CRC Press; 2009.
57. Wang C, Chen T, Chen G, Hill DJ. Deterministic learning of nonlinear dynamical systems. Int J Bifurcation Chaos. 2009;19(4):1307–28.
58. Farrell J. Stability and approximator convergence in nonparametric nonlinear adaptive control. IEEE Trans Neural Netw. 1998;9(5):1008–20.
59. Tsepis E, Giakas G, Vagenas G, Georgoulis A. Frequency content asymmetry of the isokinetic curve between ACL deficient and healthy knee. J Biomech. 2004;37(6):857–64.
60. Kaplan Y. Identifying individuals with an anterior cruciate ligament deficient knee as copers and non-copers: a narrative literature review. J Sci Med Sport. 2015;19:e26.
61. Iliopoulos E, Galanis N, Iosifidis M, Zafeiridis A, Papadopoulos P, Potoupnis M, Kirkos J. Anterior cruciate ligament deficiency reduces walking economy in copers and non-copers. Knee Surg Sports Traumatol Arthrosc. 2017;25(5):1403–11.

8

Changes in heart rate variability with respect to exercise intensity and time during treadmill running

Kenneth J. Hunt[1] and Jittima Saengsuwan[2,3]*

*Correspondence:
sjittima@kku.ac.th
[2] Department
of Rehabilitation Medicine,
Srinagarind Hospital
and Faculty of Medicine,
Khon Kaen University, Khon
Kaen, Thailand
Full list of author information
is available at the end of the
article

Abstract

Background: Heart rate variability (HRV) arises from the complex interplay of sympathetic and parasympathetic autonomic regulation of heart rate. Ultra-low frequency (ULF) and very-low frequency (VLF) components of HRV play a crucial role in automatic HR controllers, but these frequency bands have hitherto largely been neglected in HRV studies. The aim of this work was to investigate changes in ULF and VLF heart rate variability with respect to exercise intensity and time during treadmill running.

Methods: RR intervals were determined by ECG in 21 healthy male participants at rest, and during moderate and vigorous-intensity treadmill running; each of these three tests had a duration of 45 min. Time dependence of HRV was investigated for moderate and vigorous running intensities by dividing the constant-speed stages into three consecutive windows of equal duration (~ 14 min), denoted w_1, w_2 and w_3. ULF and VLF power were computed using Lomb-Scargle power spectral density estimates.

Results: For both the ULF and VLF frequency bands, mean power was significantly different between the resting, moderate and vigorous intensity levels (overall $p < 0.001$): mean power was lower for moderate vs. rest ($p < 0.001$), for vigorous vs. rest ($p < 0.001$), and for vigorous vs. moderate ($p < 0.001$). For both ULF and VLF and moderate intensity, mean power was significantly different between the three time windows (overall $p < 0.001$ for ULF, overall $p = 0.041$ for VLF): for ULF, mean power was lower for w_2 vs. w_1 ($p = 0.031$) and for w_3 vs. w_1 ($p = 0.001$); for VLF, mean power was lower for w_3 vs. w_1 ($p = 0.007$). For ULF and vigorous intensity, there was no significant difference in mean power between the three time windows (overall $p = 0.12$). For VLF and vigorous intensity, mean power was significantly different between w_1, w_2 and w_3 (overall $p < 0.001$): mean power was lower for w_2 vs. w_1 ($p = 0.001$) and for w_3 vs. w_1 ($p < 0.001$).

Conclusions: The degree of HRV in terms of ULF and VLF power was found to decrease with increasing intensity of exercise. HRV was also observed to decrease over time, but it remains to clarify whether these changes are due to time itself or to increases in HR related to cardiovascular drift. For feedback control applications, attention should be focused on meeting performance targets at low intensity and during the early stages of exercise.

Keywords: Heart rate control, Heart rate variability, Spectral analysis, Treadmills

Background

Heart rate variability (HRV) is ordinarily characterised by beat-to-beat variations in the time between peaks in the QRS complex of the ECG wave, i.e. by variations in the RR interval [1]. A formal set of signal analysis standards for measurement, interpretation and clinical application of HRV has been established [2, 3]; these standards comprise time and frequency-domain methods and combinations thereof.

In the frequency domain, HRV analysis has classically been described for four distinct bands [2, 3]:

1. Ultra-low frequency (ULF), where $f < 0.003$ Hz;
2. Very-low frequency (VLF) with $0.003 \leq f < 0.04$ Hz;
3. Low frequency (LF), $0.04 \leq f < 0.15$ Hz; and
4. High frequency (HF), $0.15 \leq f \leq 0.4$ Hz.

Since the frequency $f = 0.003$ Hz at the border between the ULF and VLF bands corresponds to a time period of 333 s, short-term recordings of duration < 5 min are restricted to analysis of VLF, LF and HF characteristics, while the ULF band requires longer-term recording [3]; clinically, portable ECG monitors are employed which typically record for up to 24 h (Holter monitor). It is also noted that HR signals may contain power at frequencies above 0.4 Hz, the upper bound of the HF range: consider a notional HR of 180 beats/min, which is 3 beats/s or $f = 3$ Hz.

It has been widely believed that HF power primarily reflects parasympathetic cardiac drive, that LF power has a predominantly sympathetic component [4], and that the LF/HF ratio can thus be used as a measure of sympatho-vagal balance [5, 6], i.e. the relative contributions of sympathetic and parasympathetic activity. However, this delineation of the different compartments of autonomic nervous system activity and the supposed correspondence with HRV power in the different frequency bands has recently been challenged [7, 8]; it now seems that HRV is the result of more complex sympathetic-parasympathetic interactions that are not yet fully understood [9]. It has furthermore been proposed that ULF and VLF power might be predictors of cardiac health [10], but it has also been pointed out that understanding of the mechanisms involved is presently limited [4].

Although it remains to fully elucidate the complex neural mechanisms of HRV and the associated implications for health, it is clear that HRV is an important phenomenon to be considered in the design of engineering systems employed in support of prescription and implementation of exercise training programmes: contemporary recommendations for exercise duration and intensity use HR for delineation of training regimes [11, 12].

In this regard, the focus in the present work is on the frequency-domain characteristics of HRV during treadmill running; concomitantly, it is intended to apply the knowledge gained to the design and analysis of feedback systems for automatic control of HR during treadmill exercise. It has previously been noted that the principal challenge in the design of controllers for HR is to ensure that the feedback system maintains acceptable performance in the face of disturbances to HR caused by physiological HR variability [13]. To this end, attention is focused here on the ULF and VLF bands: this is because, firstly, HR controllers are usually designed with low-pass characteristics and with a crossover region lying within the VLF band (typically, closed-loop bandwidths are around 0.01 Hz, [13]); secondly, disturbances in the ULF and VLF bands caused by HRV can excite the control signal, i.e. the treadmill speed command, to a degree that would be perceptible to, and possibly unacceptable to, the runner.

Previous investigations of feedback controllers for HR have noted that HRV appears to decrease over time during moderate-to-vigorous intensity exercise of duration 45 min, albeit these observations were obtained indirectly using time-domain measures of closed-loop performance rather than from direct analysis of RR intervals: there were substantial and significant decreases in root-mean-square HR tracking error and in average control signal power [13, 14].

A recent review of HRV responses during exercise concluded twofold: that the primary effect is a pronounced reduction in HRV with increasing exercise intensity, up to a moderate intensity corresponding approximately to the first ventilatory threshold; and, secondarily, that HRV decreases over time, but only during low-to-moderate intensity exercise and when accompanied by cardiovascular drift [15]. However, in terms of frequency-domain analysis, the studies included in the review reported only LF, HF and total power components, presumably due to the relatively short duration of exercise bouts that were investigated. It therefore remains an open question as to how ULF and VLF components of HRV are affected by intensity and duration of exercise.

The aim of the present work was to directly investigate changes in ULF and VLF heart rate variability with respect to exercise intensity and time during treadmill running. For this purpose, a recording duration of 45 min was chosen which is within the recommended range for development and maintenance of cardiorespiratory fitness and which is sufficiently long to capture ULF components, and exercise intensities were studied in accordance with levels used in training-intensity prescription (recommended duration is on the range 20–60 min, and recommended intensity is from 'moderate' to 'vigorous', based on heart-rate reserve [11, 12]).

Methods

Participants

Twenty-one males on the range 24–51 years participated. Inclusion criteria applied during the selection of this cohort were: male, age between 18 and 60 years, able-bodied and physically healthy. Exclusion criteria were known cardiovascular, pulmonary or musculoskeletal problems that might have interfered with or contraindicated moderate to vigorous intensity treadmill exercise.

Procedures

For each participant, three ECG measurements were made: 45 min of rest while supine; 45 min of running on a treadmill at moderate intensity; and 45 min running at vigorous intensity. These are referred to in the sequel as test conditions 'r', 'm' and 'v', respectively.

Exercise intensity levels were defined using heart rate reserve (HRR) according to formal guidelines [11, 12]: HRR is the difference between an individual's maximum and resting heart rates, $HRR = HR_{max} - HR_{rest}$: moderate intensity is 40–59% of HRR; vigorous intensity is 60–89% of HRR. Maximal heart rate was taken as the age-related prediction given by $HR_{max} = 220$-age [16].

The resting measurement was carried out on the same day as, but prior to, the first treadmill test and was also used to determine resting HR: this was taken as the mean HR during the 6th min of the resting measurement. The intensity for each participant's first treadmill test (i.e. m or v) was randomly selected. The second treadmill test was conducted on a separate day at the other intensity.

Each treadmill test was conducted as follows:

1. 10-min warm up running on the treadmill while speed was manually adjusted to find the lower end of the selected intensity range (i.e. 40% of HRR for m, where $HR = 0.4 \cdot HRR + HR_{rest}$; or 60% of HRR for v, where $HR = 0.6 \cdot HRR + HR_{rest}$); in this stage, HR was monitored using a chest belt.
2. 10-min of rest and fitting of the ECG electrodes.
3. 5-min of recorded rest while standing quietly on the treadmill.
4. 45-min of constant-speed running at the speed determined above for intensity m or v.
5. Up to 10-min cool down at a comfortable walking pace of 3.5 km/h.

Participants were instructed to avoid strenuous exercise, alcohol consumption and smoking in the 24 h preceding each test, to refrain from caffeine consumption in the preceding 12 h, and from partaking of a heavy meal in the 4 h before testing.

From the total of 21 participants, some data sets could not be included in the data analysis due to ECG measurement problems: complete data records were obtained for all three conditions, r, m and v, from 15 participants (thus, $n = 15$ for the analysis of intensity dependence); for condition m from 18 participants ($n = 18$ for analysis of time dependence at intensity m); and for condition v from 15 participants ($n = 15$ for analysis of time dependence at intensity v).

Measurement instruments

A computer-controlled treadmill was employed (model Venus, h/p/cosmos Sports and Medical GmbH, Nussdorf-Traunstein, Germany). During the warm up stage of each test, HR was monitored using a chest belt (T34, Polar Electro Oy, Kempele, Finland). For ECG measurements while running, the treadmill was controlled directly from the ECG software (below) according to a pre-specified speed profile as described above.

RR intervals were obtained using a wireless ECG recording system with up to 12 leads (custo cardio 100 BT ECG system and the associated custo diagnostic professional software, version 4.3; custo med GmbH, Ottobrunn, Germany). In the present study, a 7-lead subset of the full ECG configuration was employed to allow derivation of RR intervals and heart rate. This involved the application of five electrodes, placed according to the manufacturer's guidelines: right arm and left arm (RA, LA, positioned just below the collarbones); right leg and left leg (RL, LL, positioned just below the rib cage on each side); and one chest electrode (V_1, placed at the 4th intercostal space at the right sternal border).

ECG data were recorded at a sample frequency of 1 kHz. Following each measurement, raw RR intervals were exported with a resolution of 1 ms.

Outcomes and data processing

From the RR time series, power in the ULF and VLF frequency bands was computed using the Lomb-Scargle least-squares spectral analysis method for spectral density estimation. This method was chosen as it is specifically designed and optimised for non-uniformly-spaced data sets such as RR time series.

Prior to spectral analysis, data sets were processed for artefact detection and replacement, then trend removal, and finally low-pass filtering. Artefacts mainly occur due to the running motion and the effect of this on the ECG electrodes. Artefact detection and replacement was performed using an impulse rejection filter which was proposed for spectral analysis of biomedical signals in general [17], and which was previously applied specifically for the preprocessing of RR time series [18]. In this method, artefact detection is based on a Gaussian test statistic and a user-defined threshold (this was set to 5 in the present study). Artefact replacement uses a median filter with a specified window length (set to the value 10 here). For trend removal, a 3rd-order polynomial fit was estimated and subtracted from the data. Low-pass filtering was performed using a 12th-order Butterworth filter with cutoff frequency 0.4 Hz.

All of the above data preprocessing and spectral analysis was carried out using a custom-designed software tool for HRV analysis implemented in Matlab (The Mathworks, Inc., Natick, USA).

For investigation of the dependence of HRV on intensity (r, m and v), power values were calculated using the 45-min resting measurements and the steady-state portions of the constant-speed-running phases of duration 45 min of the tests at moderate and vigorous intensities.

The dependence of HRV on time was investigated for moderate and vigorous running intensities by dividing the constant-speed stages into three consecutive windows of equal duration ($\sim$ 14 min), denoted as windows w_1, w_2 and w_3.

Statistical analysis

One-way repeated-measures analysis of variance (ANOVA) was applied to investigate possible differences in mean ULF and VLF power under the different test conditions described above, i.e. differences in HRV with respect to intensity and time. Normality of the data sets was checked using a Kolmogorov-Smirnov test with Lilliefors correction; sphericity was checked using Mauchly's test.

The significance level for all tests was set to 5% ($\alpha = 0.05$). Whenever the ANOVA indicated the existence of a significant difference (i.e. overall $p < 0.05$), post-hoc pairwise comparisons were carried out using Bonferroni correction.

Prior to statistical analysis, all data sets were log-transformed ($\log_{10}$) to preserve normality: since the average power of a signal $x(t)$ is related to $x^2(t)$, it was anticipated that the data would follow an approximately log-normal distribution. It transpired that, for the transformed data sets, and using the tests noted above, no data set deviated significantly from normality or sphericity.

Statistical analysis was carried out using SPSS software (IBM Corp., USA).

Results

For illustration, data records for a single participant are provided (Figs. 1 and 2): these show the raw RR intervals recorded for the resting, moderate and vigorous intensity levels (Fig. 1a) and the corresponding Lomb-Scargle periodograms focused on the ULF and VLF bands (Fig. 1b); also shown are the power spectral density (PSD) estimates for this participant for the three time windows w_1, w_2 and w_3 at moderate (Fig. 2a) and vigorous (Fig. 2b) intensities.

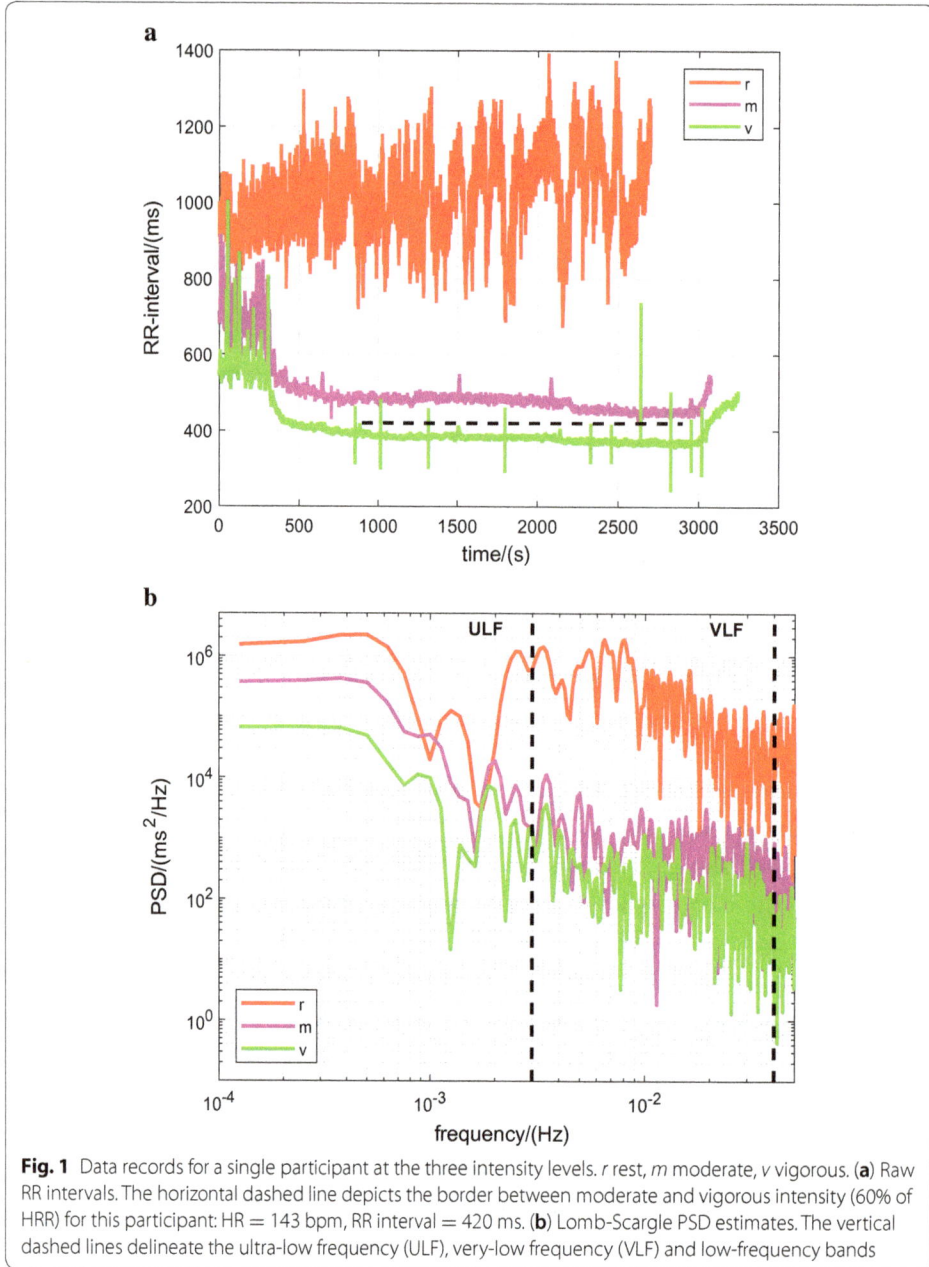

Fig. 1 Data records for a single participant at the three intensity levels. *r* rest, *m* moderate, *v* vigorous. (**a**) Raw RR intervals. The horizontal dashed line depicts the border between moderate and vigorous intensity (60% of HRR) for this participant: HR = 143 bpm, RR interval = 420 ms. (**b**) Lomb-Scargle PSD estimates. The vertical dashed lines delineate the ultra-low frequency (ULF), very-low frequency (VLF) and low-frequency bands

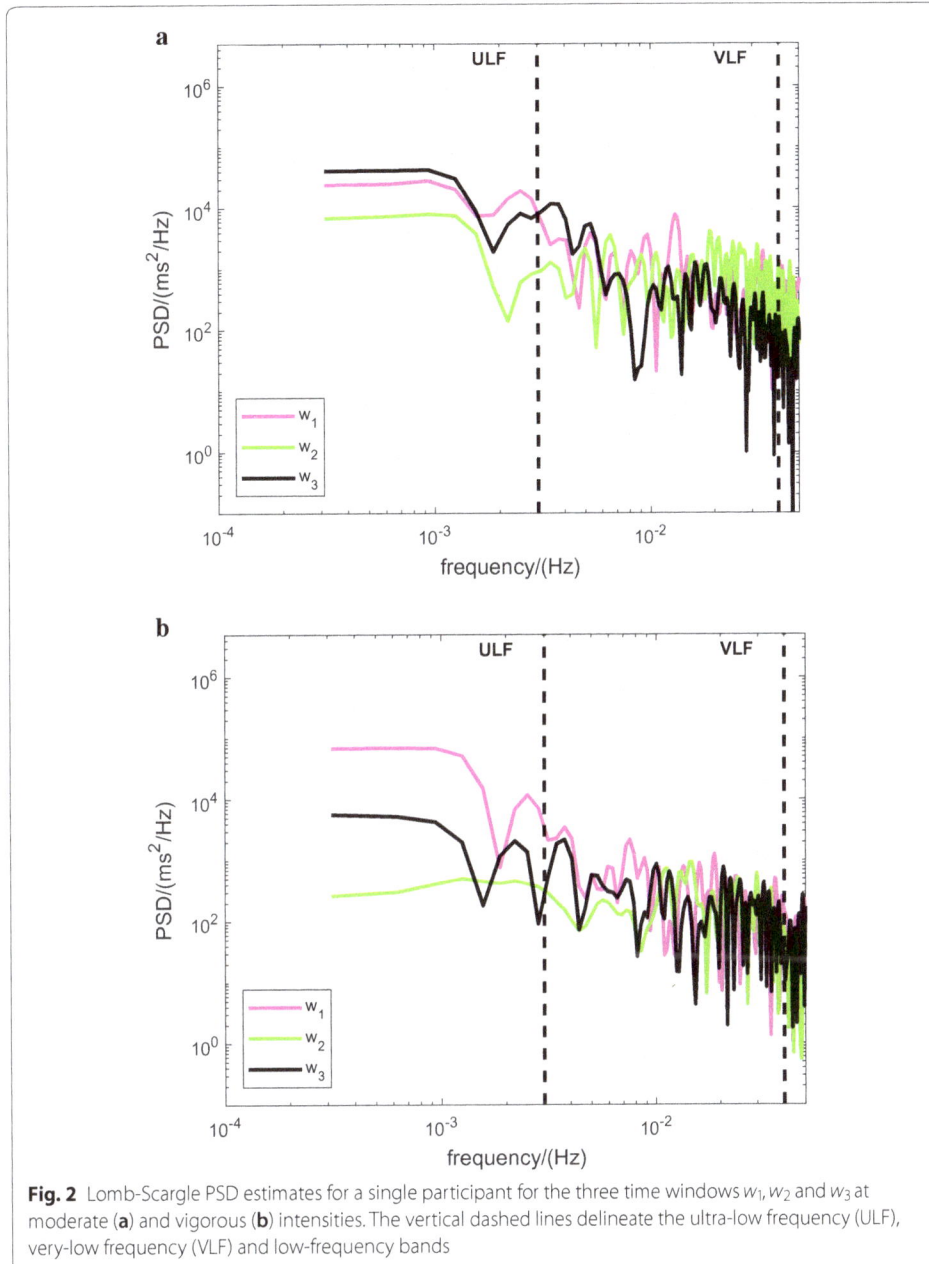

Fig. 2 Lomb-Scargle PSD estimates for a single participant for the three time windows w_1, w_2 and w_3 at moderate (**a**) and vigorous (**b**) intensities. The vertical dashed lines delineate the ultra-low frequency (ULF), very-low frequency (VLF) and low-frequency bands

Dependence of HRV on intensity

For both the ULF and VLF frequency bands, mean power was found to be significantly different between the resting, moderate and vigorous intensity levels (overall $p < 0.001$; Table 1). Paired comparisons showed that mean power was lower for the conditions moderate vs. rest ($p < 0.001$), for vigorous vs. rest ($p < 0.001$), and for vigorous vs. moderate ($p < 0.001$) (Table 1; Fig. 3a, b).

The reduction in power with increasing intensity is clearly reflected in the single-participant data records in both the time domain (lower dispersions in the respective RR intervals, Fig. 1a) and in the frequency domain (smaller areas under the PSD curves, Fig. 1b).

Table 1 Power in the ULF and VLF frequency bands: intensity dependence (see also Fig. 3a, b)

	r	m	v	overall p-value
ULF:	2.29 ± 0.35	0.79 ± 0.36	0.35 ± 0.31	< 0.001

$p < 0.001$ → r vs. m

$p < 0.001$ → m vs. v paired comparisons

$p < 0.001$ → r vs. v

	r	m	v	overall p-value
VLF:	3.00 ± 0.28	1.49 ± 0.27	1.07 ± 0.27	< 0.001

$p < 0.001$ → r vs. m

$p < 0.001$ → m vs. v paired comparisons

$p < 0.001$ → r vs. v

$n = 15$

Values are mean $\pm$ standard deviation, given as $\log_{10}([power/(ms^2)]/[1/(ms^2)])$

r rest, *m* moderate, *v* vigorous, *ULF* ultra-low frequency, *VLF* very-low frequency

Paired comparisons: *p*-values adjusted using Bonferroni correction

Table 2 Power in the ULF and VLF frequency bands: time dependence at moderate and vigorous intensities (see also Fig. 3c–f)

	w_1	w_2	w_3	overall p-value	p-values, paired comparisons		
					w_1 vs. w_2	w_1 vs. w_3	w_2 vs. w_3
m							
ULF	0.85 ± 0.48	0.50 ± 0.57	0.20 ± 0.53	< 0.001	0.031	0.001	0.255
VLF	1.87 ± 0.19	1.80 ± 0.35	1.74 ± 0.27	0.041	0.778	0.007	0.704
v							
ULF	0.30 ± 0.59	0.13 ± 0.36	-0.12 ± 0.60	0.12	–	–	–
VLF	1.57 ± 0.32	1.31 ± 0.32	1.22 ± 0.33	< 0.001	0.001	< 0.001	0.292

$n = 18$ (m); $n = 15$ (v)

Values are mean $\pm$ standard deviation, given as $\log_{10}([power/(ms^2)]/[1/(ms^2)])$

w_1 first window, w_2 second window, w_3 third window, *m* moderate, *v* vigorous, *ULF* ultra-low frequency, *VLF* very-low frequency

Paired comparisons: *p*-values adjusted using Bonferroni correction (presented only where overall *p*-value < 0.05)

Dependence of HRV on time, moderate intensity

For the moderate-intensity running condition, ULF mean power was found to be significantly different between the three time windows w_1, w_2 and w_3 (overall $p < 0.001$; Table 2). Similarly, at moderate intensity, VLF mean power was significantly different between the three time windows (overall $p = 0.041$; Table 2).

For ULF, paired comparisons showed that mean power was lower for the time windows w_2 vs. w_1 ($p = 0.031$) and for w_3 vs. w_1 ($p = 0.001$) (Table 2; Fig. 3c).

For VLF, paired comparisons showed that mean power was lower for the time window w_3 vs. w_1 ($p = 0.007$) (Table 2; Fig. 3d).

The observed differences can be discerned in the single-participant data record for moderate intensity (Fig. 2a, ULF and VLF bands).

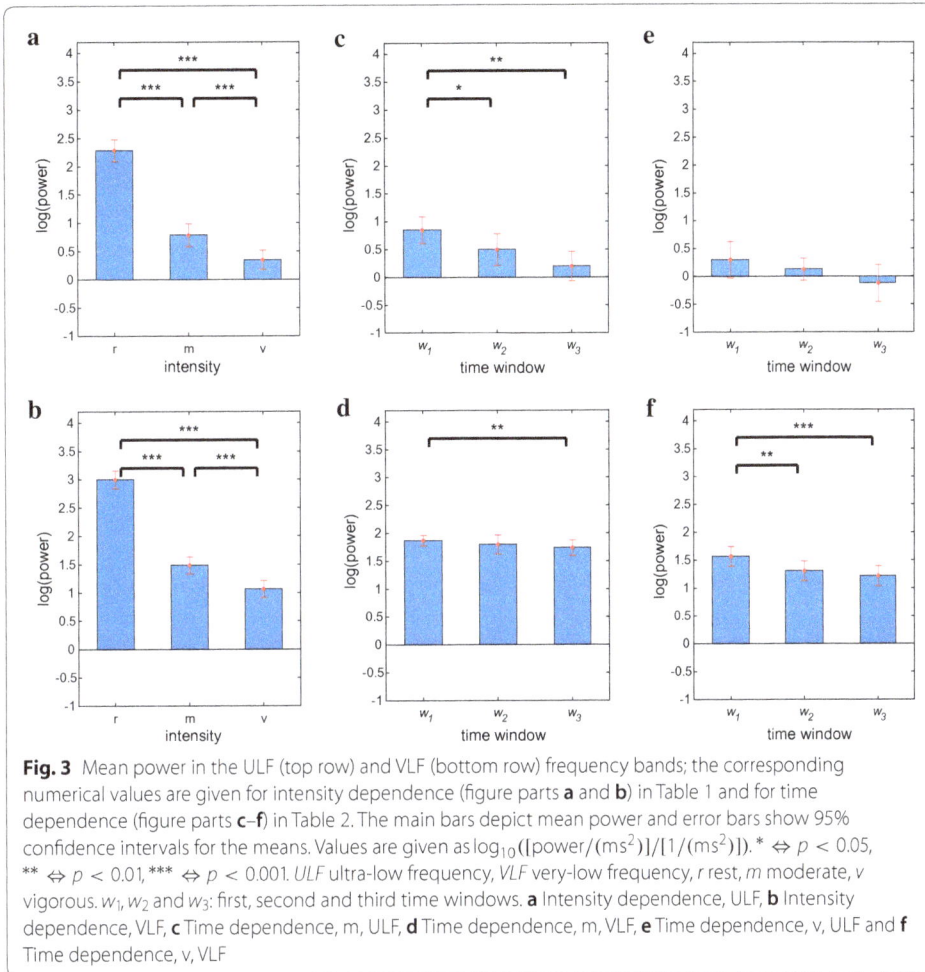

Fig. 3 Mean power in the ULF (top row) and VLF (bottom row) frequency bands; the corresponding numerical values are given for intensity dependence (figure parts **a** and **b**) in Table 1 and for time dependence (figure parts **c–f**) in Table 2. The main bars depict mean power and error bars show 95% confidence intervals for the means. Values are given as $\log_{10}([\text{power}/(ms^2)]/[1/(ms^2)])$. $* \Leftrightarrow p < 0.05$, $** \Leftrightarrow p < 0.01$, $*** \Leftrightarrow p < 0.001$. *ULF* ultra-low frequency, *VLF* very-low frequency, *r* rest, *m* moderate, *v* vigorous. w_1, w_2 and w_3: first, second and third time windows. **a** Intensity dependence, ULF, **b** Intensity dependence, VLF, **c** Time dependence, m, ULF, **d** Time dependence, m, VLF, **e** Time dependence, v, ULF and **f** Time dependence, v, VLF

Dependence of HRV on time, vigorous intensity

For the vigorous-intensity running condition, ULF mean power did not differ significantly between the three time windows w_1, w_2 and w_3 (overall $p = 0.12$; Table 2; Fig. 3e).

For the VLF frequency band, and while running at vigorous intensity, mean power was found to be significantly different between the three time windows w_1, w_2 and w_3 (overall $p < 0.001$; Table 2). Paired comparisons then showed that mean power was lower for the time windows w_2 vs. w_1 ($p = 0.001$) and for w_3 vs. w_1 ($p < 0.001$) (Table 2; Fig. 3f).

For the VLF band, the observed differences between the time windows can be discerned in part in the single-participant data record for vigorous intensity (Fig. 2b, VLF band).

Discussion

The aim of this work was to investigate changes in ULF and VLF heart rate variability with respect to exercise intensity and time during treadmill running.

Substantial and significant decreases in both ULF and VLF power were observed as intensity increased from rest to moderate-intensity running and then to vigorous running. These findings of decreasing HRV are generally consistent with previous studies included in the review by Michael et al. [15], albeit those studies focused only on LF,

HF and total power components. For example, the study of Tulpo et al. [19] reported a nonlinear decay of HRV as a function of exercise intensity: subjects had a decrease of LF and HF power from rest to light and moderate exercise intensity on a cycle ergometer. The study found a very small but significant decrease in LF power from moderate to vigorous exercise (80% of maximal oxygen uptake) and no significant decrease in LF and HF power thereafter. However, similar to most physiological studies, this article mainly focused on investigation of cardiac sympathetic and parasympathetic nervous system responses using relatively short-term recordings, so there was no report on ULF and VLF power.

Other studies have previously observed that VLF power increased during rhythmic activity (alternating rest and mild exercise) and normal/random activity when compared to rest [20], and that ULF power was higher during activity (ADL—activities of daily living) than at rest [21]. These apparent anomalies might be explained by the exercise conditions studied being of very low intensity when compared to the moderate running condition investigated in the present work: the intensity of the exercise conditions employed in those studies is not likely to have been substantially different from the resting intensity.

Significant decreases were observed over time in ULF and VLF power for the moderate-intensity running condition. At vigorous intensity, VLF power decreased significantly, while the observed decrease in ULF power was not significant. The latter might be explained by the fact that absolute ULF power levels at vigorous intensity were already very low during the first time window, thus making subsequent changes more difficult to detect. These observations are consistent with the results of previous studies which noted decreases over time in time-domain measures of HR-related signal intensity while using feedback control of HR during moderate-to-vigorous treadmill exercise of similar duration [13, 14].

Attenuation of HRV during exercise was also found in a previous study where subjects ran on a treadmill at an intensity of 60% of peak oxygen uptake: the LF and HF components were significantly higher at rest compared with exercise. Additionally, LF was significantly higher at the beginning of exercise (25–30 min of exercise) compared to near the end of exercise (85–90 min of exercise) [22].

It was suggested elsewhere that the mechanism for these time-related findings may not be the direct effect of exercise duration itself on HRV, but it may be due to dehydration, thus leading to reduced stroke volume and increased heart rate (cardiovascular drift) [15]. This concept is however challenged by the findings of Hunt et al. [13, 14], where heart rate was maintained constant over an exercise duration of 45 min by means of feedback control and automatic adaptation (reduction) in treadmill speed: despite the HR intensity staying constant, time-domain measures of HR tracking error and average control signal power were still seen to decrease. Since the latter studies used indirect time-domain measures of HRV, further investigations are recommended which combine feedback control of HR with direct frequency-domain analysis of RR intervals to more precisely elucidate the mechanisms underlying time-dependent changes in HRV.

The results of the present study have important implications for the design and analysis of feedback systems for automatic control of HR during treadmill exercise. HRV can be considered as a disturbance term that enters the feedback loop. This results in two main effects: HR will tend to deviate from the target HR level and this must be corrected by feedback action; and the HRV disturbance might, as a consequence of the feedback, cause unacceptable changes in the treadmill speed command [13]. The main challenge in feedback design is to achieve an acceptable tradeoff between accuracy of target HR tracking and control signal intensity.

The focus in the present work on ULF and VLF heart rate variability is important also for the design of exercise programmes and strategies. This is because these components of HRV dominate during long-duration exercise, and must be accounted for by appropriate design of the feedback control system for heart rate. LF and HF components, on the other hand, are less critical for two reasons: LF and HF power are lower than ULF and VLF power; disturbances within the LF and HF frequency bands will generally be attenuated in a natural way due to the low-pass characteristics of typical feedback designs.

The present work shows that it is most critical to meet these demands for low intensity exercise, and during the initial stages of any exercise bout, because the degree of HRV is higher under these conditions; for higher levels of exercise intensity, and/or as exercise progresses over time, reduction in HRV might make it possible to adaptively increase the feedback bandwidth to improve the accuracy of HR tracking performance but without necessarily provoking unacceptable levels of control signal activity.

Conclusions

The degree of HRV in terms of ULF and VLF power was found to decrease with increasing intensity of exercise. HRV was also observed to decrease over time, but it remains to clarify whether these changes are due to time itself or to increases in HR related to cardiovascular drift.

With regard to design of feedback controllers for HR, these results suggest that attention should be focused on meeting performance targets at low intensity and during the early stages of exercise. Further studies of HR control are warranted in order to systematically investigate this feedback-design issue.

Abbreviations
ANOVA: analysis of variance; bpm: beats per minute; ECG: electrocardiogram; HF: high frequency; HR: heart rate; HR_{max}: maximal heart rate; HR_{rest}: resting heart rate; HRR: heart rate reserve; HRV: heart rate variability; LF: low frequency; PSD: power spectral density; QRS complex: the combination of the Q, R and S wave deflections in the ECG; RMS: root mean square; RR interval: time between peaks in the QRS complex of the ECG wave; ULF: ultra-low frequency; VLF: very-low frequency.

Authors' contributions
KJH and JS conceived and designed the study, contributed to the analysis and interpretation of the data, wrote the manuscript and revised it for important intellectual content. Both authors read and approved the final manuscript.

Author details
[1] Division of Mechanical Engineering, Department of Engineering and Information Technology, Institute for Rehabilitation and Performance Technology, Bern University of Applied Sciences, 3400 Burgdorf, Switzerland. [2] Department of Rehabilitation Medicine, Srinagarind Hospital and Faculty of Medicine, Khon Kaen University, Khon Kaen, Thailand. [3] Exercise and Sport Sciences Development and Research Group, Khon Kaen University, Khon Kaen, Thailand.

Acknowledgements
Jonas Egger did the data acquisition. Christian Kummer contributed to the analysis and interpretation of the data.

Competing interests
The authors declare that they have no competing interests.

Funding
Not applicable.

References

1. Billman GE. Heart rate variability—a historical perspective. Front Physiol. 2011;2:86.
2. Malik M, et al. Heart rate variability. Standards of measurement, physiological interpretation, and clinical use. Eur Heart J. 1996;17(3):354–81.
3. Sassi R, Cerutti S, Lombardi F, Malik M, Huikuri HV, Peng CK, Schmidt G, Yamamoto Y. Advances in heart rate variability signal analysis. Europace. 2015;17(9):1341–53.
4. Berntson GG, Bigger JT, Eckberg DL, Grossman P, Kaufmann PG, Malik M, Nagaraja HN, Porges SW, Saul JP, Stone PH, van der Molen MW. Heart rate variability: origins, methods, and interpretive caveats. Psychophysiology. 1997;34:623–48.
5. Pagani M, Lombardi F, Guzzetti S, Sandrone G, Rimoldi O, Malfatto G, Cerutti S, Malliani A. Power spectral density of heart rate variability as an index of sympatho–vagal interaction in normal and hypertensive subjects. J Hypertens Suppl. 1984;2:383–5.
6. Pagani M, Lombardi F, Guzzetti S, Rimoldi O, Furlan R, Pizzinelli P, Sandrone G, Malfatto G, Dell'Orto S, Piccaluga E. Power spectral analysis of heart rate and arterial pressure variabilities as a marker of sympatho-vagal interaction in man and conscious dog. Circ Res. 1986;59:178–93.
7. Billman GE. The LF/HF ratio does not accurately measure cardiac sympatho-vagal balance. Front Physiol. 2013;4:26.
8. von Rosenberg W, Chanwimalueang T, Adjei T, Jaffer U, Goverdovsky V, Mandic DP. Resolving ambiguities in the LF/HF ratio: LF-HF scatter plots for the categorization of mental and physical stress from HRV. Front Physiol. 2017;8:360.
9. Goldstein DS, Bentho O, Park M-Y, Sharabi Y. Low-frequency power of heart rate variability is not a measure of cardiac sympathetic tone but may be a measure of modulation of cardiac autonomic outflows by baroreflexes. Exp Physiol. 2011;96:1255–61.
10. Hadase M, Azuma A, Zen K, Asada S, Kawasaki T, Kamitani T, Kawasaki S, Sugihara H, Matsubara H. Very low frequency power of heart rate variability is a powerful predictor of clinical prognosis in patients with congestive heart failure. Circ J. 2004;68:343–7.
11. Garber CE, Blissmer B, Deschenes MR, Franklin BA, Lamonte MJ, Lee I-M, Nieman DC, Swain DP. American college of sports medicine position stand. Quantity and quality of exercise for developing and maintaining cardiorespiratory, musculoskeletal, and neuromotor fitness in apparently healthy adults: guidance for prescribing exercise. Med Sci Sports Exerc. 2011;43(7):1334–59.
12. Pescatello LS, Arena R, Riebe D, Thompson PD. ACSM's guidelines for exercise testing and prescription. 9th ed. Philadelphia: Lippincott Williams and Wilkins; 2014.
13. Hunt KJ, Fankhauser SE. Heart rate control during treadmill exercise using input-sensitivity shaping for disturbance rejection of very-low-frequency heart rate variability. Biomed Signal Process Control. 2016;30:31–42.
14. Hunt KJ, Gerber S. A generalised stochastic optimal control formulation for heart rate regulation during treadmill exercise. Syst Sci Control Eng. 2017;5(1):481–94.
15. Michael S, Graham KS, Davis GM. Cardiac autonomic responses during exercise and post-exercise recovery using heart rate variability and systolic time intervals - a review. Front Physiol. 2017;8:301.
16. Shargal E, Kislev-Cohen R, Zigel L, Epstein S, Pilz-Burstein R, Tenenbaum G. Age-related maximal heart rate: examination and refinement of prediction equations. J Sports Med Phys Fitness. 2015;55(10):1207–18.
17. McNames J, Thong T, Aboy M. Impulse rejection filter for artifact removal in spectral analysis of biomedical signals. In: Proceeding of conference of the IEEE engineering medicine and biology society, vol. 1. San Francisco, CA, USA; 2004, p. 145–8.
18. Thuraisingham RA. Preprocessing RR interval time series for heart rate variability analysis and estimates of standard deviation of RR intervals. Comput Methods Programs Biomed. 2006;83:78–82.
19. Tulppo MP, Mäkikallio TH, Takala TE, Seppänen T, Huikuri HV. Quantitative beat-to-beat analysis of heart rate dynamics during exercise. Am J Physiol. 1996;271:244–52.
20. Bernardi L, Valle F, Coco M, Calciati A, Sleight P. Physical activity influences heart rate variability and very-low-frequency components in holter electrocardiograms. Cardiovasc Res. 1996;32:234–7.
21. Serrador JM, Finlayson HC, Hughson RL. Physical activity is a major contributor to the ultra low frequency components of heart rate variability. Heart. 1999;82:9.
22. Moreno IL, Pastre CM, Ferreira C, de Abreu LC, Valenti VE, Vanderlei LCM. Effects of an isotonic beverage on autonomic regulation during and after exercise. J Int Soc Sports Nutr. 2013;10:2.

9

Effects of disparity on visual discomfort caused by short-term stereoscopic viewing based on electroencephalograph analysis

Xiao Wang, Liuye Yao, Yuemei Zhao, Lidong Xing, Zhiyu Qian*[iD], Weitao Li and Yamin Yang

*Correspondence:
zhiyu@nuaa.edu.cn
College of Automation
Engineering, Nanjing
University of Aeronautics
and Astronautics, No. 29
Jiangjun Avenue, Jiangning
District, Nanjing 211106,
China

Abstract

Background: Discomfort evoked by stereoscopic depth has been widely concerned. Previous studies have proposed a comfortable disparity range and considered that disparities exceed this range would cause visual discomfort. Brain activity recordings including Electroencephalograph (EEG) monitoring enable better understanding of perceptual and cognitive processes related to stereo depth-induced visual comfort.

Methods: EEG data was collected using a stereo-visual evoked potential (VEP) test system by providing visual stimulus to subjects aged from 21 to 25 with normal stereoscopic vision. For each type of visual stimulus, data were processed using directed transfer function (DTF) and adaptive directed transfer function (ADTF) in combination with subjective feedbacks (comfort or discomfort). The topographies of information flow were constructed to compare responses stimulated by different stereoscopic depth, and to determine the difference in comfort and discomfort situations upon stimulation with same stereoscopic depth.

Results: Based on EEG analysis results, we found that the occipital P270 was moderately related to the disparity. Moreover, the ADTF of P270 showed that the information flows at frontal lobe and central-parietal lobe changed when stimulation with different stereoscopic depth applied. As to the stereo images with same stereoscopic depth, the DTF outflows at the temporal and temporal-parietal lobes in δ band, central and central-parietal lobes in α and θ bands, and the comparison of inflows in these three bands could be considered as discriminated indexes for matching the stereoscopic effect with viewers' comfort or discomfort state impacted by disparity. The subjective feedbacks indicated that the comfort judgments remained as a result of cumulative effect.

Conclusions: This study proposed a short-term stereo-VEP experiment that shorted the duration of each stimulus in the experimental scheme to minimize the interference from other factors except the disparity. The occipital P270 had a mid-relevance to the disparity and its ADTF showed the affected areas when viewers are receiving stimulations with different disparities. DTF could be considered as discriminated indexes for matching the stereoscopic effect with viewers' comfort or discomfort state induced by disparity. This study proposed a preferable experiment to observe the single effect of disparity and provided an intuitive and easy-to-read result in a more convenient manner.

Keywords: Stereoscopic depth, Visual discomfort, Visual evoked potential, Directed transfer function/adaptive directed transfer function

Background

Visual discomfort in stereo viewing is noted as a result of inappropriate binocular disparity. Depth features of stereoscopic images have thus been widely investigated. Lambooij et al. [1] claimed that disparity beyond one degree could cause noticeable visual discomfort. Cho et al. [2] found that the increase of binocular disparity gave rise to human fatigue level. To obtain more in-depth understanding of stereoscopic depth-induced visual discomfort, the relevance between stereo imagery and potential adverse effects have been studied under a wide variety of situations. Shibata et al. [3] reported that large crossed disparity and small uncrossed disparity led to a marked drop in comfort ratings during stereo viewing based on subjective questionnaires. Although such questionnaire-based survey presents a simple and practical method, individual differences in uncertainty tolerance may influence the results. Functional magnetic resonance imaging (fMRI) was also employed to explore the changes in human cortex during visual stereo stimulation study. Based on fMRI, Liu et al. [4] indicated that stereoscopic depth perception was correlated with several regions (hV3A, LG, hMT/V5, LOS and VIPS), which stands for the precise positions on human cortex in stereoscopic vision processing. Jung et al. [5] also used fMRI to locate the activated areas on the cortex when people felt uncomfortable during 3D viewing. fMRI remains a precise yet expensive method which also requires readers to have professional knowledge of the brain structure. For the detection of stereo visual discomfort in daily life and taking the cost-effectiveness into consideration, a relatively simple, easy-operating, labor- and money-saving method needs to be developed. Recently, electroencephalograph (EEG) has been widely used as an effective way to assess stereoscopic visual fatigue. Li et al. [6] found the power of the high frequency band in EEG became stronger in 3D viewing and the peak difference in P700 at 3D oddball paradigm might be an effective indicator for revealing 3D visual fatigue. Mun et al. [7] found significantly reduced P600 amplitudes and delayed P600 latencies appeared in accordance with 3D visual fatigue, and significant fatigue effects were also observed at P4 and O2 sites during the 8.57 Hz attended task. Zou et al. [8] evaluated visual fatigue level via random dot stereogram (RDS) with six different disparities and found that EEG could be employed as a useful tool to predict visual fatigue caused by vergence-accommodation conflict. Malik et al. [9] compared the EEG absolute power differences, coherence and complexity, then concluded that 3D viewing is more attractive than 2D and may cause high attention and involvement of working memory manipulations. Other researches also demonstrated the relevance between visual fatigue and disparity in stereo viewing based on EEG analysis [10–13]. Kang et al. [14] developed a platform to facilitate comfortable stereo video viewing using EEG-based visual discomfort evaluation technology. Frey et al. used EEG and event related potential (ERP) to assess the visual discomfort in stereoscopic displaying and established a prediction model to assess visual discomfort based on their ERP results and reached an accuracy rate more than 62% on average [15, 16]. Although the above work demonstrated the capability of distinguishing uncomfortable stereoscopic viewing conditions from comfortable ones based on EEG analysis, each stimulus in their experimental trial lasted for seconds. In the majority of previous studies, overall level of visual discomfort was recorded according to the subjective questionnaires and was determined after certain experimental session. However, it might be expected that the time course of viewing could also cause possible cumulative effects and thus affect discomfort judgments.

Visual evoked potential (VEP), which refers to the evoked potential caused by a visual stimulus, is commonly used for vision check in clinics. VEP measures the functional integrity of the visual pathways from retina via the optic nerves to the visual cortex when retinas receive stimuli, and can be induced by a stimulus repeated at a higher rate [17]. Therefore, in order to capture the fast responses upon short-term visual stimuli (less than 1 s) with various stereoscopic depth, the present pilot study described an experimental protocol which acquires both EEG signals and subjective feedbacks without interrupting the viewing.

Thanks to precise stereoscopic parameters and real-time measures of our stereo-VEP setup, the effects of disparity and to which degree the disparity would evoke visual discomfort immediately over short viewing sequences were systematically investigated in this study. The effects of disparity over short viewing sequences were systematically investigated by directed transfer function (DTF) and adaptive directed transfer function (ADTF) methods. The study will provide more intuitive and easy-to-read results to the public to help them understand their physical status in a more convenient manner and pave the way for better estimating stereoscopic discomfort and optimization of stereoscopic display parameters in the future.

Methods

Experimental protocol

Stereo-VEP experimental system

Figure 1 shows the schematic setup of our experimental system which composed of a stimulation part, a recording part and a feedback part. The simulation part is consisted of an active shutter 3D-TV (LED46XT39G3D, Hisense) and a laptop installed with E-Prime 2.0. The visual stimuli were provided by a visual paradigm written based on E-Prime 2.0. During experiment, the paradigm running on the laptop would be synchronously displayed on the 3D-TV through a high definition multimedia interface (HDMI) cable. Participants watched the stereo stimuli series through a pair of shutter glasses (FPS3D02, Hisense). The detailed information of the paradigm will be introduced in the following section.

EEGs were acquired in real-time with NuAmps electroencephalograph (Australia, Neuroscan) during experiment. Signal recording was performed according to the expanded international 10–20 montage system. Thirty of total thirty-seven electrodes were placed along the scalp for EEG recording. As alternative references, two mastoid electrodes (M1 and M2) at bilateral were used during recording. Four electrodes were used for real-time horizontal and vertical electrooculogram (EOG) recording and the ground was set at FPz. The sampling rate was 1 kHz. The contact resistance of each electrode was less than 5 kΩ. The acquisition process was monitored by the other laptop installed with Curry 7. For each trial, participants stimulated upon corresponding visual cues were asked to click the left button of the mouse as soon as they felt uncomfortable.

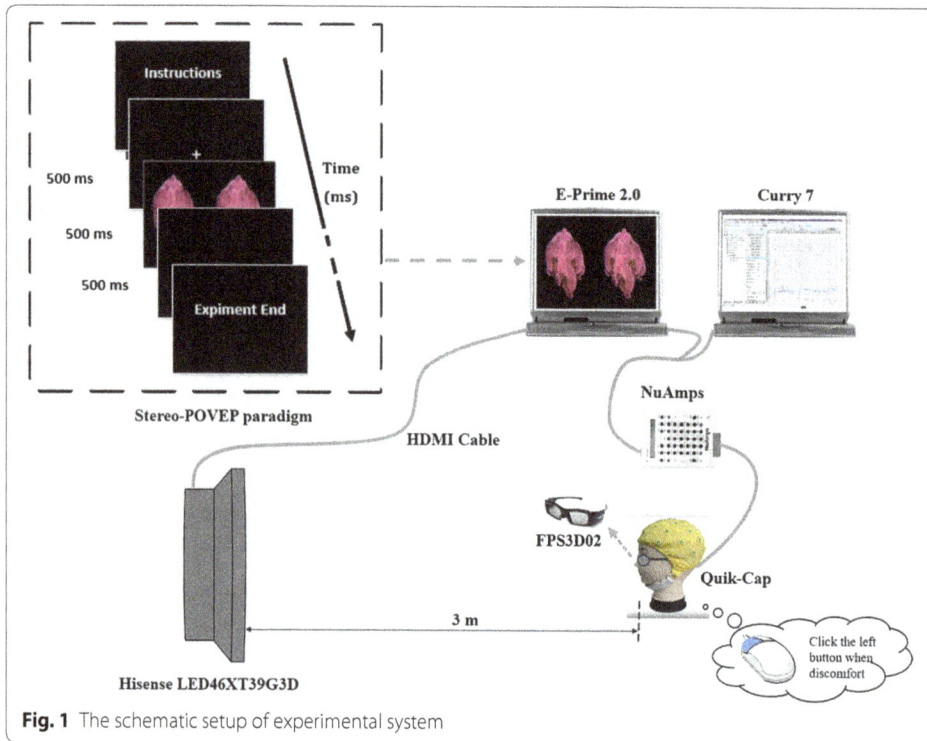

Fig. 1 The schematic setup of experimental system

The entire experiment was conducted in a quiet room and the temperature was kept at 24 °C. Ten right-handed subjects (male: 9, female: 1, aged from 21 to 25) with normal stereoscopic vision participated in our experiment. All subjects were informed about the general experiment information and signed the consent form before the experiment. All experiments were carried out in accordance with institutional guidelines of Nanjing University of Aeronautics and Astronautics (NUAA). All experimental protocols were approved by the Ethics Committee of NUAA.

The paradigm

The pattern onset/offset visual evoked potential (POVEP), including both flash visual evoked potential (F-VEP) and pattern visual evoked potential (P-VEP), was applied to explore the relevance between stereoscopic depth perception and human discomfort in this study. The stereo pattern onset/offset visual evoked potential (Stereo-POVEP) paradigm was shown in the dashed box in Fig. 1. After the subject read through the instructions for experiment description, he or she would tap the space key to start the experiment. A cross would appear at the center of the screen for 5 s to draw the participant's attention to the screen center. One of four images with different disparities would randomly appear and displayed for 500 ms for each, followed by a black background for 500 ms. A complete set of stimulation series presented 240 trials (60 trials for each image stimulus) and it costed less than 5 min in total. Each subject performed two sets of above stimulation trials.

The disparity parameters of the stereo images are illustrated in Fig. 2a. The image stimulus with zero disparity includes neither uncrossed disparities nor crossed disparities, and in other words, it is a 2D image stimulus (abbreviated as '2D'). The "±" sign in the

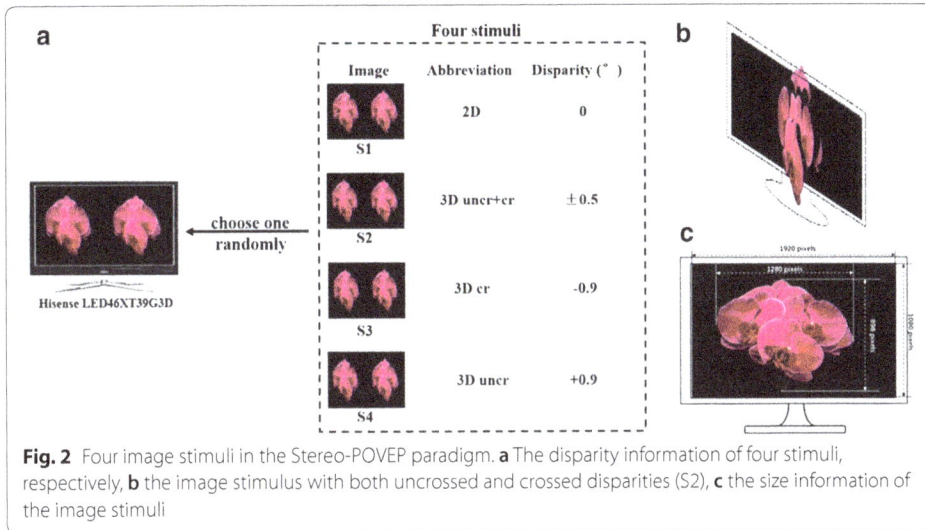

Fig. 2 Four image stimuli in the Stereo-POVEP paradigm. **a** The disparity information of four stimuli, respectively, **b** the image stimulus with both uncrossed and crossed disparities (S2), **c** the size information of the image stimuli

"disparity" column (abbreviated as '3D uncr+cr') means that, besides the 2D (zero disparity) part of the image, the image stimulus includes uncrossed (orchid bud appeared in front of the screen) and crossed disparities (orchid bud appeared behind the screen) (Fig. 2b). The "+" sign means the uncrossed disparity (abbreviated as '3D uncr'), and the "−" sign means the crossed disparity (abbreviated as '3D cr'), respectively. For convenience, we numbered these four types of stimuli '2D', '3D uncr+cr', '3D cr' and '3D uncr' as 'S1', 'S2', 'S3' and 'S4'. 'S2' has both negative and positive disparities, and the absolute value of its positive disparity or negative disparity are both in the suggested comfortable range (| positive (or negative) disparity of S2| $= 0.5 < 1$). The size of the image in left–right format was 1920×1080 and the object in the middle of the image was 1280×898 (Fig. 2c). The visual angle was approximately $4.08° \times 6.67°$. All images were provided by Prof. Qiu and his group from the School of Art in Peking University.

Data processing

The reference was switched to Cz during offline processing. After baseline correction, 50 Hz notch filtering, 0.01–30 Hz band-pass filtering, eye movement artifacts removal and bad blocks removal, VEPs were averaged by the time-locked and phase-locked EEGs. Valid data used for each averaging was over 90 trials. The processed VEPs were saved for following computations.

Brain connectivity estimators

DTF and ADTF

Granger causality is used to compute the causal relationship between two time series. Kaminski and Blinowska successfully applied DTF into neurobiological system computations, which expanded the application of Granger causality based on multivariate autoregressive (MVAR) model [18]. A short description of this algorithm is as follows.

$\mathbf{X}(t) = [X_1(t), X_2(t), X_3(t), ..., X_k(t)]^T$ is defined as a k-channel signal at time point t. The superscript T represents the matrix transposition. Then the MVAR model can be described as:

$$\mathbf{X}(t) = \sum_{n=0}^{p} \mathbf{A}(n)\mathbf{X}(t-n) + \mathbf{E}(t) \tag{1}$$

with

$$\mathbf{A}(0) = \mathbf{I} \tag{2}$$

where $\mathbf{E}(t)$ is the white noise and $\mathbf{A}(1)$, $\mathbf{A}(2)$,...,$\mathbf{A}(p)$ are $k \times k$ coefficient matrices. The order p could be derived from Schwarz Bayesian Criterion (SBC) [19]. Then it is transformed into the frequency domain by Fourier transforms:

$$\mathbf{A}(f)\mathbf{X}(f) = \mathbf{E}(f) \tag{3}$$

where

$$A(f) = \sum_{n=0}^{p} \mathbf{A}(n)e^{-2\pi ftn} \tag{4}$$

So $\mathbf{X}(f)$ could be rewritten as:

$$\mathbf{X}(f) = \mathbf{A}^{-1}(f)\mathbf{E}(f) \tag{5}$$

Define $\mathbf{H}(f) = \mathbf{A}^{-1}(f)$, and then

$$\mathbf{X}(f) = \mathbf{H}(f)\mathbf{E}(f) \tag{6}$$

$\mathbf{H}(f)$ is the transfer matrix of the system, which equals to the inverse of frequency-transformed coefficient matrix. The casual information from channel j to channel i is:

$$\theta_{ij}^2(f) = \left|\mathbf{H}_{ij}\right|^2 \tag{7}$$

The DTF measurement $\gamma_{ij}^2(f)$ is the normalization of $\theta_{ij}^2(f)$, which describes the directional causal information from channel j to channel i. The value of $\gamma_{ij}^2(f)$ is between 0 and 1.

$$\gamma_{ij}^2(f) = \frac{\left|H_{ij}(f)\right|^2}{\sum_{m=1}^{k}\left|H_{im}(f)\right|^2} \tag{8}$$

ADTF was proposed by Wilke in 2008, which was a time-varying multivariate method aiming to estimate rapidly changing connectivity relationship of the brain [20]. Some studies have confirmed that ADTF is suitable for the analysis of short duration signals [19, 21]. Similarly, the signal $\mathbf{X}(t)$ could be described as:

$$\mathbf{X}(t) = \sum_{n=1}^{p} \mathbf{A}(n,t)\mathbf{X}(t-n) + \mathbf{E}(t) \tag{9}$$

where $\mathbf{X}(t)$ is a vector that changes with time, $\mathbf{A}(n, t)$ is the time-varying coefficient matrices. Other parameters are the same as their definitions in DTF. ADTF measure $\gamma_{ij}^2(f)$ could be similarly defined as:

$$\gamma_{ij}^2(f) = \frac{\left|\mathbf{H}_{ij}(f, t)\right|^2}{\sum_{m=1}^{k} \left|\mathbf{H}_{im}(f, t)\right|^2} \tag{10}$$

Matlab toolbox eConnectome (Biomedical Functional Imaging and Neuroengineering Laboratory, University of Minnesota, Minneapolis, http://econnectome.umn.edu/) was used to compute the DTF and ADTF out-to-in of these signals. The out-to-in information indicates the causal information flows from one channel to another [22]. From out-to-in information we could compute the outflow information (or outflows) and inflow information (or inflows) of each flow. The outflow information of an electrode is the sum of the information from this electrode to all the other electrodes while its inflow information is the sum of all the other electrodes' information flow into this electrode [22]. The outflow or inflow information value represents the ability of the electrode to affect other electrodes or to be affected from other electrodes. In this paper, the electrode with strong outflows was defined as a key cause node and the electrode with sensitive inflow information was considered as a key result node.

Surrogate data

Surrogate data, a time series which fits well with the linear-dynamics null hypothesis, could assess the significance of DTF and ADTF connectivity measures [19, 23]. It demonstrated that this method is suited for DTF and ADTF analysis that are the measurement of frequency-specific causal interactions [19]. The significance setting in this study was $P < 0.05$.

Results

Visual comfort comparison among different disparities measured by Stereo-VEPs

Waveforms and time–frequency analysis

According to the subjective feedback, EEG signals in the cases of visual comfort were firstly averaged among subjects. Figure 3 depicts a representative planform of Stereo-VEP from one subject and averaged Stereo-VEP over all subject evoked by four types of stimuli at O1, Oz, O2 electrodes, respectively. In the planform, the magnitude of each amplitude was represented by colors. Colors from blue to red depict the amplitude of EEG signal from low to high. Obviously, the most distinct peak presented in the occipital lobe. In the grand average waves of Stereo-VEP, the P3 component was extremely obvious at nearly 270 ms after the onset of the stimulus (in the following contents, we use P270 for convenience). Table 1 lists the means and standard deviations of P270 from 10 participants. Compared with VEP evoked by 'S1', P270 evoked by other stimuli had a delay no less than 10 ms. As shown in Table 1, 'S2' caused nearly the same latencies at both the left and the right occipital lobe. In the case of 'S3', the latency of P270 at the left occipital lobe (O1) is slightly greater than that at the right part (O2), while 'S4' led to the contrary result. It is also quite clear that 'S3' evoked the most significant peak in the amplitude of P270.

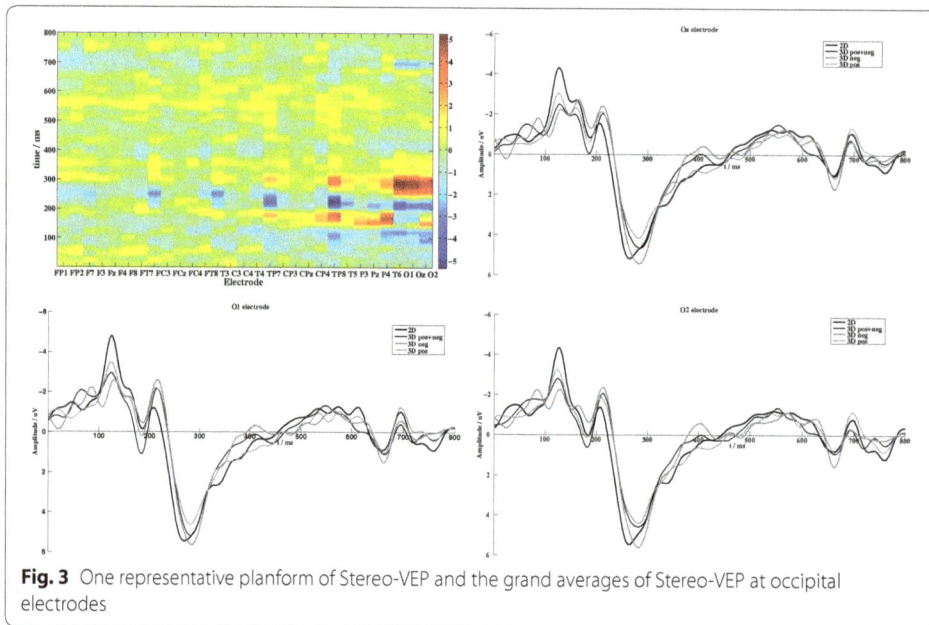

Fig. 3 One representative planform of Stereo-VEP and the grand averages of Stereo-VEP at occipital electrodes

Table 1 Means and standard deviations on the latency and peak amplitude of P270 over 10 subjects

Image	Latency (ms)			Peak (µV)		
	O1	Oz	O2	O1	Oz	O2
S1	269±15	267±12	271±14	6.23±1.93	6.02±2.01	6.33±2.86
S2	288±21	289±20	288±22	5.84±1.42	5.52±1.67	5.52±1.83
S3	284±22	281±21	281±20	6.80±1.60	6.77±1.71	6.97±2.05
S4	279±20	285±24	281±21	5.24±1.81	4.90±1.90	5.15±2.21

The Pearson correlation coefficient showed that peak amplitudes of P270 had a mid-relevance with the disparity (O1: Pearson correlation coefficient$=-0.474$, $P=0.006<0.01$; Oz: Pearson correlation coefficient$=-0.480$, $P=0.005<0.01$; O2: Pearson correlation coefficient$=-0.459$, $P=0.008<0.01$). The Paired T test between the occipital P270 from any two different types of visual stimuli was shown in Table 2. It is shown that the difference can be significantly observed by P270 component from O1, Oz, and O2 electrodes at the occipital lobe. Although between-group differences cannot be illustrated using latency and amplitude information from one single electrode, P270 based on the integrated results from three electrodes located at occipital lobe could be used as an effective indicator for differentiating different type of stimuli.

In addition, when the stimulation disappeared, an off-response at about 660 ms in the grand average showed that 'S3' caused the most significant changes in amplitude (Fig. 3). Previous study [24] suggested that off-responses had close relationship with visual persistence. It may infer that large crossed disparities would contribute to more off-responses. Another negative potential showed at around 700 ms, which could be due to the open issue of visual N700 [25, 26].

Table 2 The Paired T-test of P270

Pairs	Latency (ms)			Amplitude (µV)		
	O1	Oz	O2	O1	Oz	O2
S1–S2	T=−4.862 P=0.001	T=−7.735 P=0.000	T=−4.235 P=0.002	T=3.110 P=0.013		T=3.233 P=0.01
S1–S3	T=−2.648 P=0.027	T=−2.776 P=0.022				
S1–S4	T=−2.454 P=0.037	T=−3.995 P=0.003	T=−2.669 P=0.026		T=4.125 P=0.003	T=4.302 P=0.002
S2–S3					T=−3.363 P=0.008	T=3.423 P=0.008
S2–S4			T=2.504 P=0.034			
S3–S4				T=3.324 P=0.009	T=4.703 P=0.001	T=3.887 P=0.004

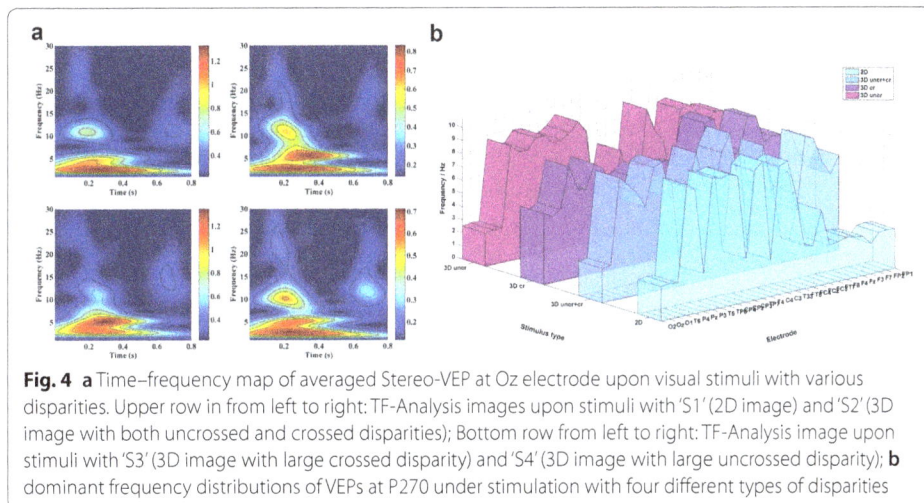

Fig. 4 a Time–frequency map of averaged Stereo-VEP at Oz electrode upon visual stimuli with various disparities. Upper row in from left to right: TF-Analysis images upon stimuli with 'S1' (2D image) and 'S2' (3D image with both uncrossed and crossed disparities); Bottom row from left to right: TF-Analysis image upon stimuli with 'S3' (3D image with large crossed disparity) and 'S4' (3D image with large uncrossed disparity); **b** dominant frequency distributions of VEPs at P270 under stimulation with four different types of disparities

Figure 4a describes the time–frequency analysis (TF-Analysis) of averaged Stereo-VEP at Oz electrode within 800 ms after the onset of various visual stimuli. These images showed that P270 component evoked by 3D stereo stimuli includes wider frequency bandwidth (δ, θ and a few α bands) at occipital lobe as compared with the 2D stimulus. Figure 4b represents the dominant frequency distribution of P270, indicating that although the frequency band of Stereo-VEPs became wider upon 3D stereo stimuli, most of leading frequency were still at δ band (about 2.6 Hz). An exception was that the dominant frequency of Stereo-VEP evoked by 'S3' was 5.4 Hz, which was in θ band.

ADTF information flows

δ, θ, α bands are related to signal matching and decision-making process, attentive processing and sensory processing, respectively [27]. Figure 5 shows the ADTF information flow topographies of P270 at δ, θ, α and δ~α (δ to α) bands (from bottom to top). In Fig. 5a, 2D stimulus evoked strong outflows at F3 electrode at δ~α bands, while 3D stimuli evoked that at F7, CP4, CP3, P4, Pz, T3 and Oz electrodes. It was easy to observe

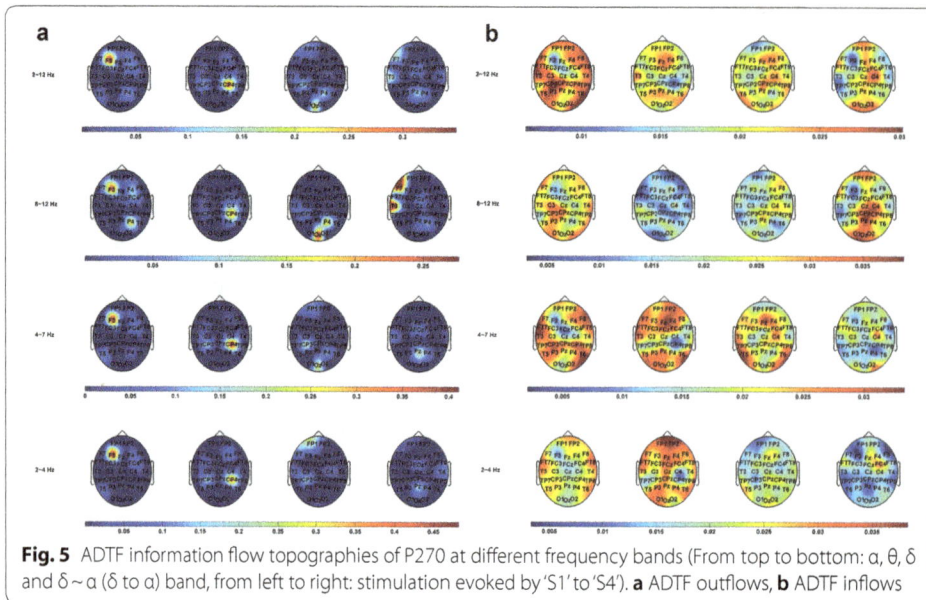

Fig. 5 ADTF information flow topographies of P270 at different frequency bands (From top to bottom: α, θ, δ and δ ~ α (δ to α) band, from left to right: stimulation evoked by 'S1' to 'S4'). **a** ADTF outflows, **b** ADTF inflows

that 2D stimulus caused more powerful outflows in the left frontal lobe in the decision-making process (δ band), while 3D stimuli caused significant influence in the central lobe and the central-parietal lobe (CP4, CP3, Pz electrodes). In addition, 'S3' caused strong outflows at FP1 electrode as well. Although the θ band topographies were almost similar to the δ band topographies, Oz electrode at θ band became the key cause node instead of FP1 electrode in the δ band under 'S3' stimulation. In α band, P4 became one of the key cause nodes under 'S1' and 'S3' stimulation, while the CP3 electrode was never the key cause node under 'S2' or 'S3' stimulation. F7 and T3 were the key cause nodes under 'S4' stimulation. All the nodes mentioned above were marked as bright color in Fig. 5a. Generally, the ADTF outflows of P270 evoked by 2D stimulus mainly centered on the frontal lobe, while upon 3D stimuli the outflows of P270 centered on the posterior brain areas.

The ADTF inflow topographies of P270 were shown in Fig. 5b. The number of disparity types included in one stimulus was defined as depth complexity (D-complex). The D-complex of 'S2' was higher than that of 'S3' or 'S4' for the reason that it includes two types (uncrossed and crossed) of disparities. The result showed that the stimulus with high D-complex ('S2') caused stronger inflows at θ and δ bands (especially in the pre-frontal area). Large crossed disparity evoked slightly stronger inflows when comparing the inflows between the two stimuli with low D-complex ('S3', 'S4'). 'S1' and 'S4' caused relatively strong inflows in α band from the central area to the occipital lobe. The difference was that 'S1' acted on the left part while 'S4' acted on the right. 'S3' also slightly influenced the right frontal and central-parietal lobe at α band. The δ ~ α band topography for each stimulus showed that 'S1' owns the strongest inflows.

Comparison between visual discomfort and comfort by DTF
Subjective feedbacks
The subjective feedbacks (Table 3) showed that no discomfort was reported during stimulation by 'S1', while 'S3' caused 93 feedbacks reporting discomfort. That is to say,

Table 3 Subjective feedbacks

Image	Abbreviation	Feedback	Total number of stimuli
S1	2D	0	1200
S2	3D uncr+cr	39	1200
S3	3D cr	93	1200
S4	3D uncr	27	1200

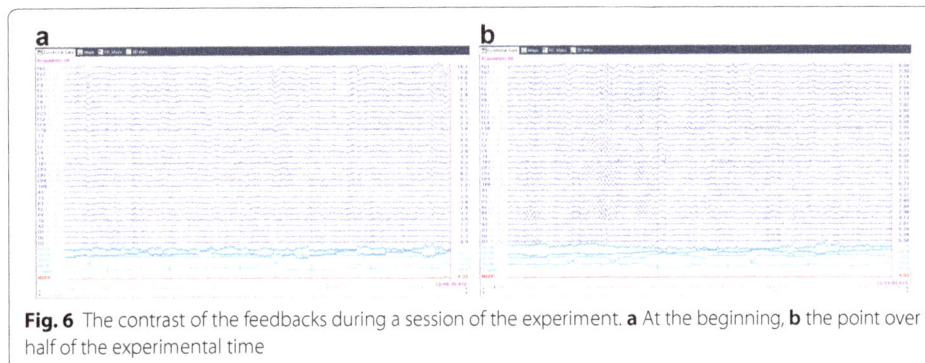

Fig. 6 The contrast of the feedbacks during a session of the experiment. **a** At the beginning, **b** the point over half of the experimental time

observers are more sensitive to crossed disparity than to uncrossed ones, especially the large ones. This result is in good accordance with our previous study about stereoscopic depth in watching 3D films [28]. Furthermore, we found that most participants did not report feedbacks at the beginning of the experiment. As the experiment continues, discomfort feedbacks were frequently marked. Typically, as shown in Fig. 6a, there was no discomfort feedbacks (marked as '1') recorded at the very beginning of the experimental session upon stimulation by 'S3' (marked as '30'), while discomfort feedbacks firstly appeared after half of the experimental process (Fig. 6b). Other labels '10', '20', '40' in Fig. 6 represented 'S1', 'S2', and 'S4' respectively. The electrodes named A1 and A2 represented the references M1 and M2. It implied that, although disparity might easily lead to visual discomfort, there could still be an accumulative process.

Comparison between visual discomfort and comfort by DTF

DTF information flows

Based on the subjective feedbacks, Stereo-VEPs were further divided into seven groups, dependent on the types of stimuli and subjective feelings (comfort and discomfort). Because in some cases uncomfortable feedbacks upon certain stimulation were rarely received, differences between comfort and discomfort at a specific time point could not be accurately represented. Therefore, the DTF method rather than the ADTF method was chosen to distinguish the difference between visual comfort and discomfort. Figure 7 depicted the DTF information flow topographies at α, θ, δ bands (from top to bottom). Data from 0 to 500 ms after the onset of the stimulus were analyzed. It was obvious that 3D stereo stimuli activated more and stronger outflows than the 2D stimulus did in posterior brain areas. In Fig. 7a, as compared to that in 2D condition, outflows became weaker in the right frontal lobe (F8 electrode)

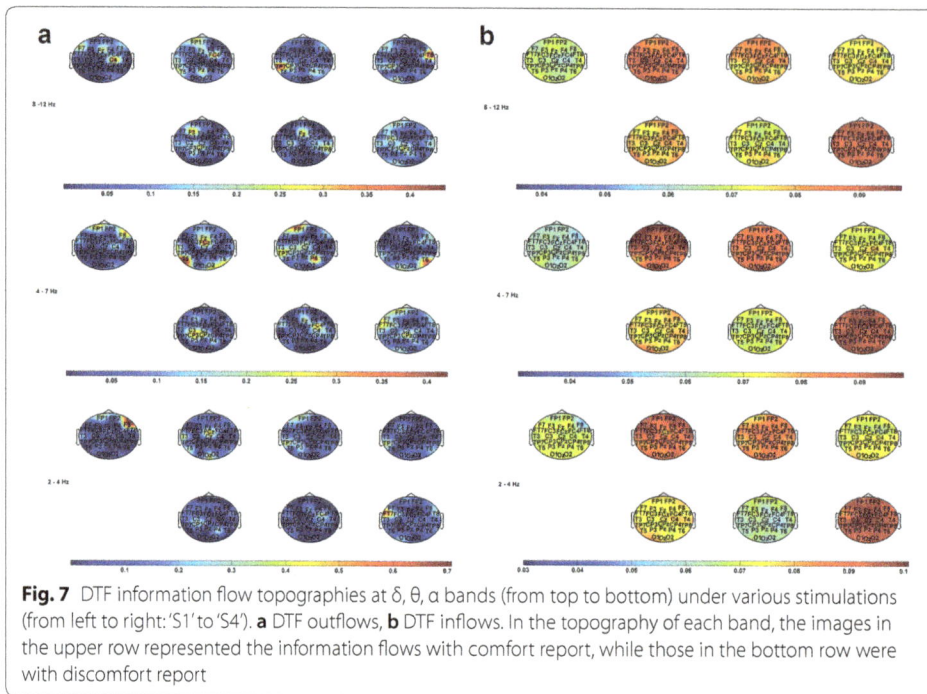

Fig. 7 DTF information flow topographies at δ, θ, α bands (from top to bottom) under various stimulations (from left to right: 'S1' to 'S4'). **a** DTF outflows, **b** DTF inflows. In the topography of each band, the images in the upper row represented the information flows with comfort report, while those in the bottom row were with discomfort report

when participants received 3D stimuli no matter whether they reported visual discomfort or not. The comparison of comfort in this finding was in consistent with what has been illustrated in Fig. 5a. In 'S2', 'S3' and 'S4' cases where no discomfort was reported, the outflows in the left temporal lobe (T3 electrode) or in the left temporal-parietal lobe (TP7 electrode) became stronger, while the outflows in the same lobes became weaker if discomfort was marked in the feedback. The central lobe and the central-parietal lobe showed that the outflows at α and θ bands became relatively strong when participants received 'S2' and 'S4' stimuli and felt uncomfortable. However, the phenomenon was opposite under 'S3' stimulation.

As shown in Fig. 7b, the DTF inflows throughout the whole brain were obviously stronger under 3D stimulations than that under 2D stimulation. Consistent with Fig. 5b, the stimulus with high D-complex ('S2') caused stronger DTF inflows when participants felt comfortable. In terms of DTF inflows caused by low D-complex, compared 'S3' with 'S4', the large crossed disparity rather than uncrossed disparity evoked stronger DTF inflows in comfortable situation. It was also noticeable that the DTF inflows became weaker when subjects received 'S2' and 'S3' stimulation and felt uncomfortable, but the result was opposite in the 'S4' case.

Discussion

Considering the safety issue, the largest disparity value used in this experiment was within the comfortable range proposed by Lambooij [1]. However, it is very close to the superior limit of the range and the effect was distinct when comparing with other stimuli in this paradigm. The uncomfortable feelings were frequently reported by participants when they received the stimulus with large crossed disparity.

P270 and other Stereo-VEP components in stereo viewing

As VEP suggests the electrical activity of occipital lobe when the retina receives stimuli [17], VEPs in occipital region indicate that the change of disparity could influence the brain. In the overall averaged Stereo-VEPs, five onset response components C1, C2, C3, N2 and P3 were found. Early components of N75 component (or N1), P100 component (or P1) and N135 component (or N2) are often regarded as a set called the NPN component [29, 30]. Shawkat et al. verified the relevance between pattern reversal visual evoked potential (PRVEP) and POVEP, that is, P100 corresponds to C1, N135 corresponds to C2, and the following positive component may correspond to C3 [31, 32]. Sometimes, there is a small negative wave called C0, which corresponds to N75 [32]. In our study, it is shown in Fig. 3 that 'S2' and 'S3' caused distinct NPN component.

P1 (C1 in present study), N2 and P3 are regarded as neural correlate of visual consciousness and are usually used to study attention and visual processes [33]. Our results are consistent with previous studies which confirmed that C1 increases due to the depth perception [34, 35]. Compared to the 2D stimulus, a significant increase in the amplitude of C1 component in the 3D conditions was noticed. The difference in C1 amplitude between different stimuli can be explained by the difference between conditions, as the amplitude of C1 is known to be modulated by depth perception of 3D stimulus. As shown in Fig. 3, although 'S3' and 'S2' both evoked a clear C1 component at around 100 ms, more significant increases in C1 amplitude were found in 'S3' condition, which explains that the larger crossed disparity ('S3') may result in more distraction of visual attention and lead to more discomfort feedback. Although the result shown in Fig. 3 was the grand averaged amplitude of VEPs, there is no any statistically significance between different individuals participated in our experiment.

In order to attract participants' attention and to avoid the decrease of attention intensity during stimulation, we used POVEP paradigm and forced participants to give feedbacks whenever there is any visual discomfort. P270 in this paper was P3b which appears in voluntary attention [36]. The TF-Analysis inferred that the P270 component at occipital lobe correlated closely to the decision-making process and attentive processing. As shown in Table 1, the latency of P270 appears to be longer in the case of 'S2', indicating that the visual cortex may need longer period of time for fusion and stereopsis upon processing images with increased disparity complexity. Similar to C1, the amplitude of P270 also increases with the depth perception by reflecting enhanced neural activity upon stimuli with larger disparity. Therefore, the occipital P270 could be chosen as an index to distinguish the effect of stereoscopic depth under comfortable feelings in stereo viewing.

The ADTF analysis showed that outflows of P270 caused by 2D stimulus were mainly on the frontal lobe, while 3D stereo stimuli evoked outflows of P270 on the posterior brain areas. It suggests that ADTF could help to judge whether the visual stimuli include any stereoscopic depth information under the comfortable state. Furthermore, the key cause nodes showed no significant change in 'S1' and 'S2' condition at δ, θ and α bands. However, in the large disparity cases, they changed obviously. According to the topographies of ADTF inflows, the result could be helpful to distinguish which kind of depth the stimulus may contain.

Comparison between comfort and discomfort caused by stereoscopic depth

Researchers have demonstrated that the posterior parietal cortex participated actively in processing depth based on fMRI and functional near infrared spectroscopy (fNIRS) methods [37, 38]. In our study, changes of DTF and ADTF outflows were also found at the central-parietal lobe when stimulated by 3D images with different stereoscopic depth (Figs. 5 and 7). The key cause nodes appeared in posterior brain areas under 3D condition, while the nodes were in the frontal lobe under 2D condition. The subjective feedbacks indicated that 'S3' caused the most significant visual discomfort. DTF outflows in Fig. 7 shows that visual discomfort is accompanying with the decrease of the importance of the cause node in left temporal lobe. Based on literatures, temporal lobe relates to visual memory, language comprehension, and emotion association [39]. The ventral part of the temporal cortices also participates in high-level visual processing of complex stimuli. Anterior parts of the ventral stream are involved in object perception and recognition during visual processing [40]. Thus, it implies that discomfort from depth perception may diminish the connection strength between temporal lobe and other parts of the brain. On the other hand, the parietal lobe relates to attention and the central area is correlated with information processing. It was shown that the outflows at θ bands in the central lobe and the central-parietal lobe became relatively strong in participants under 'S2' and 'S4' stimulation accompanied with visual discomfort. However, the phenomenon evoked by 'S3' was opposite. It infers that the discomfort caused by the uncrossed disparity may incline to catch participants' attention but that caused by the crossed disparity may reduce the attention. Figure 7b shows high depth complexity and large crossed disparity made DTF inflows weaker when participants felt uncomfortable, but it was contrary in the case with large uncrossed disparity. That is to say, the high depth complexity and the large crossed disparity would reduce the connectivity strength but the large uncrossed disparity would increase the connectivity strength when participants felt uncomfortable. This result indicated that DTF information flows from EEG signals could be considered as an index to distinguish the comfort level in stereo viewing.

Generally, this study proposed a Stereo-VEP experiment that can capture the fast responses upon short-term visual stimuli with various stereoscopic depth to minimize the interference from other factors. Compared to previous experiments, it focused more on the comfort level caused by disparity itself. ADTF and DTF results in this study illustrated that the information flows of EEG electrodes could be as effective as other methods to show the effect of disparities and comfort level. For the next step, based on the experimental and EEG analysis results in this paper, we intend to further explore the disparity-induced visual discomfort in stereo viewing in the cerebral cortex. Besides the findings reported herein, we have also found some strongly activated gyri at frontal and temporal lobes when viewers receiving stimuli with visual discomfort, which correlated well with previous studied using fMRI analysis.

Conclusions

In this paper, we established a short-term experimental system which combined viewer-interactive subjective feedback towards better understanding of visual comfort or discomfort impacted by different stereoscopic depth. The results showed that the occipital P270 had a mid-relevance to the disparity of the stimuli and its ADTF showed the strongly activated areas when viewers are receiving stimulations with different disparities. DTF of P270 during the presence of the stimuli helped understanding the difference between comfort and discomfort stimulated by the same disparity. The change impacted along with the presence of discomfort could be found from the DTF outflows at the temporal and temporal-parietal lobes in δ band, central and central-parietal lobes in α and θ bands, and the inflows in these three bands. The subjective feedbacks showed the discomfort situations remained as a result of cumulative effect. Overall, the study provided a preferable experiment to observe the effects of disparity based on the Stereo-VEP experiment. Reconstruction of the information flow following a visual stimulus could be helpful to match the stereoscopic effect with viewers' state (comfort or discomfort status) induced by disparity and provided an intuitive and easy-to-read result in a more convenient manner.

Abbreviations
ADTF: adaptive directed transfer function; D-complex: the depth complexity; DTF: directed transfer function; ECG: electrocardiograph; EEG: electroencephalograph; EOG: electro-oculogram; fMRI: functional magnetic resonance imaging; fNIRS: functional near infrared spectroscopy; F-VEP: flash visual evoked potential; MVAR: multivariate autoregressive; NIRS: near infrared reflectance spectroscopy; POVEP: pattern onset/offset visual evoked potential; PRVEP: pattern reversal visual evoked potential; P-VEP: pattern visual evoked potential; RDS: random dot stereogram; SBC: Schwarz Bayesian Criterion; Stereo-POVEP: stereo pattern onset/offset visual evoked potential; Stereo-VEP: stereo visual evoked potential; TF-Analysis: time-frequency analysis; VEP: visual evoked potential; 2D or S1: zero disparity, i.e. the image stimulus includes neither uncrossed disparities nor crossed disparities; 3D uncr+cr or S2: the image stimulus includes both uncrossed and crossed disparities; 3D cr or S3: the image stimulus only includes the crossed disparity; 3D uncr or S4: the image stimulus only includes the uncrossed disparity.

Authors' contributions
XW was involved in performing experiments, data analyzing and writing the first draft of the manuscript. LY and YZ did the data analysis. LX and ZQ was responsible for the guidance and providing suggestions during the research. WL and YY wrote and modified this paper. All authors read and approved the final manuscript.

Acknowledgements
We are thankful to those helped with the experiment and gave suggestions during the research. We also thank Prof. Qiu and his team from the School of Art in Peking University for providing the image stimuli.

Competing interests
The authors declare that they have no competing interests.

Funding
The research work was supported by Funding of Jiangsu Innovation Program for Graduate Education (KYLX_0248) and the Fundamental Research Funds for the Central Universities. National Major Scientific Instruments and Equipments Development Project Funded by National Natural Science Foundation of China (81277804, 81827803), National Natural Science Foundation of China (61875085, 81601532), Natural Science Foundation of Jiangsu Province (BK20160814), Jiangsu Science and Technology Support Plan (Social Development) (BE2016759).

References
1.   Lambooij M, Fortuin M, Heynderickx I, et al. Visual discomfort and visual fatigue of stereoscopic displays: a review. J Imaging Sci Technol. 2009;53(3):30201-1-30201-14(14).
2.   Cho S H, Kang H B. The measurement of eyestrain caused from diverse binocular disparities, viewing time and display sizes in watching stereoscopic 3D content. In: Computer Vision and Pattern Recognition Workshops (CVPRW). 2012; https://doi.org/10.1109/cvprw.2012.6238904.
3.   Shibata T, Kim J, Hoffman DM, et al. The zone of comfort: predicting visual discomfort with stereo displays. J Vis. 2011;11(8):11.
4.   Liu J, Wang F, Wei H, et al. Preliminary fMRI research of the human brain activation under stereoscopic vision. In: IEEE International Conference on Medical Imaging Physics and Engineering. 2014; https://doi.org/10.1109/icmip e.2013.6864518.
5.   Yong JJ, Kim D, Sohn H, et al. Towards a physiology-based measure of visual discomfort: brain activity measurement while viewing stereoscopic images with different screen disparities. J Disp Technol. 2015;11(9):730–43.
6.   Li H C O, Seo J, Kham K, et al. Measurement of 3D visual fatigue using event-related potential (ERP): 3D Oddball Paradigm. 2008; p. 213–216.
7.   Mun S, Park MC, Park S, et al. SSVEP and ERP measurement of cognitive fatigue caused by stereoscopic 3D. Neurosci Lett. 2012;525(2):89.

8.   Zou B, Liu Y, Guo M, et al. EEG-based assessment of stereoscopic 3D Visual fatigue caused by vergence-accommoda-
     tion conflict. J Disp Technol. 2017;11(12):1076–83.
9.   Malik AS, Khairuddin RN, Amin HU, et al. EEG based evaluation of stereoscopic 3D displays for viewer discomfort.
     BioMed Eng OnLine. 2015;14(1):21.
10. Yue K, Wang D, Hu H, et al. P-34: compare and model multi-level stereoscopic 3D visual fatigue based on EEG. Sid
     Symp Digest Tech Pap. 2017;48(1):1359–62.
11. Shen LL, Sun WP. Using EEG for assessment of stereoscopic visual fatigue caused by motion-in-depth. Chin J Eng.
     2017;39(09):1421–7.
12. Chen C, Wang J, Li K, et al. Assessment visual fatigue of watching 3DTV using EEG power spectral parameters.
     Displays. 2014;35(5):266–72.
13. Yin J, Jin J, Liu Z, et al. Preliminary study on EEG-based analysis of discomfort caused by watching 3D images. In:
     Liljenström H, editor. Advances in cognitive neurodynamics (IV). Springer: Netherlands; 2015. p. 329–35.
14. Kang MK, Cho H, Park HM, et al. A wellness platform for stereoscopic 3D video systems using EEG-based visual
     discomfort evaluation technology. Appl Ergon. 2017;62:158–67.
15. Jérémy F, Aurélien A, Fabien L, et al. Classifying EEG signals during stereoscopic visualization to estimate visual
     comfort. Comput Intell Neurosci. 2016;2016(2):7.
16. Frey J, Pommereau L, Lotte F, et al. Assessing the zone of comfort in stereoscopic displays using EEG. CHI '14
     Extended Abstracts on Human Factors in Computing Systems, ACM. 2014; 2041–2046.
17. Creel DJ. Visually evoked potentials—webvision—NCBI Bookshelf. Salt Lake City: University of Utah Health Sciences
     Center; 2012.
18. Kamiński M, Ding M, Truccolo WA, et al. Evaluating causal relations in neural systems: granger causality, directed
     transfer function and statistical assessment of significance. Biol Cybern. 2001;85(2):145–57.
19. He B, Dai Y, Astolfi L, et al. eConnectome: a MATLAB toolbox for mapping and imaging of brain functional connectiv-
     ity. J Neurosci Methods. 2011;195(2):261–9.
20. Wilke C, Ding L, He B. Estimation of time-varying connectivity patterns through the use of an adaptive directed
     transfer function. IEEE Trans Biomed Eng. 2008;55(11):2557–64.
21. Wilke C, Drongelen WV, Kohrman M, et al. Identification of epileptogenic foci from causal analysis of ECoG interictal
     spike activity. Clin Neurophysiol. 2009;120(8):1449–56.
22. Jung YJ, Kang HC, Choi KO, et al. Localization of ictal onset zones in Lennox-Gastaut syndrome using directional
     connectivity analysis of intracranial electroencephalography. Seizure. 2011;20(6):449.
23. Ding L, Worrell GA, Lagerlund TD, et al. Ictal source analysis: localization and imaging of causal interactions in
     humans. Neuroimage. 2007;34(2):575–86.
24. Li XJ, Jiang Z, Wang Y. The temporal responses of neurons in the primary visual cortex to transient stimuli. Prog
     Biochem Biophys. 2012;39(12):1190–6.
25. Bender S, Hellwig S, Resch F, et al. Am I safe? The ventrolateral prefrontal cortex 'detects' when an unpleasant event
     does not occur. Neuroimage. 2007;38(2):367.
26. Bender S, Oelkers-Ax R, Hellwig S, et al. The topography of the scalp-recorded visual N700. Clin Neurophysiol.
     2008;119(3):587–604.
27. Zhao L. ERPs experimental tutorial. Revised edition. Nanjing: Southeast University Press; 2010 **(in Chinese)**.
28. Shuai J, Xing LD, Qian ZY, et al. The association between depth information of 3D movie and viewer fatigue. Chin J
     Biomed Eng. 2014;33(3):306–12.
29. Song WQ, Tan Q. Spatial characteristics of flash visual evoked potential in normal subjects. Int J Ophthalmol.
     2006;6(2):387–9.
30. Nolan K. Monitoring the nervous system for anesthesiologists and other health care professionals. Can J Anesth.
     2013;60(2):216–7.
31. Liu W, Zhang X, Liu J, et al. Study on the comparison between pattern-reversal and pattern onset/offset visual
     evoked potentials. Chin J Forensic Med. 2015;30(02):128–31.
32. Shawkat FS, Kriss A. A study of the effects of contrast change on pattern VEPS, and the transition between onset,
     reversal and offset modes of stimulation. Doc Ophthalmol. 2000;101(1):73–89.
33. Yuan J, Fu S. Brief review of event-related potential correlates of visual consciousness and related studies. Chin Sci
     Bull. 2012;57(35):3336.
34. Zhang W, Luck SJ. Feature-based attention modulates feedforward visual processing. Nat Neurosci. 2009;12:24–5.
35. Wilenius ME, Revonsuo AT. Timing of the earliest ERP correlate of visual awareness. Psychophysiology.
     2007;44:703–10.
36. Zhu W, Zhao L, Zhang J, et al. The influence of Mozart's sonata K.448 on visual attention: an ERPs study. Neurosci
     Lett. 2008;434(1):35–40.
37. Hanazawa A, Kawashima R, Nakamura K, et al. The human posterior parietal cortex participates in stereoscopic
     depth perception. An fMRI study. Neuroimage. 2000;11(5):S694.
38. Paggetti G, Leff DR, Orihuelaespina F, et al. The role of the posterior parietal cortex in stereopsis and hand-eye coor-
     dination during motor task behaviors. Cogn Process. 2015;16(2):1–14.
39. Smith EE, Kosslyn SM. Cognitive psychology: mind and brain. Upper Saddle River: Pearson/Prentice Hall: Pearson
     Education International; 2007.
40. Schacter DL, Gilbert DT, Wegner DM, et al. Psychology. Basingstoke: Palgrave Macmillan; 2012.

10

DEDICATE: proposal for a conceptual framework to develop dementia-friendly integrated eCare support

Sara Marceglia[1,2]*⦿, Michael Rigby[3], Albert Alonso[4], Debbie Keeling[5], Lutz Kubitschke[6] and Giuseppe Pozzi[7]

*Correspondence:
smarceglia@units.it
[1] Dipartimento di Ingegneria e Architettura, Università degli Studi di Trieste, Via A. Valerio 10, 34127 Trieste, Italy
Full list of author information is available at the end of the article

Abstract

Background: Evidence shows that the implementation of information and communication technologies (ICT) enabled services supporting integrated dementia care represents an opportunity that faces multi-pronged challenges. First, the provision of dementia support is fragmented and often inappropriate. Second, available ICT solutions in this field do not address the full spectrum of support needs arising across an individual's whole dementia journey. Current solutions fail to harness the potential of available validated e-health services, such as telehealth and telecare, for the purposes of dementia care. Third, there is a lack of understanding of how viable business models in this field can operate. The field comprises both professional and non-professional players that interact and have roles to play in ensuring that useful technologies are developed, implemented and used.

Methods: Starting from a literature review, including relevant pilot projects for ICT-based dementia care, we define the major requirements of a system able to overcome the limitations evidenced in the literature, and how this system should be integrated in the socio-technical ecosystem characterizing this disease. From here, we define the DEDICATE architecture of such a system, and the conceptual framework mapping the architecture over the requirements.

Results: We identified three macro-requirements, namely the need to overcome: deficient technology innovation, deficient service process innovation, and deficient business models innovation. The proposed architecture is a three level architecture in which the center (data layer) includes patients' and informal caregivers' preferences, memories, and other personal data relevant to sustain the dementia journey, is connected through a middleware (service layer), which guarantees core IT services and integration, to dedicated applications (application layer) to sustain dementia care (formal support services, FSS), and to existing formal care infrastructures, in order to guarantee care coordination (care coordination services, CCS).

Conclusions: The proposed DEDICATE architecture and framework envisages a feasible means to overcome the present barriers by: (1) developing and integrating technologies that can follow the patient and the caregivers throughout the development of the condition, since the early stages in which the patient is able to build up preferences and memories will be used in the later stages to maximise personalization and thereby improve efficacy and usability (technology innovation); (2) guaranteeing the care coordination between formal and informal caregivers, and giving an active yet supported role to the latter (service innovation); and (3) integrating existing infrastructures and care models to decrease the cost of the overall care pathway, by improving system interoperability (business model innovation).

Keywords: Dementia, Integrated care, Integrated eCare, Patient-centred, Family-centred, Care infrastructure, eHealth, ICT architecture, Socio-technical ecosystem

Background

Dementia is recognised as a growing societal health and care problem due to the increasing survival of older people in society [1, 2]. Not only does dementia cause serious problems for the individual concerned, but it also severely disrupts and stresses the lives of family carers, neighbours, and others inevitably involved in supporting a person with dementia [3]. At the same time, services in most countries and localities across Europe are fragmented and far from adequately developed [4]. However, economic and personnel constraints are such that re-engineering of service delivery into smarter services is at least as important as increased investment. As such, the state of the art of support for dementia in the majority of locations in Europe is seriously unsatisfactory [5]. Alongside fragmentation and silos within and between care agencies, there are weak interfaces with societal support, and the generally limited support available is dispensed to carers according to availability. Yet, there is a strong recognition of the potential and value of use of information and communication technologies (ICTs) to improve the quality of life of patients with dementia and Alzheimer's, and, in turn, that of their carers [3, 6]. Evidence from earlier pilot projects points to ICT-based integration of care services for older people as resulting in improved outcomes, client satisfaction and efficiency gains [7]. In particular, it is envisaged that ICTs will enable future systems to recognize and support the needs of caregivers, who take care of mostly (highly) dependent dementia patients, thus reducing stress and depression with concomitant long term benefits [3].

To enable such potential, there is the need to define a clear reference model and conceptual framework including all the main requirements targeted to dementia patients, and including formal and informal carers. For this reason, in this paper we propose a conceptual framework and a possible architecture that could be used to develop effective ICT-based solutions for dementia patients, that we called DEDICATE. Starting from a state-of-the-art research aimed to identify the main factors hampering a full exploitation of ICT-based solutions for dementia; we ideated a socio-technical model designed for addressing the need for user-supporting and user-driven ICT-enabled services to coordinate integrated patient-centric health and care to those with dementia, from the early stages of diagnosis to full severity. Starting from the socio-technical model, we propose an architectural approach aimed to guide the implementation of future systems supporting the full formal and informal care team, as well as the patients.

It may be unusual to public a complete conceptual model ahead of prototyping and piloting. However, we feel there are deficiencies in the current normal developmental models—namely, either proprietary product development protected against widespread early review; or research grant proposals which are secret, and which normally cannot be amended between submission for consideration and final completion of a grant-funded project. Both these approaches limit constructively critical discourse, lock development into one route, and lead to defensive attitudes to evaluation at the end point [8]. They also make difficult and expensive end-stage modification. This paper aims to enable open discussion ahead of construction, to enable a better outcome and putting societal gain above solution protection.

Methods

Since the purpose of this paper is to propose a conceptual framework and a possible reference architecture for ICT-based solutions for dementia care, we first conducted a literature research, aimed to drive the requirement definition.

The literature search was based on two sources: first, the Pubmed/Medline medical bibliographic database, second, the European Commission Portal and the Google search engine. This choice allowed opening the research not only to scientific literature but also to ongoing and past projects addressing dementia care. In the case of the Google-based general search, the results were included if they:

- Fully described the implementation of a project (or a set of projects) related to dementia care or providing a report on the state-of-the-art of dementia care (minimum date: 2013).
- Referred to the European area, in order to focus on a more homogeneous environment.

In the case of literature search, we included only reviews published after 2013 and up to 2016, in order to guarantee up-to-date the conclusions.

We identified from the literature a list of the main factors that acted as barriers to the development of effective infrastructures. From these factors, we extracted the list of requirements and the reference socio-technical ecosystem needed for dementia care [3–5, 9–12]. From them, we derived the reference architecture and the conceptual framework for its dynamic evolution. Finally, we provided a preliminary evaluation of available tools that could be used to implement a system based on the proposed architecture.

Results

Barriers to development of effective infrastructures

From our literature research we found several projects developed in Europe for the purposes of dementia care: ISISEMD (http://vbn.aau.dk/en/projects/isisemd(e5e57 897-c1e6-4993-a668-b758d4a503cc).html), Dem@Care (http://www.demcare.eu), InMINDD (http://www.inmindd.eu/), InnovateDementia, and STAR. These initiatives focus on assisted living, prevention regimes and neurodegenerative disease, which, though promising, only address individual aspects of the many problems, and do not provide an integrated delivery of all forms of care to cognitively impaired patients and their carers.

We then identified seven reviews/reports specifically addressing the problem of integrated dementia care across Europe, and reviewing the current status of development [3–5, 9–12]. These analyses underlined that the dangers of closed silo service provision have been recognised in some countries at the policy level, and that steps were taken to spread responsibility more widely and to introduce cooperative structures, including third sector and citizens' groups [4].

From the searched documents we identified the following factors that hamper the mainstreaming of ICT-based solutions:

- Deficient technology innovation: recent development of ICT platforms for more flexible care delivery to older people (e.g., UniversAAL) and those having dementia (e.g., ISISMED) lack technology features and functionalities that enable integrating the full spectrum of formal and informal caregivers into a single information loop with a view to enabling truly joined-up support, including people with dementia themselves as co-producers of well-being and independent living. Long-term care for people with dementia also comprises health and social care services–diagnostic and continuing care services. However, these services are today split into organisational clusters, separately managed, delivered and recorded. At best, people with dementia and their families tend to be surrounded by uncoordinated 'Islands of Excellence', when what is needed is person-centered coordinated care [9].

- Deficient service process innovation: in the care domain, ICT-based services tend to be delivered within socio-technical systems, and value is frequently achieved by people (e.g., care professionals, family carers) utilizing technology for their purposes (i.e., delivering people services) not by technology alone [10]. As such, ICT-based solutions for integrated dementia care require technology innovation and service process innovation to be pursued in parallel. This includes the need to recognize co-contributions by informal carers alongside formal carers as part of one system, with due identification of the role of each and for sharing of selected appropriate information and recording. The latter aspect has, however, not as yet enjoyed sufficient attention in most pilot projects in the field of ICT-based elderly care [9]. The family always has a central role, supported to a greater or lesser extent by formal professional or para-professional care services [11]. Informal family care, representing the cornerstone in almost every country, is sometimes supported or supplemented by paid home caregivers, respite opportunities and palliative end-of-life care. Strengthening the capabilities of family carers to enable better coping with the challenges of their experience is essential [12]. Caregiver interventions, such as education and counselling, have been proven to reduce or delay transition from home into a care home [3].

- Deficient business model innovation: when it comes to the desired up-scaling of successfully piloted ICT-based solutions in the field of dementia care, evidence shows that—apart from benefits shift occurring between different parties involved in joined-up care delivery—current care systems are not always favourable to the financing of integrated eCare solutions [5, 10].

Requirements and socio-technical ecosystem definition

Following the identification of the main barriers, we identified the following requirements, mapped to specifically address them:

Requirements for overcoming deficient technology innovation

- Adopting a clearly user-driven, choice-giving approach and avoiding all technology 'push', aiming to increase respect, quality and also efficiency and effectiveness of ICT solutions. Lifestyle choices are paramount at any age, yet hitherto these have been difficult to accommodate. A person's preferences for mealtime patterns, foods, time of rising and going to bed, morning or evening shower or bath, music preferences and television viewing patterns are fundamentally important to sustaining confidence and independence. There is currently a strong likelihood that the pattern of care becomes provider-driven. Thus, at the very time that the person is vulnerable and needing help, their lifestyle is changed around them. This both reduces respect and increases confusion, even in the person's own home.

- Utilising components to support persons with dementia, particularly linked to earlier diagnosis and active support to enable continued living at home and increase quality of life and comfort. ICT solutions for coordination and communication should play an effective part right from the beginning. Timely diagnosis, at a point where the person with dementia can make informed lifestyle choices, is essential but a consequence is that early support services are necessary to support the person and their family in these decisions [13]. If no early services are readily deployed, there will be no encouragement towards early diagnosis, perpetuating the current problem that diagnosis only occurs once the effects are beginning to take deeper hold. Also, denial of the diagnosis, or turning away from facing it, is not uncommon amongst people who are experiencing significant cognitive changes [14, 15].

- Including elements of telecare and telehealth where needed, respecting the wishes and preferences of the person with dementia, especially in the early stages when they should be able to express informed choices on the way they view their future care. Then, record and information presentation mechanisms need to be in place that keep knowledge available and actively refreshed, in a way that is in line with the individual's persona and expectations. Institutional and personal memory is important for efficiency but also for sustaining independence and well-being. Over time, patients experience a loss of understanding and consistency on the small issues that facilitate acceptance of help and quality of life. This is also due to changes in their carers: informal carers may change as a result of the burden of caring, and formal carers may change through staff attrition, replacement and rosters.

Requirements for overcoming deficient service process innovation

- Utilising a shared planning and communication record and function across the formal and informal carer team for the individual, interfacing with legacy systems based on open protocols. Support services need to address the wider care challenge by including the social environment from the beginning, and enabling informal carers to play a fruitful role as part of an informal–formal virtual team, based on their needs, abilities and preferences [16].

- Utilising care coordination applications to run holistically as a virtual system on formal and informal care platforms (including mobile devices), such as scheduling, automatic messaging, and voice and video telephony support. Considerable improvements can be achieved by interfacing different ICT technologies to empower the remote carer [17]. The combination of video call contact with the frail relative, coupled with access to an ongoing log of carer comments, and the ability to input to that communication system, could still effectively bring a remote carer into the virtual team delivering real care.

- Planning of support in a shared timeline, linked to patient and carer needs and service capacity.

Requirements for overcoming deficient business models innovation

- Supporting stakeholder centric, evidence-based business case modelling. A sustainable care model for ICT-based dementia support has to be comprehensive and coordinated between the different stakeholders involved, and operated on a user-friendly platform. Integration of delivery of care services is essential for better quality, better efficiency and effectiveness, and to accommodate growing numbers within constrained resources [6, 9].
- Supporting evidence-based decision-making on the up-scaling of implemented solutions implying the inclusion, from the design phase, of appropriate information necessary for establishing the benefits and drawbacks of the solution, in line with the concept of "Evidence-Based Medical Informatics" [18].

A final unavoidable requirement is interoperability, which is considered as a prerequisite for successful service development and sustainable service operation. To this end, interoperability cannot be only associated with the ability of telecommunications and digital systems—and the processes they support—to exchange data and to enable the sharing of information and knowledge [19], but also with non-technical perspectives (i.e., organizational, policy, governance, and legal perspectives), such as that defined by the European Union eHealth Interoperability Framework [20].

Taken together, these requirements suggest that any ICT-based solution aimed at delivering care to dementia patients is part of a socio-technical ecosystem (Fig. 1) that includes all relevant stakeholders as actors, uses existing systems and services, and manages needs and tools that have to be coordinated. The center of the ecosystem is composed by the patients and their informal/social caregivers (informal co-producers), whereas the borders of the ecosystem are the other care stakeholders and the existing services, representing the domain of formal care. The integration technology (digital coordination infrastructure) acts as a "sharing bus" connecting the dedicated systems addressing all the needs of the informal co-producers to the other actors (formal care domain). Technology innovation requirements, service process innovation requirements, and business model innovation requirements concur to implement such holistic ecosystem.

Proposed architecture and conceptual framework

Figure 2 shows the proposed architecture, which relies upon a concept of integration of different applications mediated by a middleware that introduces an abstraction layer in the system architecture and thus reduces the complexity considerably.

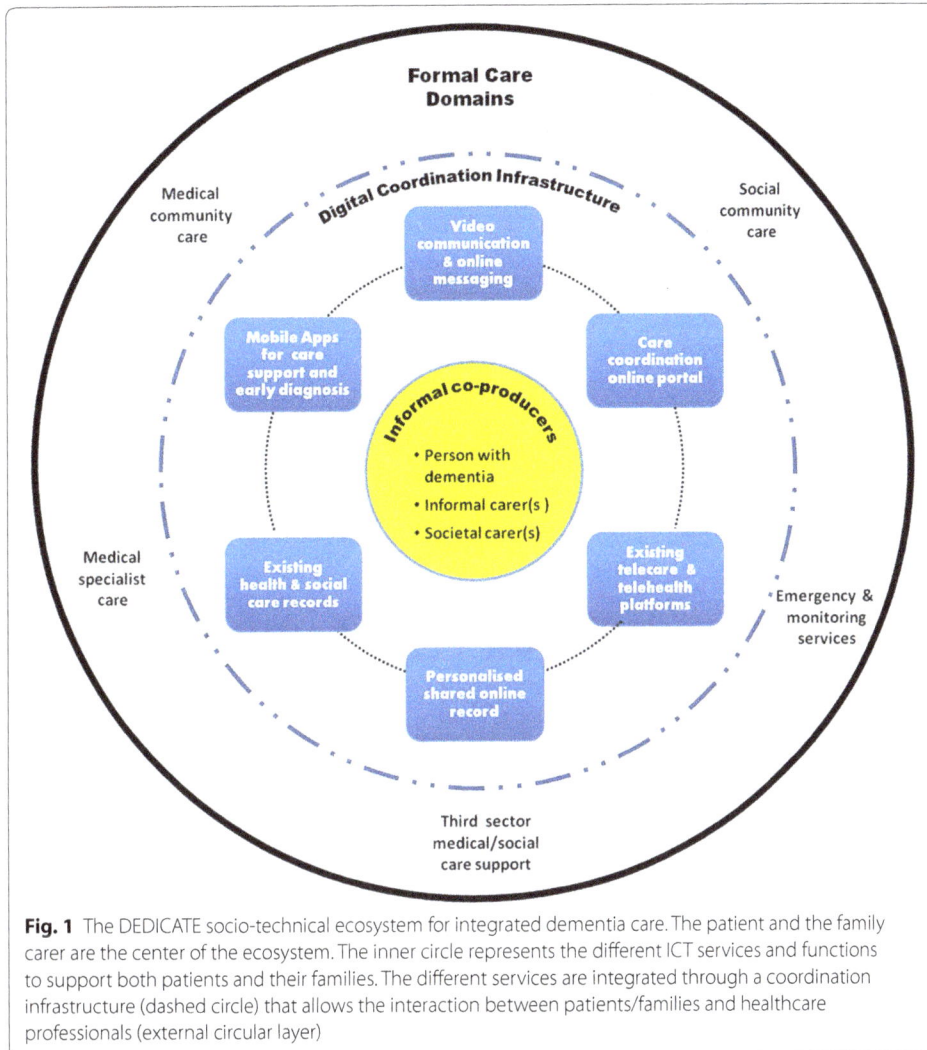

Fig. 1 The DEDICATE socio-technical ecosystem for integrated dementia care. The patient and the family carer are the center of the ecosystem. The inner circle represents the different ICT services and functions to support both patients and their families. The different services are integrated through a coordination infrastructure (dashed circle) that allows the interaction between patients/families and healthcare professionals (external circular layer)

In general, the proposed architecture is classically based on three layers, the data layer, the service layer, and the application layer. The data layer consists in a secure storage system devoted to the management of personal preferences, educational information, and all those data needed to both longitudinally follow the patient's journey with dementia and to guarantee informal caregiver inclusion. The service layer provides a set of core IT services that enables the main functionalities (e.g., access, audio/video communication, messaging, data acquisition/exchange). Acting as a middleware for integrating the various systems and applications in the final layer, a Service Oriented Architecture (SOA) design approach should be implemented for this layer. SOAs ensure that available services can be used by multiple different systems from several business domains and that the interaction among services is not affected by the service implementation. Hence, to guarantee interoperability, services should be designed with implementation-independent interfaces, and by implementing communication protocols that stress location transparency. Also, services should use explicit interfaces to define the encapsulated functions, so that either the service can

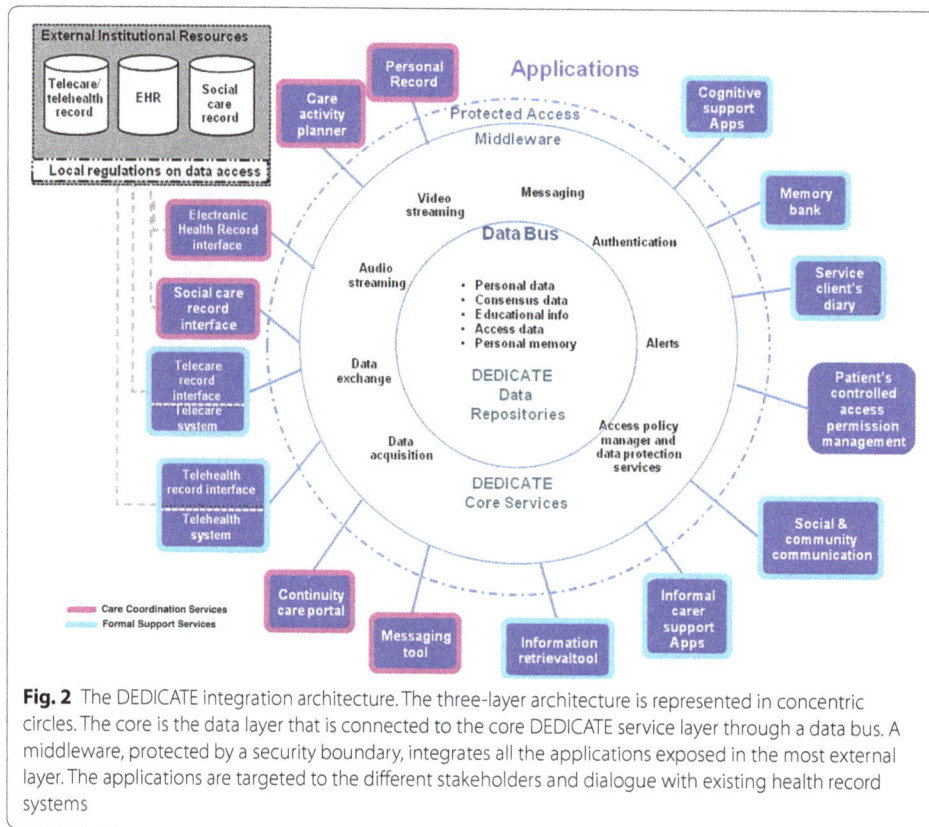

Fig. 2 The DEDICATE integration architecture. The three-layer architecture is represented in concentric circles. The core is the data layer that is connected to the core DEDICATE service layer through a data bus. A middleware, protected by a security boundary, integrates all the applications exposed in the most external layer. The applications are targeted to the different stakeholders and dialogue with existing health record systems

be used by one or more systems that participate in the architecture or the service can be easily substituted by another providing the same functionalities as declared in the interface. Finally, the application layer is composed both by existing systems, including web applications and telemonitoring and telecare systems, and newly-conceived applications (e.g., exercises for cognitive support, apps for the creation of personalized memory banks, patient's diaries, social and community applications, applications dedicated to informal carers, etc.).

The applications connected to the DEDICATE architecture can be divided into two main categories: the formal support services (FSS) and the care coordination services (CCS).

FSS aim to provide a number of support tasks to be achieved and supported throughout the various stages of the dementia journey. When dementia has been identified, there is a lot of new information to assimilate and a lot of decisions to be made. This is a period of change and upheaval both practically and emotionally for the person diagnosed and for their family and friends. At an early stage, the person themselves may become less able to recall important current information so having a reliable interface where this can be shared and not lost is crucially important (e.g., personally constructed memory banks). Also, telecare and telehealth has an increasing role to play as dementia progresses. The need for on-going support and feeling connected to favourite places and past-times remains crucial for the person living with dementia and the family carer. More specifically, FSS should:

- Provide people with information personalised to their needs at the various stages of dementia (early stage: creation of the memories; later stages: use of the memories to recall information).
- Help people with dementia and their families in adjusting to diagnosis (later stages).
- Support lifestyle changes that would help people in the longer term (later stages).
- Support people with dementia and their family in making plans and records that would help in the future (early stage).
- Support people with dementia and their families in enjoying life (all stages).
- Provide support when it comes to symptoms control and keeping fit and healthy (later stages).
- Help in dealing with crises and managing the unexpected (later stages).
- Risk management (telecare) and health monitoring (telehealth) (all stages).

CCS are, on the other hand, dedicated to the number of actors needing to coordinate their caring task(s) around the individual's needs throughout the different stages of the dementia journey. CCS support inter-organisational cooperation at the "back office" level when it comes to organisations involved in professional care (e.g., health care and social care) and those providing non-professional support (e.g., third sector organisations). Moreover, CCS also enable efficient collaboration with the family carer. Not least, they would empower persons with dementia themselves—according to their mental faculties—to take part in effective management of their condition and maintain their independence as far as possible. Such abilities have often been overlooked [13]. More specifically, CCS should provide:

- Integrated data access for care providers in different agencies and informal carers.
- Planning, scheduling and reporting of care activities enabling temporal coordination between provision steps taken by care providers in different agencies and informal carers.
- Access to analytics for home-based monitoring data (telemonitoring and/or telecare) by care providers in different agencies and informal carers.
- Real-time communication between care providers in different agencies and informal carers, for example, support to case conferences.
- Joint response to ad-hoc requests by care providers in different agencies and informal carers.
- Self-management, including links into all above cooperation mechanisms.

The architecture hypothesizes two types of access policies. One controlled by the patient that regards the access to his/her personal information for which he/she has the full responsibility (i.e., memory bank contents, cognitive supporting apps results, etc.). One depending on the local constraints and regulations regarding health-related information controlled by the care providers (i.e., therapies and treatments, care pathways, etc.).

Figure 3 illustrates the conceptual framework that shows how the proposed architecture fulfills the expected requirements. More specifically, the three macro-requirements are achieved through a threefold service integration strategy (Fig. 3).

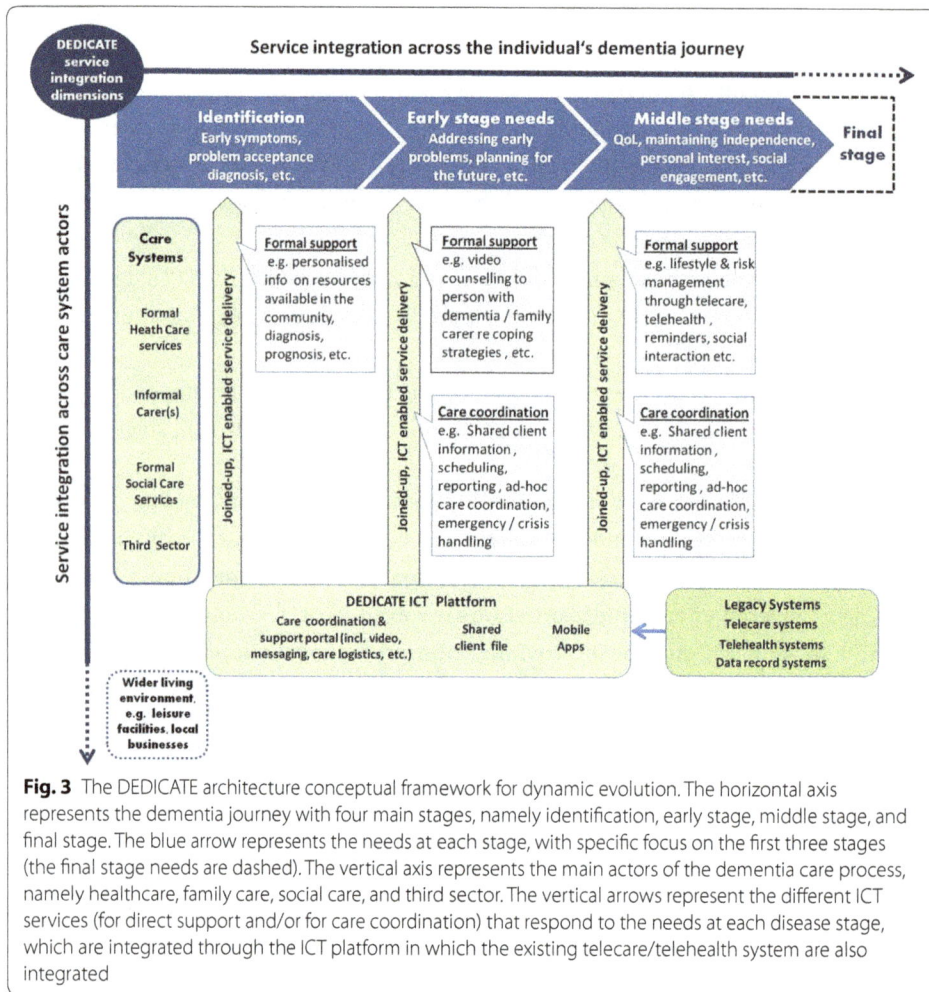

Fig. 3 The DEDICATE architecture conceptual framework for dynamic evolution. The horizontal axis represents the dementia journey with four main stages, namely identification, early stage, middle stage, and final stage. The blue arrow represents the needs at each stage, with specific focus on the first three stages (the final stage needs are dashed). The vertical axis represents the main actors of the dementia care process, namely healthcare, family care, social care, and third sector. The vertical arrows represent the different ICT services (for direct support and/or for care coordination) that respond to the needs at each disease stage, which are integrated through the ICT platform in which the existing telecare/telehealth system are also integrated

Technology innovation

The DEDICATE architecture service delivery is integrated across changing needs throughout the individual's 'dementia journey' (horizontal axis, in Fig. 3) that characterizes progressive dementia at different stages. Depending on the stage, different types of information and resources tend to be helpful to people with dementia and the surrounding living environment [13]. For example, when an individual has been diagnosed, this person and her/his immediate family and friends need to come to terms with the diagnosis and integrate it into their life. They may not be focused on the practicalities of managing the condition on a daily basis. In the DEDICATE architecture, this is achieved by guaranteeing, on the one hand, the management of personal preferences, memories, etc., together with the delivery of FSS that are mainly needed in early stages, and, on the other hand, adding CCS and new FSS based on the personal preferences previously created in the later stages. In fact, although it is difficult to predict how each person will fare with dementia, there is usually a point where the person begins to need significantly more help and assistance to function on a day-to-day basis at home, and, at this point, CCS should be added for improved care coordination.

Service innovation

The architecture guarantees integration across different key actors typically involved in dementia care (vertical axis, Fig. 3). Such integration is acknowledged as a complex process [21]; yet solutions often ignore the opportunities for incorporating the informal carer(s) into formal processes [22]. In the DEDICATE model, today's organisational care silos are overcome by enabling coordinated cross-sector delivery of support to people with dementia. In doing so, a dual role is assigned to informal family carers. On the one hand, they receive supportive services from formal health and social care providers (e.g., counselling) enabling better coping with their caring experience. On the other hand, they are enabled to coordinate their own caring activities with formal health and social care services available in the community. Recognising the value that proactive informal carers can bring to the care journey in acting as a knowledgeable and local coordinator of care (within their stated abilities and preferences).

Business model innovation

The DEDICATE architecture allows the integration of new care-coordination technology and social technologies with existing home monitoring and safety technology (DEDICATE ICT Platform, Fig. 3). If newly developed ICT-based dementia services are to be enabled for up-scaling and replication, attention should be given to the considerable diversity across countries and regions in the framework conditions within which pilot services mainstreaming occurs. This situation largely reflects today's realities of the developing market place for integrated eCare solutions [23]. Against this background, a simplistic approach to the adoption of a new model for integrated dementia care can easily be (mis)interpreted as the wholesale migration to new service processes and ICT platforms. In reality, however, such an approach poses major budgetary problems for service providers who have already invested in particular care technologies addressing at least parts of the wider spectrum of user needs arising throughout the 'dementia journey' (e.g., mainstream telecare systems), as well as feared disruption of services which are trusted even in their imperfection. Also, such an approach introduces risks in terms of system delivery and potential loss of service and data continuity. Therefore, the integration approach proposed in the DEDICATE architecture can introduce the benefits of a holistic approach while maintaining a controlled use of resources.

In Table 1 a detailed description of how the DEDICATE architecture and framework address the specific requirements is provided.

Possible implementation scenario

The main components of the DEDICATE architecture are: a platform ensuring data management and core IT services acting as a middleware for application integration and series of dedicated applications. On top of that, existing services and infrastructures can be connected through the middleware to obtain a holistic system. The core platform can be a care coordination portal that supports the different care ('people') services working together with informal carers and assisted persons with dementia;

Table 1 Detailed matching between requirements, DEDICATE architecture features, expected benefits, and proposed metrics

Macro-requirement	Specific requirement	DEDICATE architecture feature	Expected benefit	Possible example metrics
Requirements for overcoming deficient technology innovation	Adopting a clearly user-driven, choice-giving approach and avoid all technology push'	Build patient's preferences and memories in early stage of disease that will remain during the entire journey. DEDICATE also envisages the use of existing technology already available to patients/families	Improved compliance to the system and long-term use	# patients using the system along the whole dementia journey Time spent (%) using the system without caregiver help Time spent (%) by the caregiver using the system
	Utilising components to support persons with dementia	Specifically developed applications are integrated to the system through the middleware to guarantee support		
	Including elements of telecare and telehealth where needed	Existing telehealth and telecare systems are integrated in the architecture		
Requirements for overcoming deficient service process innovation	Utilising a shared planning and communication record and function across the formal and informal carer team	Electronic Health Record, Social Care Record, telehealth/telecare record are all interfaced to the system, and their data are integrated in care coordination plans	Improved inclusion of healthcare professionals in the dementia journey and facilitated disease management from a medical perspective	# of unscheduled access to the healthcare system % of healthcare documents integrated in the system # access to the system by healthcare professionals to review patients' situation
	Utilising care coordination applications to run holistically as a virtual system	Care coordination applications are part of the whole system and are supported by the core ICT services run on the service layer		
	Planning of support in a shared timeline	The architecture envisages the co-participation of all actors, from formal to informal carer as well as other stakeholder, which share the same information and plans		
Requirements for overcoming deficient business models innovation	Supporting stakeholder centric, evidence-based business case modelling	The architecture is designed so that all information is shared among stakeholders, and existing systems are integrated and not replicated	Improved sustainability and auditability of the services	# trials regarding the effectiveness of ICT interventions for dementia care based on the system % data collected by the system used for effectiveness assessment/quality assessment
	Supporting evidence-based decision-making	The integration of formal care records and personal applications will allow collecting relevant data needed to track the effectiveness of the system/interventions		

rendering the ability to access a 'CRM-type' (Customer Relationship Management) database via a browser so that users can track, monitor and request support, or simply communicate as well as highlight where interventions and support may be required.

A possible prototype to implement the care coordination portal is the Connect-I portal, developed within the EU's CIP programme INDEPENDENT project (http://independent-project.eu/home.html), and mainstreamed in the region of Milton Keynes, UK. This project has developed an integrated set of ICT-enabled services dealing with a range of threats to independent living common to older people, which could be specialized for persons with dementia according to the specific requirements of integrated dementia care.

Within the portal, the access to personalised information is only provided to selected individuals following consent by the care recipient. As a main strength, the Connect-I system is developed within the context of existing proven hardware and software technologies (standard PCs, operating systems, web cams, database software based on MS SharePoint Portal technologies and commonly available broadband services including open standard VOIP applications) that can be easily implemented in standard settings, without specific equipment required for patients and families. In addition, the Connect-I portal implements a wide range of functionalities, including: case record management functionalities (e.g., capability to take Notes on contact record, case management and workflow, broader ethnicity categories, etc.); reporting capabilities (e.g., attendance at a specified workshop/training/support group/event, history of a specific client, statistical breakdown of client groups, last contact date, etc.); self-book appointments for web-cam based support/help sessions, time booking system (for the caring processes); and, simple document sharing and key information library facility, events calendar and reminders system with internal messaging, telephone lists, service provider details, with ability to rank and rate by carers.

The second component is a set of mobile apps and telehealth systems for both patients and carers, supporting multiple devices, including mobile devices, such as industry standard tablet PCs, smartphones, and smart watches. For instance, a holistic care mobile health app should provide care support and health information both inside and outside the home, thus allowing the user to easily reach a carer or family member should they need to and giving autonomy for the person living with dementia and peace of mind for their carer (e.g., mHealthASSIST app, see https://www.hma.co.uk/work/featured-home-case-study/). Another exemplary app to be integrated is a cross-platform tablet app prototype to support people living with dementia in their leisure activities (digIT, see https://www.hma.co.uk/work/featured-home-case-study/), designed to engage the person as well as their carer.

Finally, the care coordination portal should be also open towards interlinkage—as far as possible based on available standards—with established record systems in the health care and social care arenas, to enable transferring person-specific health and social care-related data into the care coordination process as deemed appropriate for the purposes of integrated dementia care, thereby adhering to relevant ethics and data privacy rules/regulation.

Even though this is only an envisaged implementation of the proposed architecture, the technological elements described here are already available and can be integrated to create a real DEDICATE system. The use of familiar technologies enables the existing skills of users to be maintained and utilized, or at least, that use skills are within the realm of everyday experiences and thereby accelerating adaptation to such systems.

Proposed evaluation and metrics

Implementing a system based on the proposed architecture and evaluating its performance and usability with respect to existing systems would be the next steps. System evaluation needs the definition of appropriate metrics, which have to be defined according to the specific case study and the priorities and values of stakeholders. However, the proposed conceptual framework can ground the identification of the outcomes and benefits that, in turn, can be used to define the metrics [24], using, for instance, the Goal Question Metrics (GQM) approach [25].

From a high-level perspective, the requirements for overcoming deficient technology innovation are expected to improve the conformity of the patients and caregivers to the system, thanks to an improved usability, acceptability, and benefits (Table 1). These benefits can be measured, for instance, by tracking the number of patients using the system along the whole dementia journey, or the percentage of time spent by the patients using the system with and without caregiver help.

Similarly, the requirements for overcoming deficient service process innovation are expected to improve the integration of healthcare professionals in the dementia journey, as well as the disease management from a medical perspective. These benefits can be measured by monitoring the number of unscheduled accesses to the healthcare system, or the percentage of healthcare documents that are successfully integrated into the system, and accessed by healthcare professionals to review patients' situation (Table 1).

Next, the requirements for overcoming deficient business models innovation would improve the long-term sustainability of the ICT interventions, as well as their monitoring. Hence, the metrics can be used to measure the usage of the data collected through the system for evaluating the effectiveness and/or the quality of the interventions for the dementia patients (Table 1).

Finally and holistically, the core benefits can be measured in terms of patient gain, such as by length of successful living at home, reduction of adverse incidents, or reduced carer fatigue symptoms; and in terms of service provider benefit from optimized use of time, and reduced emergency calls or emergency room attendances. This matches the true aim of stakeholders for optimum quality of life, and efficient use of resources.

Discussion

This work presents a conceptual framework and architecture for supporting the development of ICT-based solutions that overcome the presently existing barriers. The proposed approach is grounded on the principles of avoiding the fragmentation of support by focusing on holism and integration, of giving continuity along the path of disease progression, and of seeking to accommodate all supporting options, thereby also seeking to delay or even avoid institutional admission.

The proposed DEDICATE architecture is based on the pursuit of: coordinated working of formal carers and informal carers with their clients, patients and families; affordable technology infrastructure, featuring volume ('high street') products where possible; interoperability of services, information and products; and, the adoption of open interface standards. It also envisages the interoperability between a range of applications that can be implemented by different systems according to the different infrastructures available. Hence, the proposed architecture can be applied to different environments with different constraints, provided that the whole set of interoperability principles are fulfilled.

The value of this work resides in the definition of a complete holistic architects' and people vision—encompassing both patients and carers—which is fundamental before building an implementation. The integration of the sociological, technical, caring, and governance issues into one vision is itself an accomplishment, and a far healthier springboard than starting with a project-led technology-based build.

The first innovative aspect introduced by the whole DEDICATE architecture is to focus on different stages of dementia in a holistic manner, rather than selected aspects emerging during the typical dementia journey at a particular stage. The real benefits of this approach will be slow to demonstrate, as they are aimed in particular at delaying—or even avoiding—permanent admission to residential facilities as well as long-term mental health issues developing amongst care groups. Meanwhile, the interim benefits should, not least, be reduced anxiety and confusion in newly diagnosed persons and their family carers, and reduced adverse incidents and emergency interventions or admissions.

Secondly, the DEDICATE architecture is innovative by introducing the idea to enable the person with dementia, with their carers and over a period of time, to build up a profile of themselves. This includes familiar events and key memories, so that these will be available to future carers to act as a means of personalisation, but it also includes a recording of preferences and wishes ranging from lifestyle preferences to wishes on patterns of care at a future more dependent stage of their journey. It also enables formal carers to document professional and institutional knowledge about the individual as a care recipient, including responses to interventions and best means of obtaining cooperation, facilitating care in the future as staff or settings change.

Thirdly, the DEDICATE architecture has been designed to move well beyond normal practice by providing a vehicle for coordination of care. There are two aspects of such coordination. One is temporal coordination, particularly in the next and later stages of the disease, enabling support to be coordinated and distributed according to need of the person and the family carers, and to times they prefer. The other one is communication and messaging facility, whereby carers can pass on messages about needs, or other issues, and have the means of communication with the formal care team, if necessary, because of an anxiety or a deterioration in the patient's condition.

As described above, the implementation of the DEDICATE architecture does not require the development of new technologies but it introduces an innovative way for its construction and adoption. Whatever the starting solution, a range of functions, from care delivery schedules to personal profiles, and from care plans to remote monitoring results, should be bundled into the deployed platform to give a holistic application. Since users have access, including informal carers by individual agreement, this platform should provide a means whereby a next-of-kin living at a distance (even in another jurisdiction) will be able to keep in touch with the care team.

Limitations

However, the applicability of the approach may face several limiting non-technological challenges. First, the DEDICATE architecture, while addressing the needs and requirements of patients and carers, does not take into account the regulatory and governance constraints that may be encountered during implementations in real settings and that can vary in different deployment contexts. Second, although social care provider organisations increasingly maintain electronic records bringing together all the relevant information for a service user into an electronic data repository, the present stage of deployment of electronic social care records (ESR) is less widespread when compared with EHRs, and the level of integration between ESRs and EHRs is still low. This would introduce the need to integrate the social and the health record without relying on recognized standards, thus limiting the generalizability of the solution deployed in a specific site. Nevertheless there is work internationally in different regions in identifying what would be needed to establish these standards as well an integrated care record (see, for instance, http://wiki.hl7.org/index.php?title=Coordination_of_Care_Services_Speci fication_Project). Harmonised overarching operational and governance supervision are essential prerequisites, but understanding and development of these is still rare. Even though new technologies are available, and the vision of a new paradigm of service benefits can be shared, the organizational and human challenges of agreeing accountability and governance are often the greatest challenges—hopefully the articulation of an achievable step change will stimulate the drive to overcome these hurdles. Lastly, we are still facing fragmented reimbursement systems to achieve integrated service financing. This applies in two dimensions. First, the funding of the services themselves, and secondly the funding and maintenance of the inter-sectoral electronic ecosystem.

Finally, there are always user issues to be considered. For example, user understanding of and trust in a secure system of access to personal information and care records. We should not underestimate the value of the information that is being committed by the individual, their family and carers into the system. Also, the emphasis on familiar, everyday technologies that build on existing or, at least, familiar skills, recognizes user skills as a pivotal issue in the successful implementation of any system and aims to accelerate skills development. For these reasons, the DEDICATE architecture is built around the concepts of family/carers/patients, and it takes into account the 'needs understanding' instead of the technological implementation of the system.

Conclusions

In conclusion, the benefit of creating and describing the DEDICATE conceptual framework and architecture is that, taking a longitudinal perspective, it puts the needs of patients and carers foremost and central with technology as the enabler of a new human approach. The DEDICATE architecture should be adopted as a framework grounding the development of new ICT-based solutions for dementia care. Putting it forward for critical debate at this stage is aimed at ensuring strengthening, gaining greater understanding from the very different stakeholders necessarily involved, as well as stimulating interest in progressing development and implementation.

Authors' contributions

SM ideated the architecture, conceptualized the infrastructure, drafted the manuscript and the artwork, MR coordinated the conceptualization of the scope and spread of the research model, ideated the socio-technical framework and the model for dementia care, and reviewed the manuscript, AA contributed to the conceptualization of the framework and of the model for integrated dementia care, DK contributed to the conceptualization of the framework and of the model for dementia care, and reviewed the manuscript and the artworks, LK coordinated the research team and supported the conceptualization of the architecture and infrastructure, GP contributed to the conceptualization of the architecture and the infrastructure, and reviewed the manuscript. All authors read and approved the final manuscript.

Author details
[1] Dipartimento di Ingegneria e Architettura, Università degli Studi di Trieste, Via A. Valerio 10, 34127 Trieste, Italy. [2] Fondazione IRCCS Ca' Granda Ospedale Maggiore Policlinico, Milan, Italy. [3] Health Information Strategy, Keele University, Keele, UK. [4] Hospital Clinic Barcelona, Barcelona, Spain. [5] University of Sussex, Brighton, UK. [6] Empirica Communications and Technology Research, Bonn, Germany. [7] Dipartimento di Elettronica, Informazione e Bioingegneria, Politecnico di Milano, Milan, Italy.

Acknowledgements
The authors wish to acknowledge all the researchers and clinicians who participated in the preparation of the preceding DEDICATE project proposal. S.M. research is supported by the grant GR-2011-02352807 from the Italian Ministry of Health, by the Roche Italia Research Award 2017 "Roche per la Ricerca", and by the "Ricerca Corrente 2018" of the Fondazione IRCCS Ca'Granda Ospedale Maggiore Policlinico.

Competing interests
The authors declare that they have no competing interests.

Consent for publication
Not applicable.

Funding
Not applicable. No funding was received for the present research.

References
1. Abbott A. Dementia: a problem for our age. Nature. 2011;475:S2–4.
2. Prince M, Wimo A, Guerchet M, Ali G-C, Wu Y-T, Prina M. World alzheimer report 2015: the global impact of dementia, an analysis of prevalence, incidence, cost and trends. London: Alzheimer's Disease International; 2015. https://www.alz.co.uk/research/world-report-2015.
3. Martínez-Alcalá CI, Pliego-Pastrana P, Rosales-Lagarde A, Lopez-Noguerola JS, Molina-Trinidad EM. Information and communication technologies in the care of the elderly: systematic review of applications aimed at patients with dementia and caregivers. JMIR Rehabil Assist Technol. 2016;3:e6.
4. Leichsenring K, Billings J, Nies H, editors. Long-term care in Europe: improving policy and practice. Houndmills: Palgrave Macmillan; 2013.
5. Meyer I, Müller S, Kubitschke L, editors. Achieving effective integrated E-Care beyond the silos. IGI Global; 2014. https://doi.org/10.4018/978-1-4666-6138-7.
6. ICTs and the Health Sector. Paris: OECD Publishing; 2013. https://doi.org/10.1787/9789264202863-en.
7. Stroetmann KA, Kubitsche L, Robinson S, Stroetmann V, Cullen K, McDaid D. How can telehealth help in the provision of integrated care?. Copenhagen: World Health Organization; 2010.
8. Michael R. Evaluation: 16 powerful reasons why not to do it- and 6 over-riding imperatives. Stud Health Technol Inform. 2001;2:1198–202.
9. Rigby M, Koch S, Keeling DI, Hill P. Developing a New Understanding of enabling health and wellbeing in Europe: Harmonising health and social care delivery and informatics support to ensure holistic care. Strasbourg: European Science Foundation; 2013.
10. Stroetmann KA. Achieving the integrated and smart health and wellbeing paradigm: a call for policy research and action on governance and business models. Int J Med Inf. 2013;82:e29–37.
11. De Cola MC, Lo Buono V, Mento A, Foti M, Marino S, Bramanti P, et al. Unmet needs for family caregivers of elderly people with dementia living in italy: what do we know so far and what should we do next? Inq J Med Care Organ Provis Financ. 2017;54:46958017713708.
12. Dam AEH, Boots LMM, van Boxtel MPJ, Verhey FRJ, de Vugt ME. A mismatch between supply and demand of social support in dementia care: a qualitative study on the perspectives of spousal caregivers and their social network members. Int Psychogeriatr. 2017;30:1–12.
13. Herron RV, Rosenberg MW. "Not there yet": examining community support from the perspective of people with dementia and their partners in care. Soc Sci Med. 1982;2017(173):81–7.
14. Joseph-Williams N, Elwyn G, Edwards A. Knowledge is not power for patients: a systematic review and thematic synthesis of patient-reported barriers and facilitators to shared decision making. Patient Educ Couns. 2014;94:291–309.
15. Higgins T, Larson E, Schnall R. Unraveling the meaning of patient engagement: a concept analysis. Patient Educ Couns. 2017;100:30–6.
16. McColl-Kennedy JR, Snyder H, Elg M, Witell L, Helkkula A, Hogan SJ, et al. The changing role of the health care customer: review, synthesis and research agenda. J Serv Manag. 2017;28:2–33.

17. Piraino E, Byrne K, Heckman GA, Stolee P. Caring in the information age: personal online networks to improve caregiver support. Can Geriatr J CGJ. 2017;20:85–93.
18. Rigby M, Ammenwerth E. The need for evidence in health informatics. In: Ammenwerth E, Rigby M, editors. Evidence-based health informatics – promoting safety and efficiency through scientific methods and ethical policy. Studies in health technology and informatics series, vol. 222. Amsterdam: IOS Press; 2016. pp. 634–637. http://ebooks.iospress.nl/volumearticle/42811.
19. European Community. European interoperability framework for pan-European eGovernment services. Luxembourg: European Community; 2004.
20. European Commission, Directorate-General for the Information Society and Media. European eHealth interoperability roadmap: final European progress report. Luxembourg: EUR-OP; 2011.
21. Spanjol J, Cui AS, Nakata C, Sharp LK, Crawford SY, Xiao Y, et al. Co-production of prolonged, complex, and negative services: an examination of medication adherence in chronically ill individuals. J Serv Res. 2015;18:284–302.
22. Ferrante S, Bonacina S, Pozzi G, Pinciroli F, Marceglia S. A design methodology for medical processes. Appl Clin Inform. 2016;7:191–210.
23. Kubitsche L, Cullen K. ICT and ageing—European study on users, markets and technologies; 2010. http://ec.europa.eu/information_society/newsroom/cf/dae/document.cfm?doc_id=952. Accessed 7 Sept 2018.
24. Marceglia S, Ferrante S, Bonacina S, Pinciroli F, Lasorsa I, Savino C, et al. Domains of health IT and tailoring of evaluation: practicing process modeling for multi-stakeholder benefits. Stud Health Technol Inform. 2016;222:63–76.
25. van Solingen R, Basili V, Caldiera G, Rombach HD. Goal question metric (GQM) approach. In: Marciniak JJ, editor. Encyclopedia of software engineering. Hoboken: Wiley; 2002. https://doi.org/10.1002/0471028959.sof142.

11

Lying position classification based on ECG waveform and random forest during sleep in healthy people

Hongze Pan[iD], Zhi Xu[*], Hong Yan, Yue Gao, Zhanghuang Chen, Jinzhong Song and Yu Zhang

*Correspondence:
azhi_2840@qq.com
China Astronaut Researching
and Training Center, Beijing,
China

Abstract

Background: Several different lying positions, such as lying on the left side, supine, lying on the right side and prone position, existed when healthy people fell asleep. This article explored the influence of lying positions on the shape of ECG (electrocardiograph) waveform during sleep, and then lying position classification based on ECG waveform features and random forest was achieved.

Methods: By means of de-noising the overnight sleep ECG data from ISRUC website dataset, as well as extracting the waveform features, we calculated a total of 30 ECG waveform features, including 2 newly proposed features, S/R and $\angle$QSR. The means and significant difference level of these features within different lying positions were calculated, respectively. Then 12 features were selected for three kinds of classification schemes.

Results: The lying positions had comparatively less effect on time-limit features. QT interval and RR interval were significantly lower than that in supine ($P \leq 0.01$). Significant differences appeared in most of the amplitude and double-direction features. When lying on the left side, the height of P wave and T wave, QRS area and T area, the QR potential difference and $\angle$QSR were significantly lower than those in supine ($P \leq 0.01$). However, S/R was significantly greater on left than those in supine ($P \leq 0.01$) and on right ($P \leq 0.05$). The height of T wave and area under T wave were significantly higher in supine than those on right ($P \leq 0.01$). For the subject specific classifier, a mean accuracy of 97.17% with Cohen's kappa statistic κ of 0.91, and AUC > 0.97 were achieved. While the accuracy and κ dropped to 63.87% and 0.32, AUC > 0.66, respectively when the subject independent classifier was considered.

Conclusions: When subjects were lying on the left side during sleep, due to the effect of gravity on heart, the position of heart changed, for example, turned and rotated, causing changes in the vectorcardiogram of frontal plane and horizontal plane, which lead to a change in ECG. When lying on the right side, the heart was upheld by the mediastinum, so that the degree of freedom was poor, and the ECG waveform was almost unchanged. The proposed method could be used as a technique for convenient lying position classification.

Keywords: ECG waveform, Sleep, Lying position, Random forest, Classification

Background

Sleep is an essential process in human life, which plays a necessary role in self-repair, self-recovery of body condition, as well as integration and consolidation of memory. It is an indispensable part of human health. About one-third of a person's lifetime is spent during sleep. Good sleep can eliminate fatigue, restore one's strength and energy, and ensure body functioning well. For healthy subjects during the overnight sleep, different lying positions appear such as lying on the left side, supine (lying on the back), lying on the right side, and prone (lying on the stomach). This may cause the skin to squeeze or stretch, and the distance between the electrodes to shorten or prolong. On the other hand, the heart is squeezed slightly, and chest is pressed so that breath is influenced. All these body changes will result in ECG (electrocardiograph) waveform changes.

As early as in 1997, in the course of clinical myocardial ischemia monitoring, Adams et al. had found that the side lying position frequently caused obvious ECG changes [1]. Shinar et al. found that the R-wave durations were significantly different in three lying positions, and thus successfully identified 90% of body position changes during sleep by calculating the R-wave duration of lead I, II, and III lead ECG, simultaneously [2]. Shinar further used these three leads to classify four positions, finding that the II lead ECG worked best and achieved 80% accuracy [3]. When comparing standing and supine positions of healthy subjects, Batchvarov et al. found that the RR interval of 12-lead ECG was significantly shorter in standing than that in supine [4]. Smit et al. investigated the changes of QRS waves in ECG after normal exhalation, maximum inspiration, and maximum exhalation. It was concluded that the three kinds of breath-holding conditions had little effect on the QRS complex and individual differences were large [5].

Existing studies have shown that body positions and chest changes could cause changes in ECG waveforms, but there's no study exploring the consistent principle of such changes in ECG waveforms, systematically. It is of great importance for researchers to consider these impact in mind from lying position changing when studying the ECG waveform changes in different sleep stages. And furthermore, these changes in waveforms can be applied to non-artificial and low-intrusion lying position supervision. Consequently, in this article, we present a method of exploring the influence of lying positions on the shape of ECG waveforms during the overnight sleep in healthy subjects, and then lying position classification based on such principle and random forest is applied.

Methods

The study presented in this article can be divided into 3 parts. Data process mainly includes ECG signal preprocessing, character points detection, data epoch segmentation, features extraction with three kinds of waveform features. Then the significant differences between lying positions of waveform features are calculated. Finally lying position classification based on ECG waveform and random forest during sleep is achieved. The workflow is shown in Fig. 1.

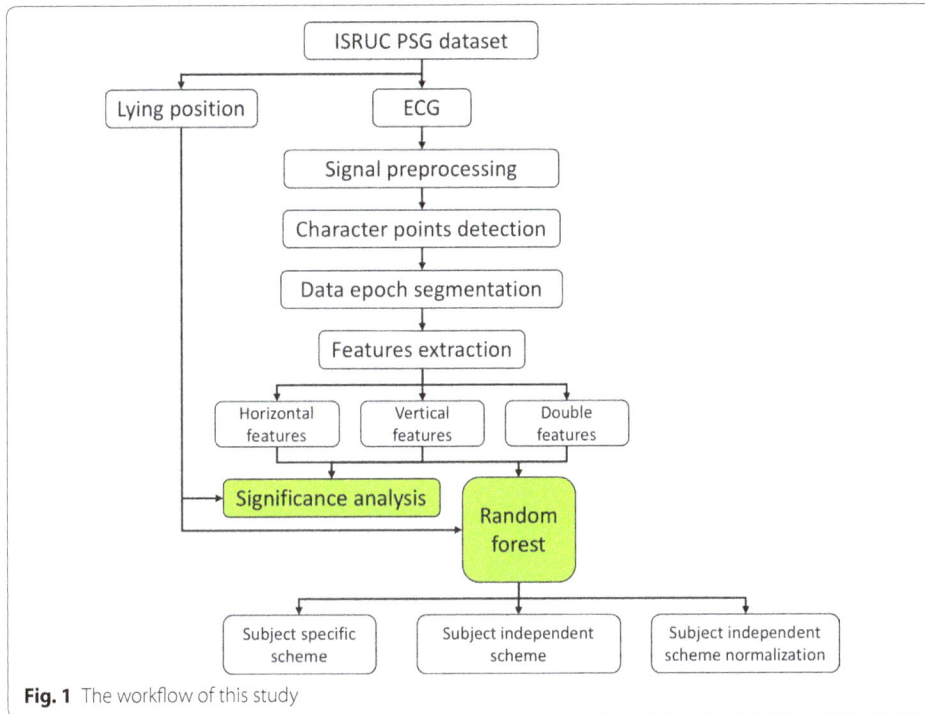

Fig. 1 The workflow of this study

Dataset

The data used in this article was from the ISRUC web sleep database, which provided a variety of physiological data from 10 healthy subjects [6]. The overnight sleep data in this database was recorded by polysomnography (PSG), which lasted for about 8 h. The experiment was finished at the Sleep Medicine Center of the University of Coimbra. For each subject, the database provided a total of 19 physiological data such as electrocardiogram (ECG) and lying position. The ECG sampling rate was 200 Hz. Because the R wave peaks morphology of No. 5 subject in the database was double-peak, the determination of the R-wave peak point's horizontal and vertical coordinates were interfered. Thus this piece of data wasn't included in this study. For the remaining 9 participants, only a small number of subjects had prone position during the overnight sleep. Therefore, this article studied the ECG waveform changes within the left, supine and right-side lying position during the overnight sleep for 9 healthy subjects.

Signal preprocessing

The ECG signal in the ISRUC database mainly contained two kinds of noises, myoelectric interference caused by muscle electrical activity with a frequency of 2 Hz–2 kHz, and baseline drift caused by human respiratory coupling. In this study, first of all, the mean filter was applied to remove the interference from AC (alternating current) in the ECG signal. Secondly, the three-layer lifting wavelet decomposition method was used to remove the high frequency myoelectric interference. Finally, the effect of baseline drift was eliminated by the function fitting method. Since this article was to explore the changes of ECG waveform features, it was necessary to acquire high accuracy point locations of P-wave, QRS-wave, and T-wave. In this study, the multi-character points

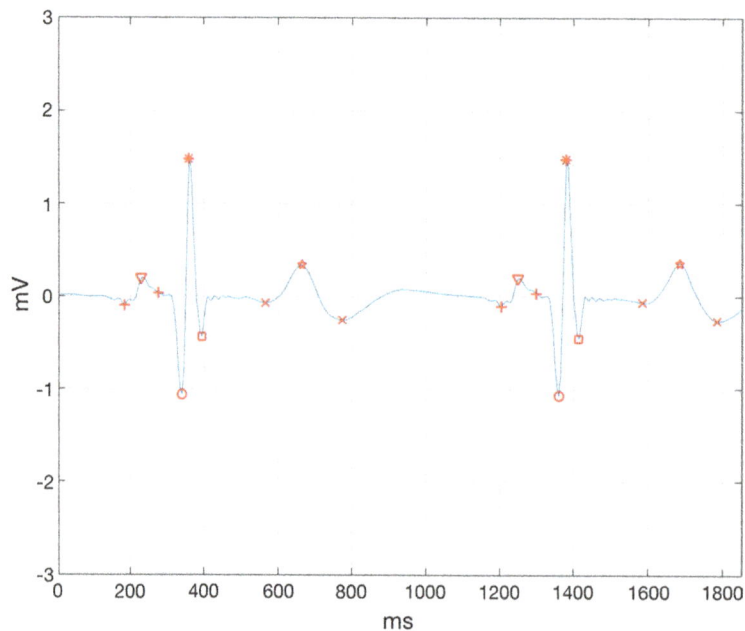

Fig. 2 The results after signal preprocessing and character points detection. From left to right, there are P wave origin, P wave peak, P wave end, Q wave peak, R wave peak, S wave peak, T wave origin, T wave peak, T wave end. This part of ECG signal was from No.1 subject, which appeared from 5 h 40 min 11 s 505 ms to 5 h 40 min 13 s 355 ms

detection algorithm of ECG signals based on wavelet transform, proposed by Yang et al. was used to decompose and de-noise the original signal, and the position of the QRS complex were obtained [7]. Then the area increment method, which was proposed by Song et al. was applied to locate the P wave end at the right side of P wave peak, and the T wave origin at the left side of T wave peak [8]. Finally, all the subject's overnight ECG character points and waveforms were manually checked. After signal preprocessing and character points detection, the results are shown as follows in Fig. 2.

Data segmentation and ECG waveform features

The ISRUC database divided the subject's overnight sleep data into 30 s epochs. Then the sleep stage of each epoch was determined and the lying position was recorded. In this study, we excluded the time segments whose lying position duration was no longer than 1 min (two epochs), and those the ECG signal waveform disturbed during the body position changing so that the character points detection could not be performed.

The characteristics of ECG waveform morphology features and their meanings are shown in Table 1. In this study, these features are divided into three classes according to their orientation in the ECG chart, which are the time-limit features (horizontal direction features), amplitude features (vertical direction features) and double-direction features (features reflecting both time and amplitude simultaneously). The time-limit features reflect the time interval between the ECG waveforms character points on the time axis. The amplitude features reflect the height of the ECG waveforms and potential

Table 1 The ECG waveform features explored in this study

No.	Features	Meaning	Orientation
1	QT interval	Interval between Q peak and T end	Horizontal
2	RR interval	Interval between contiguous R peak	Horizontal
3	PR inter	Interval between P begin and QRS begin	Horizontal
4	PR segment	Segment between P end and QRS begin	Horizontal
5	ST inter	Interval between QRS end and T end	Horizontal
6	ST segment	Segment between QRS end and T begin	Horizontal
7	RT slope	Slope of the line between R peak and T peak	Double
8	P wide	Wide between P begin and P end	Horizontal
9	QS wide	Wide between Q peak and S peak	Horizontal
10	T wide	Wide between T begin and T end	Horizontal
11	TP segment	Segment between T end and next P begin	Horizontal
12	P peak	Amplitude of P peak	Vertical
13	R peak	Amplitude of R peak	Vertical
14	T peak	Amplitude of T peak	Vertical
15	T area	Area under T wave	Double
16	Rp-Tp x	Wide between R peak and T peak	Horizontal
17	Rp-Tp y	The difference of amplitude between R and T	Vertical
18	Tp-Te	Wide between T peak and T end	Horizontal
19	QR	Difference of amplitude between Q and R	Vertical
20	RS	The difference of amplitude between R and S	Vertical
21	QRS area	Area under QRS wave	Double
22	S peak	Amplitude of S peak	Vertical
23	RS slope	Slope of the line between R peak and S peak	Double
24	S/R	Amplitude ratio of S and R	Vertical
25	T/R	Amplitude ratio of T and R	Vertical
26	Ta/QRSa	Area ratio of T area and QRS area	Double
27	QRSa–Ta	Area difference of T area and QRS area	Double
28	ST slope	Slope of the line between J point and T begin	Double
29	QTc	Corrected QT interval	Horizontal
30	Angle qsr	Angle of $\angle$QSR	Double

difference of points in the amplitude direction. The double-direction features mainly include area features, slope features and angle feature.

The calculation methods for several special waveform features are described as follows.

a. Waveform height features

The height of the waveform reflects the amplitude of the electrical signal. In actual ECG signal, the amplitude of the reference equipotential is not zero, and it fluctuates within a certain range. Therefore, the heights of P wave, R wave, S wave, and T wave cannot be directly represented by the vertical coordinates of waveform points. It is necessary to calculate the reference equipotential amplitude and the amplitude of each waveform with respect to the reference equipotential line. In TP segment all myocardial cells are at rest, so that there is no potential difference between them, and almost no electrical activity appears. TP segment is longer and more stable than PR segment, so TP segment was selected in this study to calculate the baseline equipotential line.

Firstly, the mean filter was selected with width 5 to smooth the TP segment. Then we selected 5 points (TP(i), i = 1, 2, 3, 4, 5) at equal intervals in the TP segment. The average amplitude of this 5 points was recorded as a stable point, which was used to represent the baseline equipotential of the corresponding ECG waveform before this TP segment. Finally, the potential difference between the P wave, R wave, S wave and T wave peaks and the stable point was calculated as the height of the corresponding waveforms. Take R wave height as an example, the waveform height formula is as follows:

$$R_p = R - stable$$
$$= R - \frac{1}{5}\sum_{i=1}^{5} TP(i) \tag{1}$$

b. Slope features

Slope features can reflect both time and amplitude change at the same time. The absolute value of slope features will increase with the amplitude of waveform increasing, and will decrease with the time interval increasing. Taking RT slope as an example, the formula for calculating the absolute value of the slope of the connection line between the R wave peak point and T wave peak point is as follows:

$$RT\_slope = \left| \frac{R\_y - T\_y}{R\_x - T\_x} \right| \tag{2}$$

c. Area features

In order to reduce the influence of lying position changes on the depth of Q-wave and S-wave, in this study we used the method of calculating the triangular-like area when calculating the QRS complex area. The origin of T wave might be affected by the double effect of the baseline drift and the ST segment change, resulting in different heights between the T wave start point and end point. Therefore, this method was also used when calculating the area under T wave. As shown in Fig. 3, the area of the QRS complex and the area under the T-wave should be calculated by subtracting the area of the triangle from the area obtained by summing the vertical ordinates of the ECG waveform, thereby correcting the calculation of QRS complex area and T-wave area. The formula is as follows:

$$QRSa = S_{OQRS} - S_{\Delta OQS}$$
$$= \sum_{x=Q}^{S} ecg(x) - \frac{1}{2} \times OS \times OQ \tag{3}$$

$$Ta = S_{OTsTTe} - S_{\Delta OTsTe}$$
$$= \sum_{x=Ts}^{Te} ecg(x) - \frac{1}{2} \times OTe \times OTs \tag{4}$$

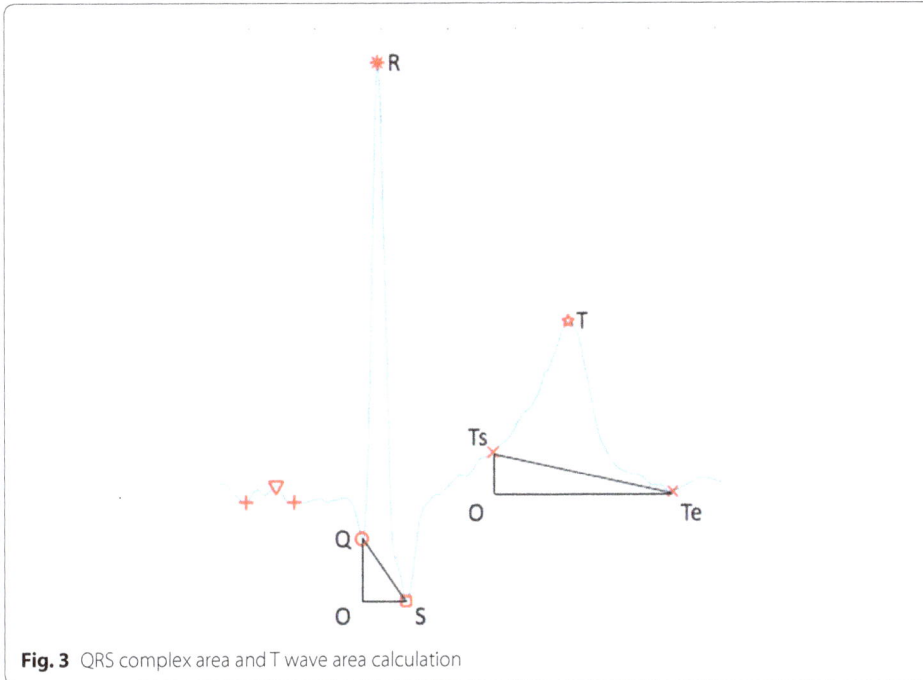

Fig. 3 QRS complex area and T wave area calculation

Among them, Q represents the Q-wave peak horizontal ordinate, S represents the S-wave peak horizontal ordinate, Ts represents the beginning of the T-wave horizontal ordinate, Te represents the end of the T-wave horizontal ordinate. The meaning of other segments in the formula is shown in Fig. 3.

d. Corrected features

QTc (corrected QT interval) is heart-rate-corrected QT interval, that reflects the entire process of cardiac depolarization and repolarization. The calculation formula is Bazetts's algorithm as follows:

$$QTc = \frac{QT}{\sqrt[2]{HRn}} \qquad (5)$$

Among the formula HRn is the standardized heart rate. It is calculated as follows:

$$HRn = \frac{60}{HR} = \frac{60}{60 \big/ \frac{RRi}{fs}} \qquad (6)$$

e. Newly proposed features

As shown in Fig. 4, further observation on the ECG waveforms in three lying positions revealed that when lying on the left side, the S wave was lower than those in supine and lying on the right side. And the waveform amplitudes of the R waves in different lying positions were obviously different. Therefore, this study proposed two new features, namely S/R and angle ∠QSR. S/R is the ratio of S wave depth and R-wave height, which can reflect the relative depth of S waves.

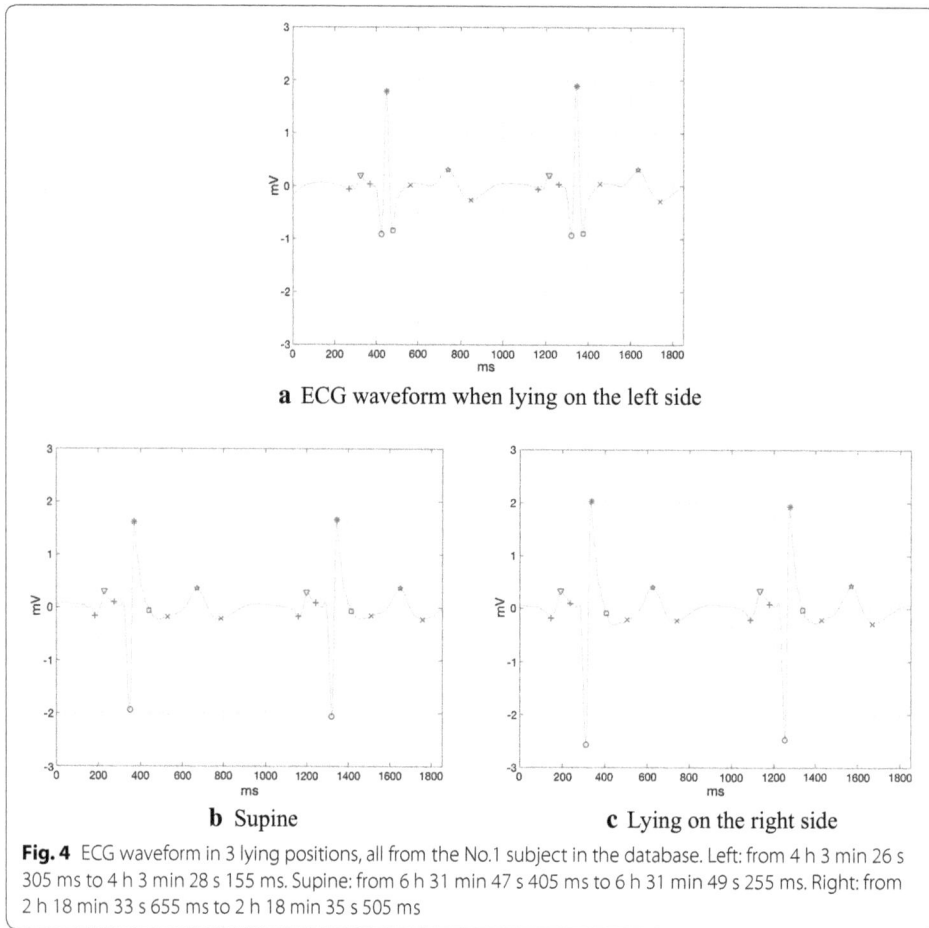

a ECG waveform when lying on the left side

b Supine   **c** Lying on the right side

Fig. 4 ECG waveform in 3 lying positions, all from the No.1 subject in the database. Left: from 4 h 3 min 26 s 305 ms to 4 h 3 min 28 s 155 ms. Supine: from 6 h 31 min 47 s 405 ms to 6 h 31 min 49 s 255 ms. Right: from 2 h 18 min 33 s 655 ms to 2 h 18 min 35 s 505 ms

$$S/R = \frac{S - stable}{R - stable} \tag{7}$$

Angle $\angle QSR$ is the angle value of the inner angle $\angle QSR$ of the triangular QRS. Firstly, the lengths of QR, RS and QS are calculated. Then according to the cosine theorem, $\angle QSR$ can be obtained. In this article, the unit of $\angle QSR$ is degree, and the formula is as follows:

$$\angle QSR = \cos^{-1}\left(\frac{QS^2 + RS^2 - QR^2}{2 \times QS \times RS}\right) \tag{8}$$

Classifier: random forest

RF (Random forest) is a novel classification method proposed by Breiman in 2001 [9]. It is a classifier that is built randomly and contains a large number of decision trees. The classification result is acquired by voting, because the output is determined by the mode of the output of each tree. Such randomness is mainly embodied in two aspects. On the one hand, a dataset of size N, which is the same as all training dataset, is selected using the bootstrapping procedure to train each decision tree. On the other hand, a subset of all features is randomly selected at each internal node. Consequently, RF can handle high-dimensional dataset (involving many features) without feature selection, and it is better at solving multiple classification problems when comparing with SVM (supporting vector machine). The decision trees are independent of each other in training procedure, so the parallel computing can be applied, which leads to fast calculation compared with ANN (artificial neural network). Besides, the structure of RF is simpler and it is easy to build, and it has strong ability to avoid over-fitting at the same time.

Because of the advantages of fast calculation, high precision, strong anti-noise ability and avoiding over-fitting when compared with other good classification method, random forest was chosen in this study. The number of trees was set as 500. After significance analysis, 12 features, including QT, RR, TP, ∠QSR, S/R, QR, P peak, R peak, T peak, T area, QRS area, T area/QRS area, were selected for classification.

When establishing each decision tree, there are two random processes to avoid over-fitting. The input data for random forest is sampled by bootstrapping procedure randomly, that is, there may be duplicate samples in the input data. Assuming N dataset, the number of input data is also N. This makes the input data of each tree not a full dataset during training, making it relatively easy to avoid over-fitting.

Then from M features, m features (m ≪ M) are randomly selected. After that, the decision tree is created by completely splitting way, so that either one leaf node of the decision tree cannot continue to split, or all the samples inside belong to the same class. Since the two random processes applied, over-fitting does not occur even without pruning. Every tree obtained by this algorithm is very weak, but they are very powerful when combined as random forest.

Each decision tree is like an expert proficient in a narrow field (because we choose m from M features to let each decision tree learn), so that there is a random forest including many experts who are proficient in different fields. When solving a new problem (new input data), they can view this from different perspective. And in the end, various experts vote to get the results. In this study, we separated the data as training data and testing data, building the RF as classifier by TreeBagger through MATLAB and the classification was achieved. We randomly selected 1–99% of the data in the database as training data, and the rest as testing data. Then the learning curves including accuracy and Cohen' k were plotted to verifying the absence of overfitting. When the proportion of training data was more than 30%, the accuracy and Cohen' k didn't increase any more. But when the proportion of training data was more than 50%, the accuracy was stable and the Cohen' k started to decrease, which meant that the overfitting existed. As we can saw in Fig. 7 in "Results" section, when the proportion was 20%, the accuracy reached a high level of 97.17% and the Cohen' k reached an acceptable level of 0.91. Besides, less training data would lead to faster calculation. Consequently, we selected 20% of the data as training data to acquire high accuracy as well as Cohen' k, and avoiding overfitting.

Performance evaluation

The performance of classifier was evaluated by accuracy, Cohen's kappa statistic κ, ROC–AUC (receiver operating characteristic curve–area under curve), Sensitivity, Specificity and F1-scores. Accuracy stands for the percentage of correctly classified epochs in the whole dataset. Statistic κ is a more effective evaluator because it takes the prior probability into account. It can be calculated as

$$\kappa = \frac{P_A - P_C}{P_{prio} - P_C} = \frac{\sum_{i=1}^{m} P_{ii} - \sum_{i=1}^{m} P_{i\cdot}P_{\cdot i}}{P_{prio} - \sum_{i=1}^{m} P_{i\cdot}P_{\cdot i}} \tag{9}$$

P_A is the proportion of correctly observed, while P_C is the proportion of randomly expected. P_{prio} is equal to 1. Such variables can be calculated by the second formula. m means the number of class. In this study m $= 3$. And P means the proportion of the corresponding sample to the entire. Statistic κ ≤ 0 means that the observed result is even worse than random expecting. And κ ≥ 0 means that all sample are classified into the correct class. A higher value of κ indicates a better classification result between our classifier and the expected results.

ROC curve is a graphical plot that presents the ability of a binary classifier system. It is created by plotting the FPR (false positive rate) and TPR (true positive rate) at various threshold. Because that the classifiers in this study are ternary classifiers, after classification results are obtained, in order to draw ROC curve and calculate the AUC, Sensitivity, Specificity and F1-scores of one lying position, the other two lying positions are combined. E.g. before drawing ROC curve and calculating such several indexes of lying on the left, epochs of supine and lying on the right are combined as not-left, then the 2×2 confusion matrix is built.

Generally speaking, a good classifier should be associated with high values of accuracy, statistic κ and AUC.

Classification scheme

In this study, we developed three kinds of classification scheme for different cases, including subject specific scheme, subject independent scheme without feature normalization and subject independent scheme with feature normalization. The result of ECG waveform features significance analysis between different lying positions will be presented in "Results" section. After significance analysis, 12 features, which showed strong significant difference between lying positions including QT, RR, TP, ∠QSR, S/R, QR, P peak, R peak, T peak, T area, QRS area, T area/QRS area, were selected for classification.

A total of 5114 epochs of the overnight sleep data from 9 subjects were included in this study. Due to the fact that most subjects did not have prone position, or only had several prone epochs in overnight sleep, the prone epochs were manually removed. Consequently, there are only three classes in classification including lying on the left, supine, and lying on the right. The details and workflow are shown in Fig. 5.

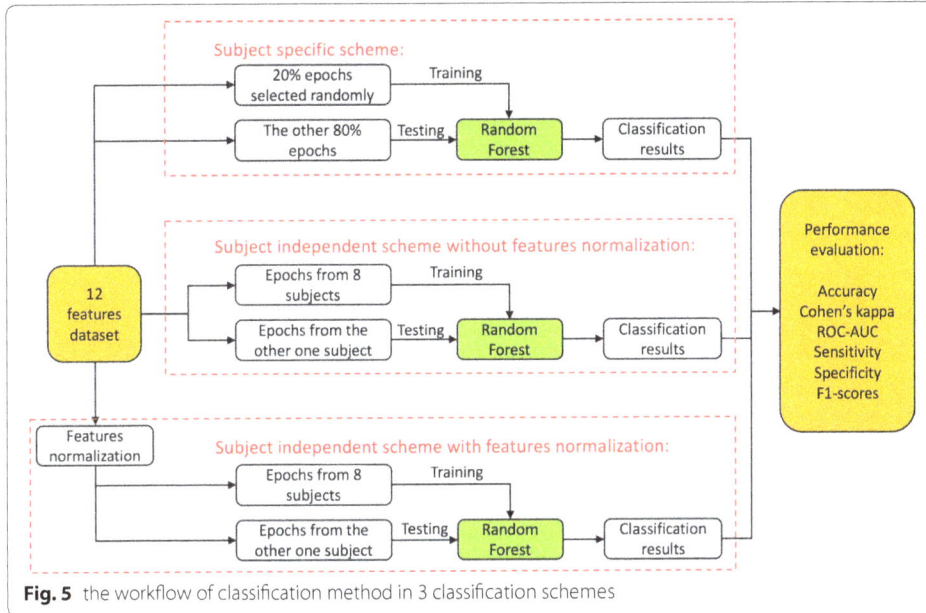

Fig. 5 the workflow of classification method in 3 classification schemes

a. Subject specific scheme

For each subject, 20% epochs of three kinds of lying positions were randomly selected for training the classifier, and the rest 80% were used as testing data. The reasons 20% for training and 80% for testing are that on the one hand, the waveforms were obviously different in 3 lying positions. Strong significant difference of waveform features appeared in "Results" section. On the other hand, we were trying to train the classifier with limited data. So that when putting into application, we could build a small database for patients, extracting ECG signals for only half an hour, to train the classifier. And then clinical automatic classification with high accuracy were achieved. In order to avoid the errors caused by selecting samples randomly, the training and classification processes were repeated for 10 times with different training data. At last, the average value and standard deviation of accuracy and κ statistic were calculated.

b. Subject independent scheme without feature normalization

For each specific subject to be analyzed, all the records from other 8 subjects were pooled together to form the training dataset. This process repeated for 9 times. Finally, the same as the specific scheme, the average value and standard deviation of accuracy and κ statistic were calculated.

c. Subject independent scheme with feature normalization

However, because of the individual differences, all features need normalization before classifier training. One of the most widely used normalization method is to transform all the features scales to a new range, such as [0,1]. But when the outliers of data appear, the transformed data scale will be unsymmetrical. To solve this problem, we developed a normalization method based on quantile. The 5% and 95% quantiles of data were selected firstly and the scale of these two samples was linearly transformed to [0,1],

Table 2 Sleep data epochs frequency distribution about sleep stages and postures

	Wake	REM	Light sleep	Deep sleep	Sum
Left	244	336	960	827	2367
Supine	84	52	583	108	827
Right	158	292	770	473	1693
Prone	21	49	35	122	227
Sum	507	729	2348	1530	5114

which covers 90% of the whole samples. The rest samples were transformed with the same linear coefficients.

Results

A total of 5114 epochs of the overnight sleep data from 9 subjects were included in this study. Table 2 shows the frequency distribution of sleep stages and lying positions for these epochs. The results part mainly includes significance analysis of features and classification performance.

Significance analysis of features

This study calculated the 30 waveform features of the overnight ECG sleep data from 9 healthy subjects in the database, and calculated the means and standard deviations according to the four lying positions. The calculation results and significant differences between the different lying positions are shown in Tables 3 and 4, respectively. Due to the fact that most subjects did not have prone, or only had several prone epochs in overnight sleep, the standard deviations of features in prone were not shown in Table 3. On the other hand, the waveform features significance level of only three conditions, including left–supine, left–right and right–supine positions, were calculated. The P values of ECG waveform features significant level among different lying positions are shown in Table 4.

Classification performance

After significance analysis, 12 features, which showed strong significant difference between lying positions including QT, RR, TP, $\angle$QSR, S/R, QR, P peak, R peak, T peak, T area, QRS area, T area/QRS area, were selected for classification. Table 5 gives the confusion matrices of all individuals for subject specific scheme and subject independent scheme without or with feature quantile normalization. The numbers in Table 5 refers to the amount of epochs of target position while classified as output position.

Table 6 shows the classification performance based on 12 features for subject specific scheme and subject independent scheme without or with feature normalization. The process repeated 10 times, and the means and standard deviation were calculated and listed in Table 6. Figure 6 shows the classifier performance of three scheme: (a–c) show the ROC curve of 3 lying positions respectively, and (d–f) show the AUC,

Table 3 Means and standard deviations of 30 ECG waveform features in 4 lying positions

Lying positions	Left	Supine	Right	Prone
QT	442.234 ± 4.763	434.916 ± 4.907	439.099 ± 5.361	427.654
QTc	446.541 ± 11.936	448.476 ± 10.903	453.259 ± 10.506	440.679
RR	985.050 ± 15.378	946.057 ± 17.053	942.002 ± 16.838	944.844
PR inter	139.201 ± 3.113	139.811 ± 3.233	139.574 ± 2.526	164.698
PR segment	21.708 ± 3.162	24.096 ± 2.731	23.382 ± 2.766	58.389
ST inter	341.285 ± 8.573	331.296 ± 8.029	327.401 ± 8.539	352.817
ST segment	77.024 ± 10.596	71.534 ± 9.446	71.020 ± 8.098	86.649
P wide	117.493 ± 2.397	115.715 ± 2.548	116.192 ± 2.580	106.308
QS wide	65.397 ± 2.820	70.183 ± 2.684	73.107 ± 2.848	54.014
T wide	264.261 ± 5.653	259.763 ± 4.260	256.381 ± 5.765	266.168
TP segment	382.120 ± 9.830	350.049 ± 11.086	342.160 ± 9.964	330.975
P peak	0.138 ± 1.414	0.233 ± 2.303	0.222 ± 1.739	0.112
R peak	1.808 ± 11.290	2.428 ± 16.243	2.291 ± 10.299	1.184
S peak	0.655 ± 4.970	0.533 ± 5.212	0.551 ± 3.110	0.206
T peak	0.547 ± 2.812	0.776 ± 3.427	0.696 ± 3.662	0.553
QRS area	266.365 ± 108.806	389.564 ± 158.961	387.565 ± 144.974	179.905
T area	165.776 ± 59.467	228.880 ± 67.021	200.339 ± 74.594	139.677
Ta/QRSa	1.129 ± 0.365	1.153 ± 0.381	0.848 ± 0.361	1.036
QRSa–Ta	100.588 ± 81.630	160.684 ± 120.856	187.226 ± 123.533	40.228
Rp-Tp x	301.607 ± 4.836	294.935 ± 4.940	295.938 ± 5.479	296.884
Rp-Tp y	21.477 ± 10.474	28.500 ± 15.368	27.279 ± 9.478	11.814
Tp-te	116.747 ± 1.705	115.795 ± 1.878	118.661 ± 2.742	108.543
QR	3.228 ± 18.543	4.661 ± 30.424	4.431 ± 20.729	2.678
RS	2.463 ± 14.825	2.961 ± 18.830	2.841 ± 12.079	1.390
RT slope	0.361 ± 0.185	0.489 ± 0.267	0.472 ± 0.180	0.199
RS slope	5.142 ± 2.698	5.600 ± 3.576	5.106 ± 3.137	3.527
ST slope1	0.106 ± 0.097	0.116 ± 0.108	0.080 ± 0.039	− 0.028
S/R	0.365 ± 0.151	0.232 ± 0.154	0.258 ± 0.122	0.174
T/R	0.245 ± 0.177	0.242 ± 0.090	0.220 ± 0.100	0.336
Angle qsr	106.798 ± 29.477	126.956 ± 25.830	127.175 ± 24.154	127.711

In this table, the time-limit features are calculated in millisecond (ms), the amplitude features are calculated in millivolt (mV), and the angle indicator is calculated in degree. Due to the fact that most subjects did not have prone, or only had several prone epochs in overnight sleep, the standard deviations of features in prone were not shown in this table

Sensitivity, Specificity and F1-scores of the classification result. The AUC of three lying positions in subject specific scheme reached at 0.9886 ± 0.0043, 0.9725 ± 0.0106 and 0.9925 ± 0.0019, respectively. While in subject independent scheme without features normalization 0.6859 ± 0.0050, 0.3570 ± 0.0035, 0.6321 ± 0.0055, and in subject independent scheme with features normalization 0.7708 ± 0.0017, 0.6646 ± 0.0047, 0.7132 ± 0.0040.

Because the results of subject specific scheme presented in Table 6 and Fig. 6 include overall accuracy of $97.17\% \pm 2.74\%$, κ 0.9121 ± 0.1010 and AUC > 0.97 in three lying position classification), we tried to decrease the proportion of training data. The results are shown below in Table 7. In order to verify the absence of overfitting, the learning curve are shown in Fig. 7. The comparison of the classification performance between RF, SVM and ANN is shown in Fig. 8. We can see that RF and ANN perform better than SVM, and the accuracies of RF and ANN are close. The

Table 4 The P value of ECG waveform features significant level among different lying positions

	Left–supine	Left–right	Supine–right
QT	0.0059**	0.1341	0.2947
QTc	0.1148	0.0229*	0.1207
RR	0.0066**	0.0206*	0.4106
PR inter	0.3549	0.0881	0.9203
PR segment	0.1287	0.0222*	0.7313
ST inter	0.1023	0.1906	0.3491
ST segment	0.2737	0.3323	0.5255
P wide	0.1912	0.1905	0.2587
QS wide	0.1157	0.0975	0.3296
T wide	0.2199	0.0609	0.3164
TP segment	0.0111*	0.0182*	0.3068
P peak*	0.0028**	0.0039**	0.1516
R peak*	0.0143*	0.0235*	0.1003
S peak*	0.0360*	0.0831	0.6562
T peak*	0.0002***	0.0180*	0.0004***
QRS area*	0.0063**	0.0228*	0.2293
T area*	0.0000***	0.0814	0.0032**
T a/QRS a*	0.4441	0.0435*	0.0379*
QRSa–Ta*	0.0562	0.0369*	0.3981
Rp-Tp x	0.0191*	0.0696	0.4688
Rp-Tp y*	0.0307*	0.0316*	0.1696
Tpte	0.3207	0.3413	0.2490
QR*	0.0066**	0.0111*	0.1585
RS*	0.0418*	0.1086	0.0651
RT slope*	0.0244*	0.0205*	0.1921
RS slope*	0.2182	0.4109	0.0197*
ST slope*	0.0126*	0.4443	0.1886
S/R*	0.0014**	0.0207*	0.3539
T/R*	0.4710	0.2809	0.2560
Angleqsr*	0.0012**	0.0275*	0.4710

The first column in this table includes 30 waveform features. To facilitate the observation, the amplitude features and double-direction features are marked by *. Columns 2, 3, and 4 show the significant level of the waveform features between two lying positions. ***$P \leq 0.001$, **$P \leq 0.01$, and *$P \leq 0.05$

Cohen' k of ANN is slightly higher than RF. However, according to Table 8, the calculation of RF is much faster. Consequently, RF performs best in general.

Discussion

Discussions of results

The reason why we developed three kinds of schemes is that firstly we tried to establish a database which could be used for many subjects. However, because of the individual difference, the results were not acceptable. Consequently, we applied the normalization method to transform all the features scales to a new range. The results

Table 5 Confusion matrices based on 12 features

Output position	Target position			
	Left	Supine	Right	Sum
(a) Subject specific scheme				
Left	1868	39	7	1914
Supine	8	603	9	620
Right	14	15	1336	1365
Sum	1890	657	1352	3899
(b) Subject independent scheme without feature normalization				
Left	1492	458	247	2197
Supine	716	203	1068	1987
Right	79	166	378	623
Sum	2287	827	1693	4807
(c) Subject independent scheme with feature normalization				
Left	1757	261	334	2352
Supine	288	377	459	1124
Right	312	176	892	1380
Sum	2357	814	1685	4856

Confusion matrices based on 12 features for (a) subject specific scheme, (b) subject independent scheme without feature normalization, (c) subject independent scheme with feature normalization

Table 6 Classification performance based on 12 features

Left	Supine	Right	Overall	κ statistic
(a) Subject specific scheme				
98.71% ± 2.03%	72.22% ± 23.41%	98.46% ± 2.34%	97.17% ± 2.74%	0.9121 ± 0.1010
(b) Subject independent scheme without feature normalization				
55.22% ± 43.25%	38.38% ± 41.36%	24.21% ± 37.28%	44.73% ± 31.61%	0.0866 ± 0.2180
(c) Subject independent scheme with feature normalization				
75.04% ± 24.10%	46.40% ± 35.61%	44.34% ± 38.14%	63.87% ± 16.32%	0.3171 ± 0.1755

Classification performance based on 12 features for (a) subject specific scheme, (b) subject independent scheme without feature normalization, (c) subject independent scheme with feature normalization

of subject independent scheme with feature normalization were much better but the accuracy was still not enough for clinical application. Finally, we developed the subject specific scheme, which was similar to building a database with the ECG data from a specific subject and then classifying the lying positions for this subject based on the database. That's why the results were acceptable and this method could be applied in clinical monitoring.

As can be seen from Table 4, the lying positions have less influence on time-limit features, because most of the time-limit features show no significant differences between different body lying positions. Compared with supine, only QT interval, RR interval, and TP segment are significantly shorter when lying on the left side. The reason needs further exploration.

It can be seen that the influence of lying position on ECG waveforms is mainly reflected in the amplitude features and double-direction features. The amplitude features include the heights of P wave, R wave, and T wave. The relative height features include QR potential difference, RS potential difference, R peak T peak potential

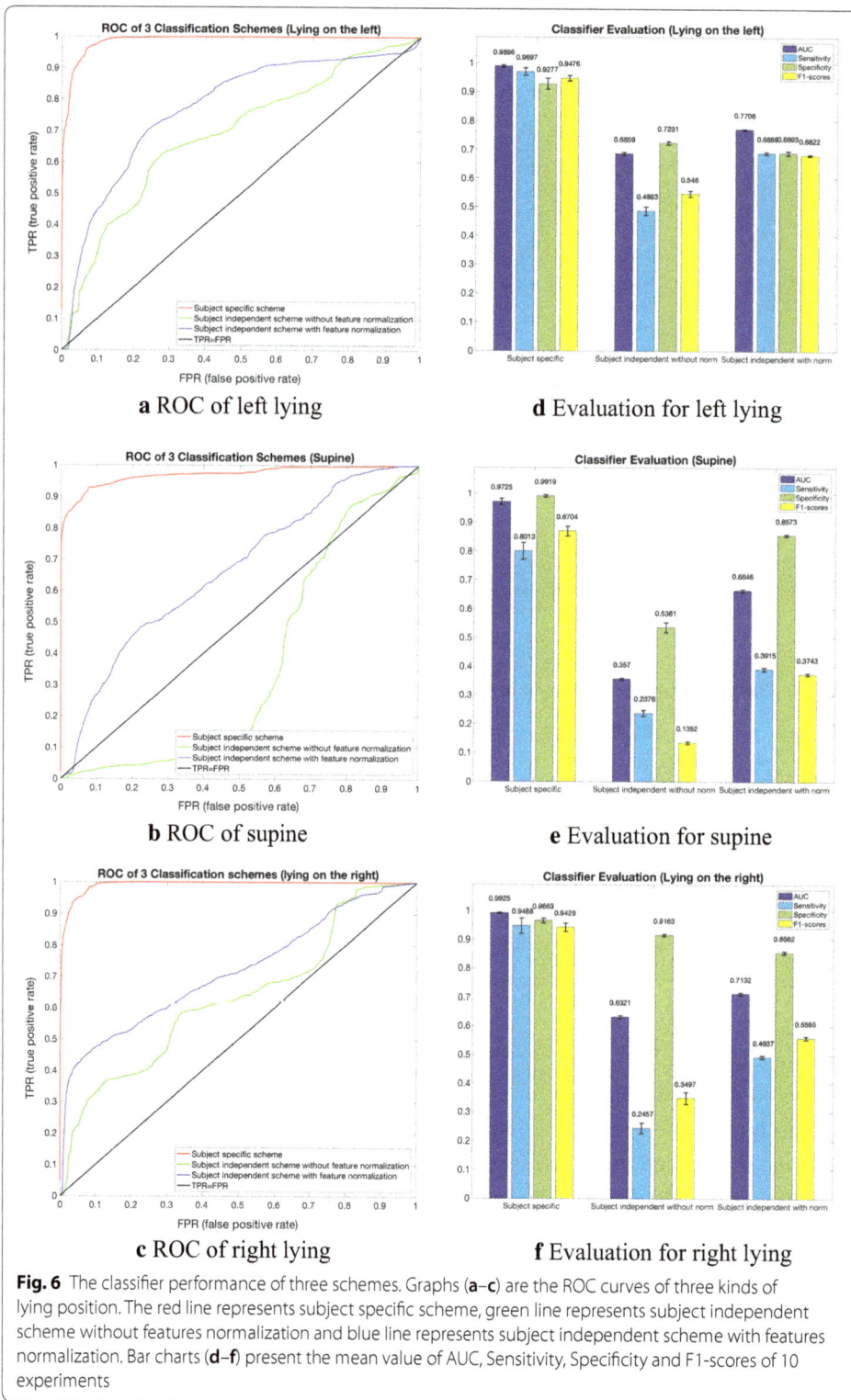

a ROC of left lying

d Evaluation for left lying

b ROC of supine

e Evaluation for supine

c ROC of right lying

f Evaluation for right lying

Fig. 6 The classifier performance of three schemes. Graphs (**a–c**) are the ROC curves of three kinds of lying position. The red line represents subject specific scheme, green line represents subject independent scheme without features normalization and blue line represents subject independent scheme with features normalization. Bar charts (**d–f**) present the mean value of AUC, Sensitivity, Specificity and F1-scores of 10 experiments

Table 7 Subject specific scheme with the training proportion of 0.2, 0.1 and 0.05

Left	Supine	Right	Overall	κ
(a) Subject specific scheme, the proportion of training is 0.2				
98.71%±2.03%	72.22%±23.41%	98.46%±2.34%	97.17%±2.74%	0.9121±0.1010
(b) Subject specific scheme, the proportion of training is 0.1				
97.99%±2.74%	59.18%±34.50%	97.44%±3.02%	95.71%±3.38%	0.8418±0.2156
(c) Subject specific scheme, the proportion of training is 0.05				
96.63%±6.41%	51.83%±37.91%	95.02%±4.55%	93.91%±4.87%	0.7902±0.2546

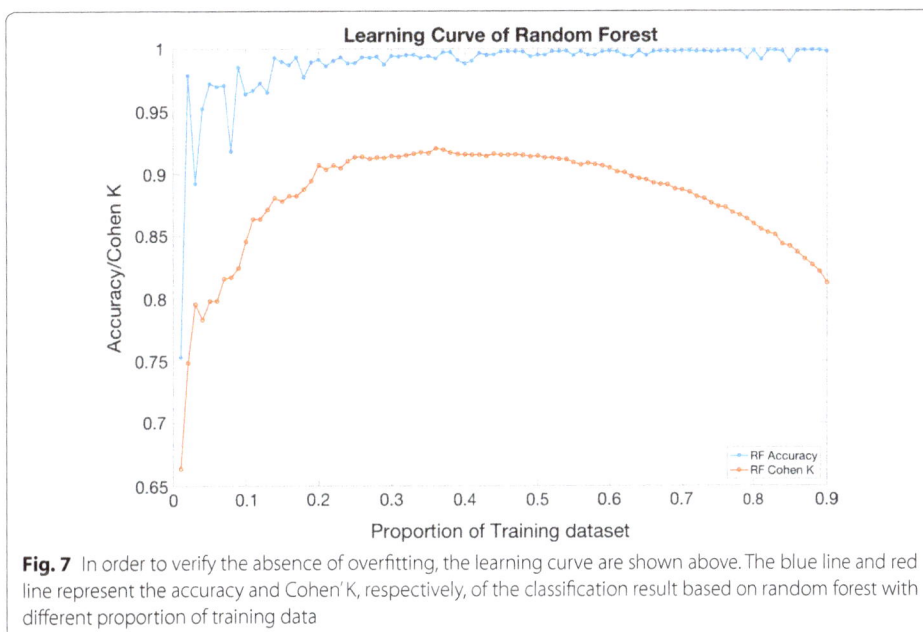

Fig. 7 In order to verify the absence of overfitting, the learning curve are shown above. The blue line and red line represent the accuracy and Cohen'K, respectively, of the classification result based on random forest with different proportion of training data

difference, and RT slope. Area features includes QRS complex area and T wave area. These three types of amplitude features were significantly smaller when lying on the left side than those in supine or right, or less than those in other two lying positions simultaneously. Only a few features show significant differences between supine and lying on the right side.

However, the S-wave-related waveform features are different. When lying on the left side, the depth of S wave is significantly greater than that in supine, and S/R is significantly greater than that both in supine and right. This feature reflects the decrease of R wave and the deepening of S wave in left-side lying. ∠QSR is significantly smaller in left than that in supine and right. This feature reflects the difference between the relative depth of the Q wave and S wave.

The influence of lying positions on ECG waveforms is mainly reflected in the amplitude features. Since the ECG waveform directly reflects the potential difference of the leads, and the signal is extracted from the electrodes on body surface, the body position changes will cause a change of relative position between the electrodes and heart. Thus ECG waveform morphology changed. This change can be embodied in two aspects. On the one hand, when the chest is under pressure, the distribution of body

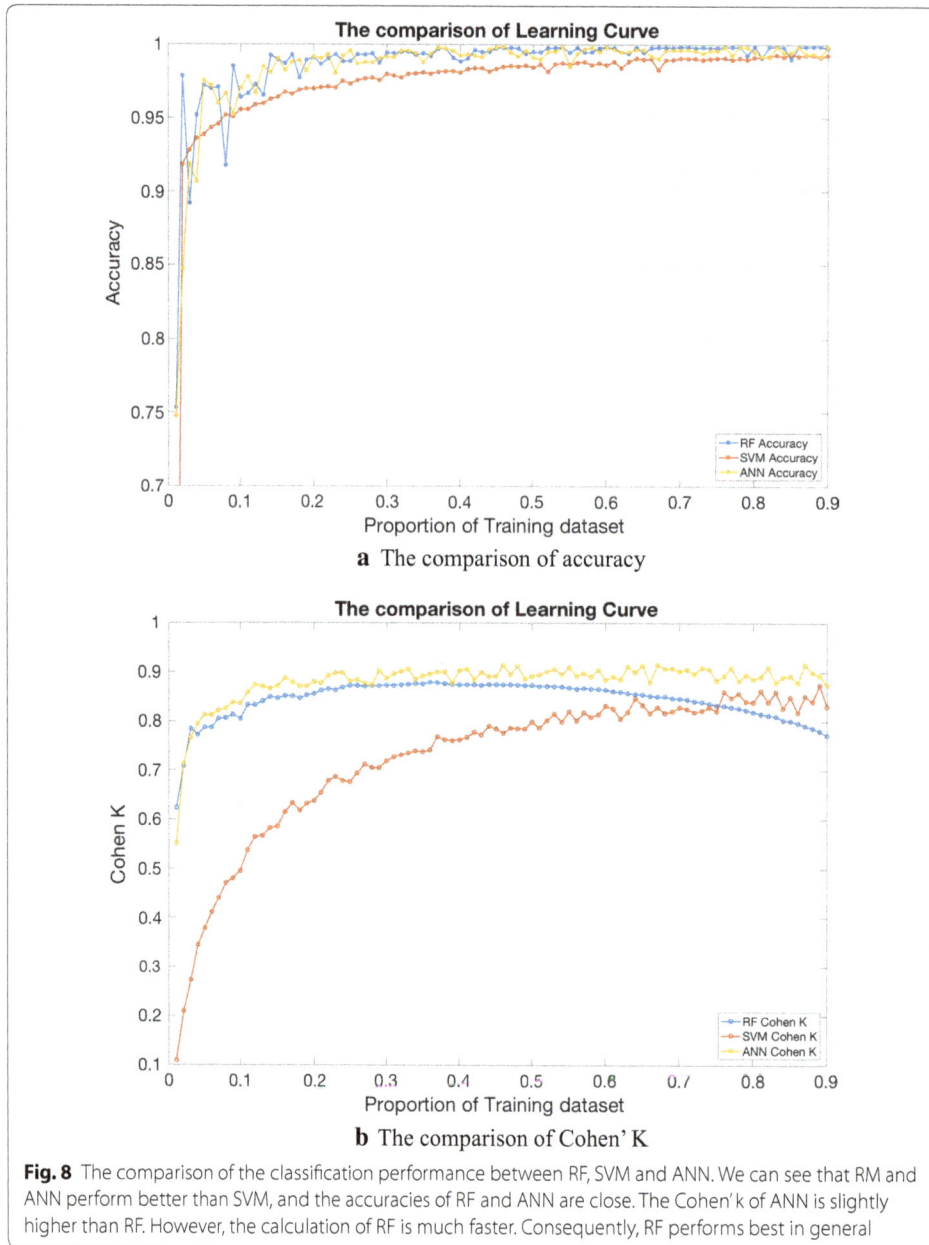

a The comparison of accuracy

b The comparison of Cohen' K

Fig. 8 The comparison of the classification performance between RF, SVM and ANN. We can see that RM and ANN perform better than SVM, and the accuracies of RF and ANN are close. The Cohen'k of ANN is slightly higher than RF. However, the calculation of RF is much faster. Consequently, RF performs best in general

fluids changes, so that the impedance of the chest changes. Also the heart is squeezed and deformed. On the other hand, the heart is affected by gravity when lying on the side. Different parts of heart have different degree of freedom, which results in heart rotation and swing.

The significant differences of ECG waveform features in 3 lying positions could be utilized for automatic lying position classification during sleep. For three kinds of schemes, the overall classification accuracy of subject specific scheme reached 97.17%, κ statistic 0.91 and AUC > 0.97, which was almost perfect. This can be used for clinical lying position monitoring after setting up a subject specific dataset. Further study in Table 7 showed that such dataset didn't need to be large, and the performance could

Table 8 The comparison of the calculation time between RF, SVM and ANN in subject specific scheme with the training proportion of 0.2

Subject no.	RF	SVM	ANN
1	3.22291 ± 0.19684	0.04193 ± 0.00423	3.34373 ± 1.16847
2	3.02385 ± 0.12278	0.03636 ± 0.00254	3.18128 ± 0.81556
3	3.13849 ± 0.15552	0.03429 ± 0.00262	4.72749 ± 1.02038
4	2.97439 ± 0.12926	0.02707 ± 0.00118	3.10264 ± 1.28542
5	3.06737 ± 0.15559	0.03440 ± 0.00151	3.76409 ± 0.76218
6	2.83269 ± 0.13927	0.00537 ± 0.00056	2.43798 ± 0.12534
7	3.18591 ± 0.12517	0.03952 ± 0.00134	5.70489 ± 0.26425
8	3.00057 ± 0.16294	0.03204 ± 0.00114	5.50479 ± 0.32688
9	2.92328 ± 0.18162	0.03034 ± 0.00091	3.57378 ± 1.14730

The mean time and standard deviation of 10 experiments were calculated in seconds. Each value represented the mean time of all epochs lying position classification of one subject

be acceptable. The results of subject independent scheme without or with feature normalization were accuracy 44.73% and 63.87%, κ statistic 0.09 and 0.32, respectively. The classification accuracy of three lying positions in subject independent scheme was much better with feature normalization when compared with the results without feature normalization. On the other hand, the classification accuracy of lying on the left side was higher than those in supine and right. This can be applied for avoiding left lying in some patients with specific diseases, clinically.

The accuracy of classification results may be influenced by the ECG quality. Firstly, in order to distinguish the horizontal features (several time features were < 0.1 s), we chose the dataset with sampling rate 200 Hz. This could make sure that the time resolution was 0.005 s. Secondly, when the subjects were turning over during sleep, the signal was disturbed severely and we had to discard this epoch. But when the subjects were not changing their lying position, the signal was stable. Thirdly, we applied signal preprocessing based on wavelet transform, and it worked well. At last, the ECG signal acquisition technology is mature in recent years. As mentioned above, the ECG signal quality was good enough for this study, which could be reflected in the accuracy of character points detection.

The structure of heart and vectorcardiogram

The bottom of heart in anatomical mainly consists of left atrium and a small part of right atrium, where the aorta and pulmonary artery cross [10]. Because of this structure, the bottom of heart in the thorax is comparatively fixed, while ventricular and the apex of heart are comparatively free. When the lying position changes or the diaphragm contracts, the heart apex will swing to a limited extent. This leads to the direction of electrocardial vector change, and so that it's projection, ECG, changes.

In a complete cardiac cycle, action potential begins from the sinoatrial node firstly, and then passes through the anterior, middle and posterior inter-nodal tract to the atrioventricular node. During this process the electrocardial vector is always from the upper right to the lower left. The process of forming the P loop is shown in Fig. 9a. Then the action potential passes through the bundle of His to the ventricle, firstly from the left bundle branch to the inter-ventricular septum, and then from the left and right bundle

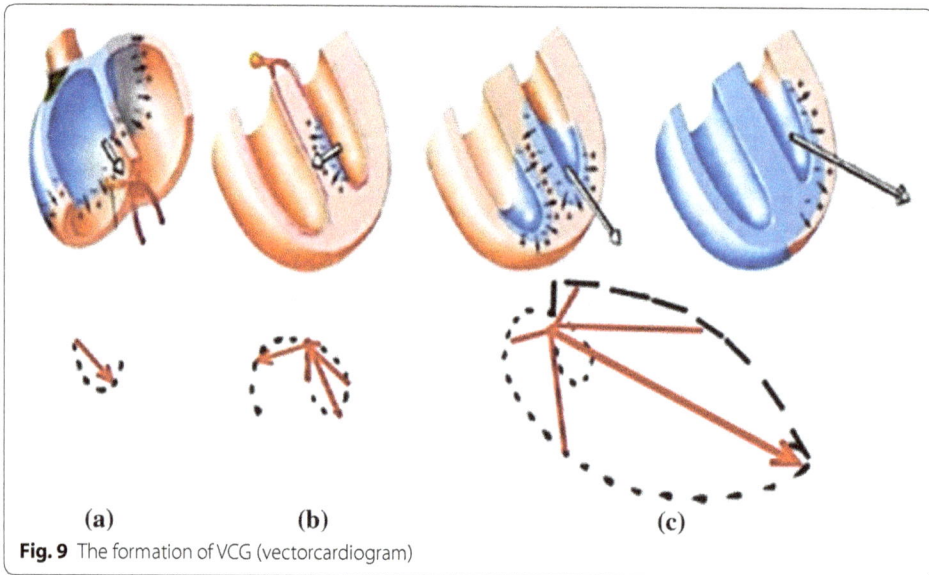

Fig. 9 The formation of VCG (vectorcardiogram)

branches to the left and right ventricular walls, respectively. Due to the left ventricular wall being much thicker than the right, the direction of the two vectors composition is to the lower left. The formation of QRS loop is shown in Fig. 9b, c. After the action potential arrives at the apex, it travels upward along the Purkinje fiber. In this process, the direction of electrocardial vector is still to the left. Finally, after a period of time, ions reflux inside and outside the cell membrane. The formation of T loop reflects the repolarization of ventricular. A complete ECG cycle ends.

The causes of this phenomenon

VCG intuitively reflects the direction and magnitude of the action potential vector in heart, and the ECG is actually the projection of the vector in different leads. The relationship between frontal VCG and limb lead, transverse VCG and chest lead are shown in Fig. 10a, b, respectively. The influence of lying positions on the heart can be reflected in VCG. Compared with the upright position, the position of the heart is in a relatively horizontal position when supine. As the heart rotates along the long axis (see this change in the direction from the apex to the bottom of heart, the heart rotates clockwise), the right atrium and right ventricle move left and slightly forward, and the left atrium and left ventricle are correspondingly shifted to the posterior position. The ventricular septum is almost parallel to the frontal plane instead of the side plane. View this from the frontal plane, the apex moves to the upper left and back, and the heart rotates anticlockwise along the long axis. So that there is a left-leaning tendency on the electric axis. When subjects are lying on the left side, because of the position of the bottom of heart fixed, the apex is swinging to the left, and the VCG in frontal plane is rotating anticlockwise. So that the projection lengths of P loop and T loop in lead II direction are reduced, that means, the heights of P wave and T wave in ECG waveform decrease. Reflected in the waveform features, P peak as well as T peak were significantly reduced. On the other hand, the projection length of huge part of QRS loop decreases while the tiny

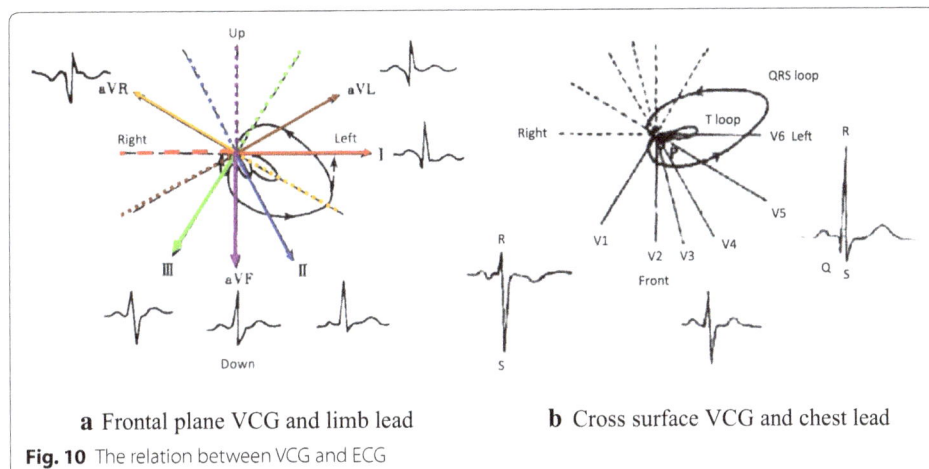

a Frontal plane VCG and limb lead   **b** Cross surface VCG and chest lead

Fig. 10 The relation between VCG and ECG

part increases, so the R wave of the ECG waveform becomes lower and S wave becomes deeper. Reflected in the waveform features, S/R increased while the $\angle$QSR decreased.

The accessible volume of heart in chest is larger when the subject is lying on the left side, because the left lung of human body is smaller than the right part and the heart is at the left side inside human chest. Therefore, the swing amplitude of heart is relatively larger. When subject is lying on the right side, the apex of the heart moves towards the mediastinum, and the heart rotates clockwise along the long axis. There shows a right-leaning tendency on the electric axis (notes: The left discussed here is the left of subject, not the left of observer). However, because the heart is upheld by the mediastinum, the range of motion is limited, so there is no obvious swing and rotation as lying on the left side. This may explain the results that waveform features rarely show significant differences between supine and lying on the right side.

Discussions of other studies

The changes of position and shape of heart in chest have drawn the researchers' attention. Mincholé et al. modeled the changes in the Karhunen–Loeve transform coefficients of the QRS complex and the ST–T waveform. It was found that the changes in body position can be reflected in the gradual changes of the two coefficients series. Then based on ECG, they determined the lying position changes of healthy people. The resulting probability of detection reached 94%, and the probability of false alarm was 0%, respectively. However, the false alarm rate in ischemia database was once per hour [11]. Since myocardial ischemia is widely judged by ST–T segment, the accuracy of lying position detection will decrease sharply, and the misjudgment as well as missed judgment of myocardial ischemia may be more severe if the influence of lying position on S wave morphology is not taken into consideration. Li et al. compared the heart morphology in supine and standing upright. When the subject was in supine, the heart rotated clockwise along the long axis. The heart apex moved to the left and back position. But it moved in the opposite direction when standing upright. When the subjects were standing upright, the diaphragm muscles moved down, and the heart remained vertical. At this time the electrical axis shifted to the right, the SNS (sympathetic nerve system) activity increased. But PNS (parasympathetic nerve system) activity increased in supine position [12]. Sahakian et al. studied changes in frontal QRS loop and P axis in standing upright, sitting, walking, supine, and two kinds of side lying conditions, and specialized the difference between left-side, supine and right-side lying positions, which confirmed the body positions' influence on VCG. They found that the change of P wave is greater [13]. Most of the results are consistent with the results in this study. By means of MRI, Mase et al. presented the frontal and horizontal cross sections images of the chest. From these images, it could be seen that when lying on the left side or left-prone side, due to the effect of gravity, the heart moved down remarkably. But when lying on the right side or right-prone side, the position of heart showed no obvious difference with that in supine [14]. Such changes can also be seen in CT imaging [15, 16]. This could confirm the fact that the ECG waveform features rarely show significant differences between supine and right in this study.

Kutbay et al. study showed that the AHI (activity apnea-hypopnea index) and average minimum oxygen saturation (SOP) were significantly lower in supine than those in other lying positions, and the heart rate as well as average awakening index were higher [17]. George et al. found that lying on either side can significantly reduce OSA (obstructed sleep apnea) [18]. Garcia et al. found that the influence of body position on ECG waveform resulted in ST segment deformation. When lying on the left side, the R waves and T waves became larger and the S waves became deeper, which caused ST pattern misjudgment, and then led to false positive error or false negative error of myocardial ischemia determination [19].

Researchers have tried to classify lying positions form ECG, but most of them can only detect body position changes without lying position classification. Shinar et al. used the R wave duration (RWD) as indicator of body position changes for healthy subjects, who were asked to rotate between four body positions (back, left, prone and right). They could identify over 90% of the changes in body position. However, they couldn't identify the exact body positions [2]. In their further study, the results showed over 90% correct identification of body position changes and up to sensitivity 79% and specificity 93% of body position classification when using any of the three leads, including leads I II and III. Lead II, which we used in this study, had the best performance for the classification of body position and correctly classified 80% of heartbeats. Classification did not improve for a combination of two leads [3]. In 2003, García et al. investigated two ECG signal processing methods for detecting body position changes. The spatial approach was based on VCG loop rotation angles and the scalar approach was based on the K–L transform coefficients. They could detected 95% of the body position changes by angle-based detector, whereas the KLT-based detector produces values of 89% [20].

The researchers also tried to classify lying position by other signals and sensors. In 2011, Zachary et al. presented a method for lying position classification using load cells placed under bed, which resulted in generalized accuracies of 0.68, 0.57, 0.69, and 0.33 for the back, right, left, and stomach positions respectively, and 0.92, 0.75, and 0.86 for the back/stomach, right, and left positions respectively [21]. The resulting accuracies, especially for left and right, were not precise enough for clinical application. In 2016, without differentiation of sitting and standing, 100% accuracy was achieved using random forest by Marcel et al. However, the signals were recorded by a gyroscope from an iPhone fixed with a belt around the torso, which was very intrusive for normal sleep. On the other hand, they couldn't classify lying on the left or right, and the number of testing data segments were only 78 (sitting and standing were not included) [22]. In 2017, Timo et al. performed sleep position classification from a depth camera using bed aligned maps. They used Convolutional Neural Networks and achieved an accuracy of 94.0%. This approach directly recorded the body positions of patients and achieved high accuracy, but the apparatuses needed were complicated, the complexity of operations and the costs were so high that may not suitable for clinical and home nursery [23].

Studies about the influence of human lying position on ECG waveform during sleep can be widely applied in different field. First of all, changes of S wave and T wave can be used to correct the shape of ST–T segment, which can improve the determination accuracy of myocardial ischemia, and warn the sudden death early and effectively. Secondly, when studying the changes of ECG waveform and the related features in different sleep stages, the influence from lying position should be taken into consideration. Furthermore, in the process of collecting body signals and studying changes in physical conditions during sleep, if we can achieve lying positions determination based on ECG, the number of signal acquisition channels and the workload of researchers in monitoring process can be reduced. Also, patients will feel more comfortable. On the other hand, lying position monitoring can also prompt the patients to adjust their lying position during sleep, consciously. So that the frequency of respiratory disorders and sleep apnea events can be reduced. The occurrence of disease symptoms can probably be avoided and finally, sleep quality can be improved.

Conclusion

In conclusion, this study explored the influence of lying positions on the shape of ECG waveform during sleep, and then lying position classification based on ECG waveform features and random forest was achieved. When subjects were lying on the left side during sleep, due to the effect of gravity on heart, the position of heart changed, for example, turned and rotated, causing changes in the VCG of frontal plane and horizontal plane, which lead to a change in ECG. When lying on the right side, the heart was upheld by the mediastinum, so that the degree of freedom is poor, and the ECG waveform is almost unchanged. The overall classification accuracy of subject specific scheme reached 97.17%, κ statistic 0.91 and AUC > 0.97, while the results of subject independent scheme with feature normalization were accuracy 63.87%, κ statistic 0.32 and AUC > 0.66, respectively. The proposed method could be used as a technique for convenient lying position classification.

Abbreviations
ECG: electrocardiograph; PSG: polysomnography; AC: alternating current; RF: random forest; VCG: vectorcardiogram; MRI: magnetic resonance imaging; AHI: activity apnea-hypopnea index; OSA: obstructed sleep apnea.

Authors' contributions
HP, ZX and HY accomplished the data processing, finished the significant analysis as well as classification, and wrote the manuscript. YG, JS, ZC and YZ revised the manuscript. All authors read and approved the final manuscript.

Acknowledgements
Not applicable.

Competing interests

The authors declare that they have no competing interests.

References

1. Adams MG, Drew BJ. Body position effects on the ECG: implication for ischemia monitoring. J Electrocardiol. 1997;30(4):285–91.
2. Shinar Z, Baharav A, Akselrod S. R wave duration as a measure of body position changes during sleep. Comput Cardiol. 1999;26:49–52.
3. Shiner Z, Baharav A, Akselrod S. Detection of different recumbent body positions from the electrocardiogram. Med Biol Eng Comput. 2003;41(2):206–10.
4. Batchvarov V N, Bortolan G, Christov I I. Effect of heart rate and body position on the complexity of the QRS and T wave in healthy subjects. In: computers in cardiology. New Jersey: IEEE; 2008. p. 225–8.
5. Smit D, Cock CCD, Thijs A, et al. Effects of breath-holding position on the QRS amplitudes in the routine electrocardiogram. J Electrocardiol. 2009;42(5):400.
6. Khalighi S, Sousa T, Santos JM, et al. ISRUC-Sleep: a comprehensive public dataset for sleep researchers. Comput Methods Programs Biomed. 2016;124:180–92.
7. Yang X, Yan H, Ren Z, Chen J. Survey of ECG based for human identification. Chin J Sci Instrum. 2010;31:541–5.
8. Song J, Yan H, Li L, et al. A squeeze approach for electrocardiogram ST-segment detection based on R-wave and T-wave. Sheng Wu Yi Xue Gong Cheng Xue Za Zhi. 2011;28(5):855–9.
9. Breiman L. Random forests. Mach Learn. 2001;45:5–32.
10. Faletra FF, Pandian NG, Ho SY. Location of the heart: body planes and axis. Anat Heart Multislice Comput Tomogr. 2009.
11. Mincholé A, Sörnmo L, Laguna P. ECG-based detection of body position changes using a Laplacian noise model. Conf Proc IEEE Eng Med Biol Soc. 2011;2011(4):6931–4.
12. Li H, Yao F. The influence of body position change on electrocardiogram. J Med Front. 2016;6(29):136.
13. Ng J, Sahakian A V, Swiryn S, et al. The effect of body position on P-wave axis. Heart Lung Circ. 2001;28.
14. Mase K, Noguchi T, Tagami M, et al. Compression of the lungs by the heart in supine, side-lying, semi-prone positions. J Phys Ther Sci. 2016;28(9):2470–3.
15. Albert RK, Hubmayr RD. The prone position eliminates compression of the lungs by the heart. Am J Respir Crit Care Med. 2000;161(5):1660–5.
16. Ball WS, Wicks JD, Jr MF. Prone-supine change in organ position: CT demonstration. AJR Am J Roentgenol. 1980;135(4):815–20.
17. Kutbay Özçelik H, Bayram M, Doğanay E, et al. Effects of body position on sleep architecture and quality in subsyndromal adults without apparent obstructive sleep apnea. Sleep Biol Rhythms. 2015;13(3):279–86.
18. George CF, Millar TW, Kryger MH. Sleep apnea and body position during sleep. Sleep. 1988;11(1):90–9.
19. García, J, Astrom, M, Laguna P, et al. Detection of body position changes on the surface ECG. In: Computers in cardiology. New Jersey: IEEE; 2003. p. 45–8.
20. García J, Aström M, Mendive J, et al. ECG-based detection of body position changes in ischemia monitoring. IEEE Trans Bio Med Eng. 2003;50(6):677.
21. Beattie ZT, Hagen CC, Hayes TL. Classification of lying position using load cells under the bed. In: International conference of the IEEE engineering in medicine and biology society. New Jersey: IEEE; 2011. p. 474–7.
22. Mlynczak M, Berka M, Niewiadomski W, et al. Body position classification for cardiorespiratory measurement. In: Engineering in Medicine and Biology Society. New Jersey: IEEE; 2016. p. 3515.
23. Grimm T, Martinez M, Benz A, et al. Sleep position classification from a depth camera using Bed Aligned Maps. In: International conference on pattern recognition. New Jersey: IEEE; 2017. p. 319–24.

12

Multimodal MRI-based classification of migraine: using deep learning convolutional neural network

Hao Yang[†], Junran Zhang[*†], Qihong Liu and Yi Wang

*Correspondence:
zhangjunran@126.com
[†]Hao Yang and Junran Zhang
contributed equally to this
work
Department of Medical
Information Engineering,
School of Electrical
Engineering and Information,
Sichuan University, Chengdu,
Sichuan, China

Abstract

Background: Recently, deep learning technologies have rapidly expanded into medical image analysis, including both disease detection and classification. As far as we know, migraine is a disabling and common neurological disorder, typically characterized by unilateral, throbbing and pulsating headaches. Unfortunately, a large number of migraineurs do not receive the accurate diagnosis when using traditional diagnostic criteria based on the guidelines of the International Headache Society. As such, there is substantial interest in developing automated methods to assist in the diagnosis of migraine.

Methods: To the best of our knowledge, no studies have evaluated the potential of deep learning technologies in assisting with the classification of migraine patients. Here, we used deep learning methods in combination with three functional measures (the amplitude of low-frequency fluctuations, regional homogeneity and regional functional correlation strength) based on rs-fMRI data to distinguish not only between migraineurs and healthy controls, but also between the two subtypes of migraine. We employed 21 migraine patients without aura, 15 migraineurs with aura, and 28 healthy controls.

Results: Compared with the traditional support vector machine classifier, which has an accuracy of 83.67%, our Inception module-based convolutional neural network approach showed a significant improvement in classification output (over 86.18%). Our data also indicate that the Inception module-based CNN performs better than the AlexNet-based CNN (Inception module-based CNN reached an accuracy of 99.25%). Finally, we also found that regional functional correlation strength (RFCS) could be regarded as the optimum input out of the three indices (ALFF, ReHo, RFCS).

Conclusions: Overall, our study shows that combining the three functional measures of rs-fMRI with deep learning classification is a powerful method to distinguish between migraineurs and healthy individuals. Our data also highlight that deep learning-based frameworks could be used to develop more complicated models or systems to aid in clinical decision making in the future.

Keywords: Deep learning, Migraine, Diagnosis, Resting-state functional MRI, Convolutional neural networks

Background

A migraine is a common, chronic, incapacitating neurovascular disorder that is characterized by attacks of severe headaches, autonomic nervous system dysfunction, and in some patients, aura-associated neurologic symptoms [1]. A recent survey by the World Health Organization highlighted that migraine ranks as the third most prevalent disorder, affecting roughly 12% of adults globally. Moreover, people who experience migraines may have an increased risk of ischemic stroke, unstable angina, and/or affective disorder [2–5]. In migraine without aura (MWoA), previously known as common migraine, attacks are usually associated with nausea, vomiting, and/or sensitivity to light, sound, or movement. The other major subtype of migraine, which is known as migraine with aura (MWA), is accompanied by early symptoms. As the diagnosis of migraines is based on a combination of features and it is relatively difficult to exclude possible causes, achieving an accurate diagnosis using traditional methods (e.g., symptoms analysis, medical tests) is not easy. In fact, according to American Migraine Studies, only 65.2% of migraineurs receive the correct diagnosis [6]. Undoubtedly, there has been substantial interest in developing automated methods with the potential to assist in the diagnosis of migraines.

In recent years, resting-state functional magnetic resonance imaging (rs-fMRI) has attracted considerable attention for studying neural mechanisms. This approach not only overcomes the potential limitation associated with task paradigms in fMRI studies, but is also a non-invasive imaging technique capable of measuring spontaneous brain activity. In this approach low-frequency fluctuations in blood oxygen level-dependent (BOLD) signals are used to identify areas of increased or decreased neuronal activity [7–9]. By using resting-state fMRI, researchers have demonstrated that migraines are related to different indices of functional brain alterations, including amplitude of low-frequency fluctuations (ALFF), regional homogeneity (ReHo), and regional functional correlation strength (RFCS). For instance, compared with the healthy subjects, migraineurs showed significant changes in ALFF within the anterior cingulate cortex and prefrontal cortex, Moreover, a correlation analysis demonstrated that in migraineurs, ReHo values changes in the prefrontal cortex [10], and orbitofrontal cortex [11]. Despite these results demonstrating that migraines might contribute to functional brain alterations due to the repetitive occurrence of pain-related processes, very few studies have considered the possibility of using these functional features to improve the classification and diagnosis of migraines.

Deep learning is a relatively new technique in the field of computational medical imaging [12, 13]. This technique automatically creates multilevel models with hierarchical representations of the input data and permits powerful automatic feature extraction. Deep learning models are being used more frequently to improve diagnostic abilities and to discover the diverse patterns in patients' data that are characteristic of a disease [14, 15]. Convolutional neural networks (CNNs) is a deep learning model that has been applied successfully in the diagnosis of diseases based on MRI. For instance, Li et al. [16] constructed a CNN with two convolutional layers and one fully connected layer to identify the Alzheimer's disease (AD). They obtained a final recognition rate of up to 92.87% when comparing AD and healthy controls (HC). Similarly, Sarraf et al. [17] were able to make early diagnosis of Alzheimer's disease based on GoogleNet [18], a successful network that is broadly used for object recognition and classification. This network based on a modern design module called "Inception", and this module was come up with to enhance the performance of classifier, which resulted in a higher accurate, reaching to 98.74% between AD and HC [19]. These deep learning-based frameworks have implications for numerous applications in classifying brain disorders in clinical trials and large-scale research studies. In light of these previous findings, it is evident that the classification accuracy for a specific disease can be further improved by choosing the proper deep learning model.

Here, we used two deep learning-based classifiers to distinguish between healthy brains, brains affected by MWA and brains affected by MWoA. The first model we used was the CNN network based on AlexNet [21], and the second model was the CNN with Google's Inception module. As a lot of information can be lost when decomposing 4D rs-fMRI data into 2D data, many fMRI studies use feature mapping instead of raw data as the original input [17, 20]. Thus, after preprocessing, we extracted three indices—ALFF, ReHo and RFCS—as inputs to improve the classification results for migraine patients. We compare the classification results obtained using a 3-way classifier (MWA vs. MWoA vs. HC) and two binary classifiers (migraines vs. HC, and MWA vs. MWoA). To the best of our knowledge, this is one of the first studies to examine the performance of different deep learning-based frameworks and to apply them specifically for migraine discrimination. Taken together, our results could help improve diagnostic accuracy for patients suffering from migraines.

Methods

Subjects

Our study population included 21 migraine patients without aura, 15 migraineurs with aura, and 28 healthy controls. The age and gender differences between the three groups were tested using the t-test and no significant difference was observed (p > 0.05). In addition, the diagnosis of migraineurs was made by neurologic practitioners according to the criteria from the Second Edition of the International Classification of Headache Disorders (ICHD-II) [22]. All subjects were right-handed, aged between 18 and 50 years, and underwent brain scans at the Huaxi MR Research Center of the West China Hospital. The patients were attack-free for at least 72 h prior to the brain scan, and 48 h after the scan. Patients with chronic migraine, other chronic or current pain disorders, a history of mental disease, or other neurological disorders were excluded from this study. The study was approved by the Medical Ethics Committee of the West China Hospital, Sichuan University, and written informed consent was obtained from each participant. Table 1 lists the demographic and clinical data of the 64 subjects.

Data preprocessing

All data were acquired using a 3.0 Tesla MRI system (Trio Tim, Siemens, Erlangen, Germany). Subjects were instructed simply to rest with their eyes closed, not to think of anything, and not to fall asleep. Imaging data were collected transversely by using an echo-planar imaging (EPI) sequence with the following settings: repetition time/echo time (TR/TE) = 1900/2.26 ms, flip angle = 9°, slice thickness/gap = 1/0 mm, field of view (FOV) = 256 × 256 mm^2, matrix = 256 × 256, and voxel size = 1 × 1 × 1 mm^3. The rs-fMRI data were also collected using an echo planar

Table 1 Demographic and clinical characteristics of the 64 participants

Variables (mean ± SD)	MWoA	MWA	HC	p-value		
				MWoA vs. HC	MWA vs. HC	MWoA vs. MWA
Sex (male/female)	7/14	4/11	13/15	0.366	0.216	0.679
Age (years)	29.67 ± 6.45	32.2 ± 7.68	31.57 ± 6.72	0.323	0.782	0.291
Education (years)	16.90 ± 4.40	15.07 ± 1.98	16.36 ± 2.87	0.601	0.129	0.141
24-HAMD	5.95 ± 6.76	8.73 ± 5.13	3.57 ± 2.41	0.012	0.002	0.190
14-HAMA	4.28 ± 5.53	7.47 ± 6.13	2.07 ± 2.38	0.031	0.005	0.113

SD standard deviation, *HAMD* Hamilton Depression Scale, *HAMA* Hamilton Anxiety Scale

imaging (EPI) sequence but with the following settings: $TR/TE = 2000/30$ ms, flip angle $= 90°$, slice thickness/gap $= 5/0$ mm, $FOV = 240 \times 240$ mm^2, matrix $= 64 \times 64$, and voxel size $= 3.75 \times 3.75 \times 5$ mm^3. Resting-state functional images were analyzed with the Statistical Parametric Mapping software (SPM8, http://www.fil.ion.ucl.ac.uk/spm) and the Data Processing Assistant for Resting-State fMRI (DPARSF, http://rfmri.org/DPARSF) toolbox. The first 10 EPI volumes of each functional data were discarded due to instability of the initial MRI signal and adaptation of the participants to the situation. The remaining volumes first underwent slices-timing correction, and were then realigned to the first volume to correct for susceptibility-by-movement interaction. None of the subjects' heads in this study had a movement in any direction that exceeded a 2-mm displacement or a 2° of rotation. Then, the resulting images were spatially normalized into a standard stereotaxic space at 3-mm isotropic voxels using the Montreal Neurological Institute (MNI) template. Next, band-pass-filtering (0.01–0.08 Hz) was performed on the time series of each voxel to reduce the effect of low-frequency drifts and high-frequency physiological noise [23]. Finally, the ALFF, ReHo and RFCS were calculated as described below.

Feature mapping

ALFF is an effective indicator of regional intrinsic or spontaneous neuronal activity in the brain [24]. The time series for each voxel was first transformed to the frequency domain using a Fast Fourier Transform, and the power spectrum was obtained. The square root of the power spectrum was computed, and subsequently averaged across a predefined frequency interval. When calculating ALFF, the normalized and resliced images were smoothed using a 4-mm FWHM Gaussian kernel. In our study, the ALFF within the 0.01–0.08 Hz frequency band was calculated for each voxel using the Resting-State fMRI Data Analysis Toolkit (REST, http://rest.restfmri.net). To reduce the global effects of variability across participants, the ALFF of each voxel for a certain subject was divided by the global mean ALFF value. ReHo was used to measure the similarity between the time series of a given voxel and that of its nearest neighbor. It is calculated using Kendall's coefficient of concordance, which produces reliable results in rs-fMRI data analysis. In our study, we used cubic cluster of 27 voxels for each normalized and resliced image and assigned the ReHo value of every cubic cluster to the central voxel. A larger ReHo value for a given voxel indicates a higher local synchronization of rs-fMRI signals among neighboring voxels. As for ALFF, the ReHo of each voxel was also divided by the global mean ReHo value for each subject. All these procedures were performed using REST software.

By measuring functional connectivity (FC) of the distinct brain regions, rs-fMRI can be used to evaluate the brain function. To reduce the effect of the region of interest (ROI) selection strategy on the FC results, we employed the RFCS method, which measures the extent of the average correlation of a given brain region with all other regions [25]. To compute resting-state FC, we repressed the spurious effects of nuisance covariates [26]. We parcellated the rs-fMRI into 116 ROIs according to the Automated Anatomical Labeling (AAL) template, and thereby obtained 116 FC results for each subject. The RFCS values were then calculated using a previously described method [27]. The RFCS was defined as:

$$S(i) = \frac{1}{N} \sum_j |R_j| \tag{1}$$

where, R_j is the result of the FC in a certain region of the AAL, S_i represents the average activation level, and the N is the number of regions.

For each subject, we therefore obtained three functional maps: ALFF, ReHo and RFCS. To some extents, three features respectively reflect the degree of activity, the degree of synchronization and the degree of global synchronization of spontaneous neuronal activity.

Convolutional neural network

We used a deep CNN that was based on AlexNet and limited by the amount of data: three convolutional layers, two pooling layers and two fully connected layers (Fig. 1). With respect to the input layer, the post-feature mapping fMRI images were used as input data, with the convolutional layer playing the most important role in the CNN architectures as it is the core building block of such networks. The parameters of the convolutional layer consist of a set of learnable filters, where very filter is spatially small but extends through the full depth of the input volume. In our study, the filters had a size of 3×3 and computed the output of the neurons that were connected to local regions in the input. Thus, each output is computed as a dot product between its weight and the region that it is connected to the input volume. The pooling layer performs a downsampling operation along the spatial dimensions, and to reduce hyper parameters and control overfitting, we exploited the max pooling layer in our framework. The normalization layer also plays an important role in CNN architecture. Here we used the rectified linear units (ReLu) layer, which applies an element-wise activation function, such as max $(0, x)$ thresholding at zero. This layer does not change the size of the image volume [28, 29]. Next, the fully connected (FC) layer, as the name implies, each neuron in this layer is connected to all the numbers in the previous volume [16], and computes the class scores, resulting in the volume of the number of classes. The main goal when using a dropout layer is to avoid overfitting [30], especially in cases of limited training data. The dropout layer randomly drops neurons with some probability, p, which in our study, is equal to 0.5. In addition, we also exploited Adam [31] as an optimizer, which is a simple

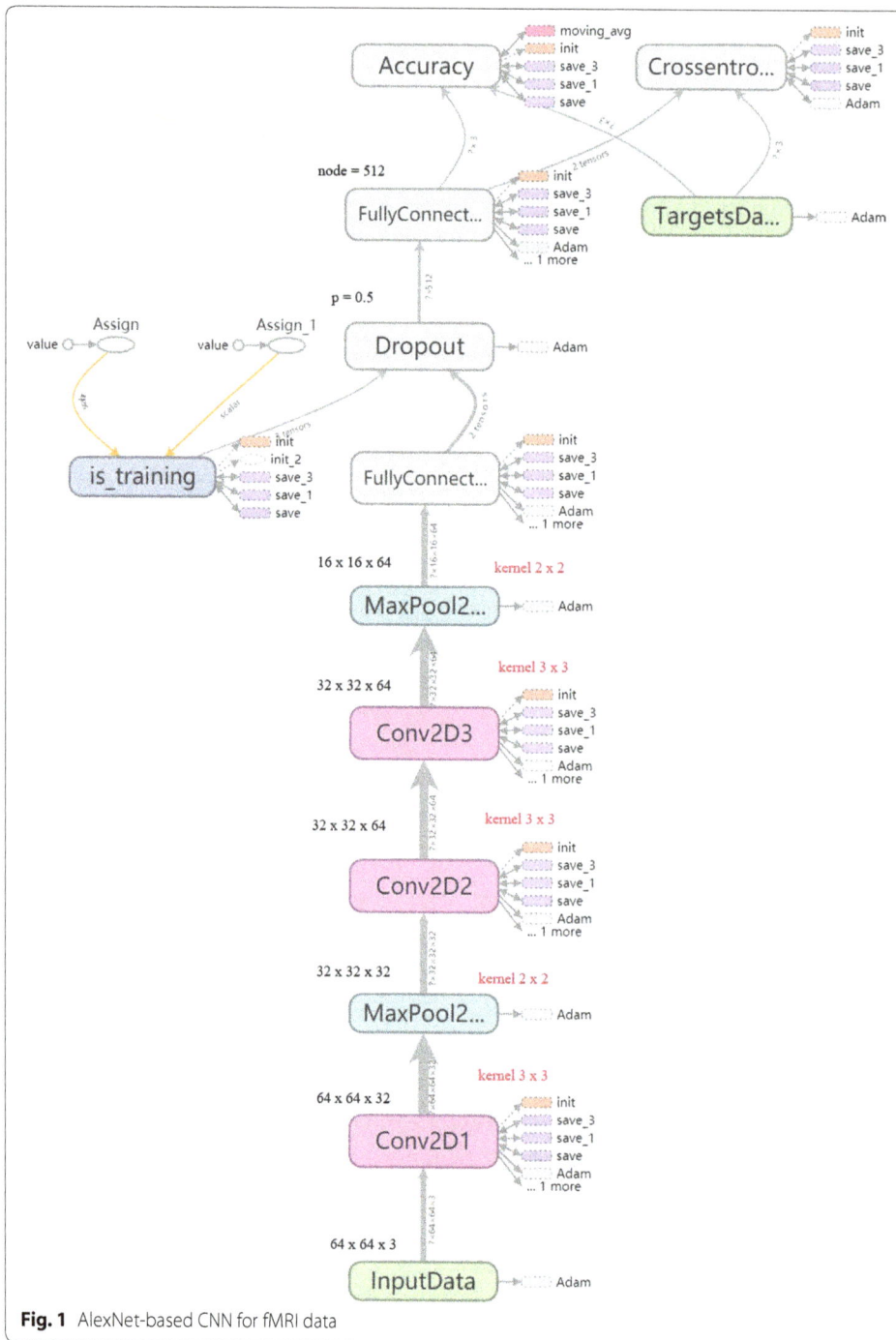

Fig. 1 AlexNet-based CNN for fMRI data

and computationally efficient algorithm for gradient-based optimization of stochastic objective functions.

Inception module

GoogleNet, which was developed by Szegedy et al. [18], is a successful network broadly used for object recognition and classification. The architecture of GoogleNet consists of a deep network based on a modern design module called Inception (Fig. 2). This module

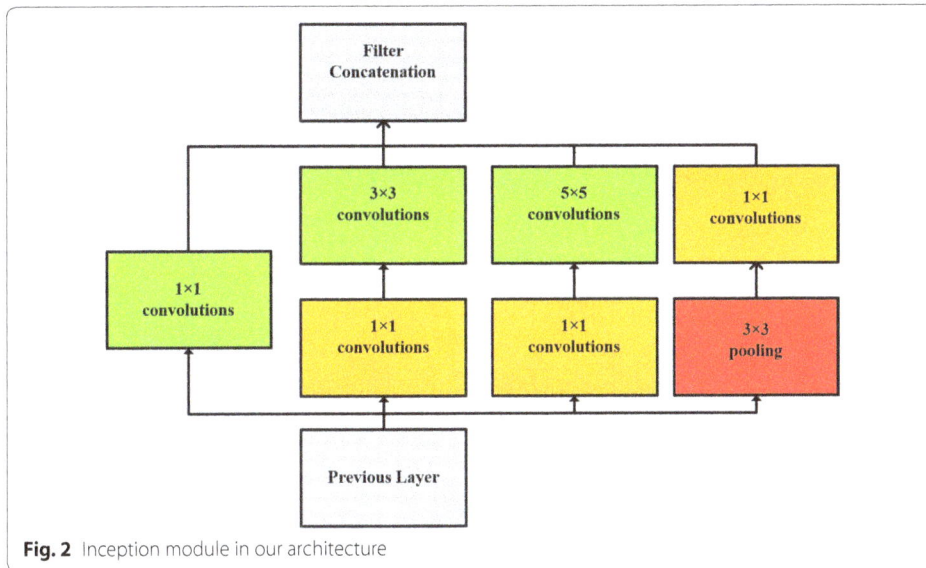

Fig. 2 Inception module in our architecture

allows for a significant increase in the number of units at each layer without increasing computational complexity until later stages. This is achieved through a global dimensionally reduction prior to costly convolutions with larger patch sizes.

In our second frames, we use this module to improve the performance of the classifier. Before development of the Inception module, the fundamental approaches used to improve accuracy of CNN architecture were to increase the size of the layers and make the network deeper. However, this straightforward solution causes two major issues. First, a large number of hyper parameters require more training data and can result in overfitting, especially in the case of limited training data. Second, uniform increases in network size dramatically increase interactions with computational resources, thereby affecting both the timing performance and cost of providing infrastructure. In consideration of our datasets, our frame as shown in Fig. 3 (Additional file 1: TableS1).

Training

The preprocessed rs-fMRI time series data were first loaded into memory using the neuroimaging package Nibabel (http://nipy.org/nibabel/) and were then decomposed into 2D (x, y) matrices along z and time (t) axes. Next, the 2D matrices were converted to lossless PNG format using the Python and OpenCV (opencv.org). To improve the performance of the classifier, we removed some slices of each time course because they included no functional information. During the training process, about 80% of the subjects from the three groups (51 individuals in total) were assigned to the training dataset, and the remaining 20% were used for testing purposes. This ensures the relative independence between the training set and the testing set. In the data conversion process, a total of 5760 images were produced, including 1890 for MWoA, 1350 for MWA and 2520 for normal control PNG samples. These training and testing images were randomly shuffled and resized to 63×63 pixels. For these images, we have exploited data augmentation, such as cropping, rotating, and flipping input images [32], to avoid overfitting. Then they were converted to h5py-formatted data through Tensorflow, which is a Deep

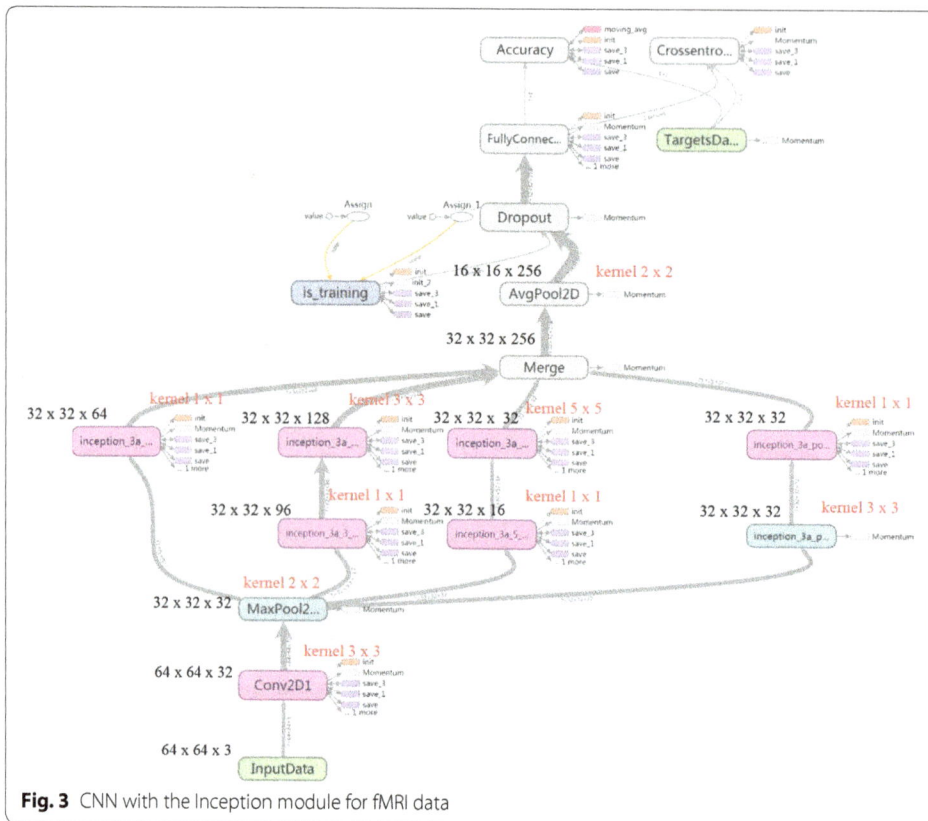

Fig. 3 CNN with the Inception module for fMRI data

learning platform (http://www.tensorflow.org) used for this classification experiment. In our study, we divided the data into three groups, migraine vs. HC, HC vs. MWoA vs. MWA, and MWA vs. MWOA, and used them as three different inputs. Then, we adopted two different deep learning frameworks and compared the identification results obtained between the two. The general flow of the classification process is depicted in Fig. 4.

The weights of hidden layers are randomly initialized, and for the output layer, we choose the SoftMax function as the classifier. The numbers of the units in the output layer represent the conditional probabilities that the input belongs to each of the classes. In the training phase, the learning rate drops every 10 epochs; we started with 0.01 and divided it by 10 every 10 epochs. We used the cross-entropy as our cost function, $J\,(w,\,b)$, as it is commonly used for classification tasks [29]. This function is:

$$-\frac{1}{N}\sum_{i=1}^{N}\sum_{j=1}^{J}\left[\mathrm{I}\left\{y^{i}=j\right\}\log\left(h_{w,b}\left(x^{i}\right)_{j}\right)\right] \tag{2}$$

where, N is the number of MRI images, j is the sum over the number of classes, $h_{w,b}$ is the function computed by the network, and $x,\,y$ are the input and label of image, respectively.

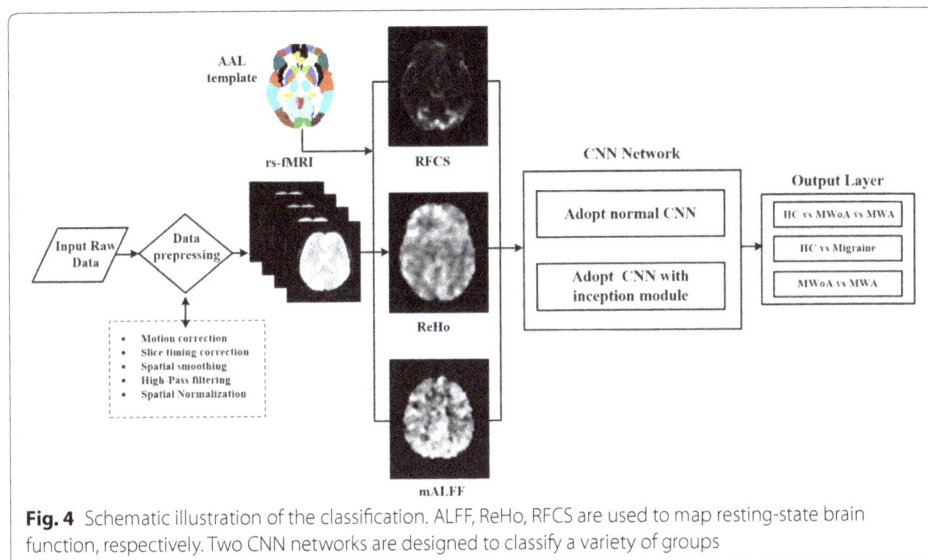

Fig. 4 Schematic illustration of the classification. ALFF, ReHo, RFCS are used to map resting-state brain function, respectively. Two CNN networks are designed to classify a variety of groups

Results

Here, we made classifications based on different feature types and used test examples (unseen examples) to evaluate the performance of the models. For robustness and reproducibility purposes, we repeated cross-validation of the deep neural networks four times (fourfold cross validation) [33]. The final accuracy of each CNN was subsequently computed by averaging the accuracies across the four runs (Tables 2, 3).

The best performer out of three different features was the RFCS, and the highest identification rate achieved was 99.25% when using the Inception module-based CNN to distinguish between the HC and migraine groups. According to the experimental data, the recognition rate of the AlexNet-based CNN was lower than that of the Inception module-based CNN, especially in terms of the ALFF feature. As expected, the Inception module-based CNN improved the classification performance in most cases. It was relatively hard for either framework to distinguish between the MWoA and MWA groups, with the AlexNet-based CNN showing an identification rate of just 86.43%. In addition, compared with other features, the RFCS feature mapping improved the classification accuracy of both of the deep learning-based models, with a noticeable difference between the two.

Table 2 The accuracy of testing dataset in CNN based on AlexNet is demonstrated

Feature	Data group	Accuracy
ALFF	HC vs. migraine	89.56% ± 0.24%
	HC vs. MWoA vs. MWA	87.31% ± 0.26%
	MWoA vs. MWA	86.43% ± 0.54%
ReHo	HC vs. migraine	93.42% ± 0.13%
	HC vs. MWoA vs. MWA	92.66% ± 0.21%
	MWoA vs. MWA	87.39% ± 0.27%
RFCS	HC vs. migraine	98.63% ± 0.28%
	HC vs. MWoA vs. MWA	98.78% ± 0.36%
	MWoA vs. MWA	96.87% ± 0.43%

Table 3 The accuracy of testing dataset in CNN with Inception module is demonstrated below

Feature	Data group	Accuracy
ALFF	HC vs. migraine	92.38% ± 0.14%
	HC vs. MWoA vs. MWA	92.31% ± 0.15%
	MWoA vs. MWA	89.77% ± 0.39%
ReHo	HC vs. migraine	95.07% ± 0.19%
	HC vs. MWoA vs. MWA	94.81% ± 0.35%
	MWoA vs. MWA	93.44% ± 0.20%
RFCS	HC vs. migraine	99.25% ± 0.36%
	HC vs. MWoA vs. MWA	98.69% ± 0.29%
	MWoA vs. MWA	96.13% ± 0.22%

ROC curve

The area under the receiver operating characteristic (ROC) curve is the most popular metric to measure the performance of a classifier [34]. It is a plot of the true positive rate against the false positive rate. Here, we calculated the true positive, true negative, false positive, and false negative rates from 0 to 1 based on the prediction results. Next, for each classifier, the sensitivity, specificity and accuracy were calculated as follows:

$$Sensitivity = True\ Positive\ Rate\ (TPR) = \frac{TP}{P} = \frac{TP}{TP + FN} \tag{3}$$

$$Specificity = True\ Negative\ Rate\ (TNR) = \frac{TN}{N} = \frac{TN}{FP + TN} \tag{4}$$

$$Accuracy = \frac{TP + TN}{TP + FN + FP + TN} \tag{5}$$

where, in the case of the HC vs. migraine group, TP denotes the number of patients correctly classified, FN denotes the number of controls correctly predicted, and FP denotes the number of controls classified as patients. To evaluate the performance of the two classifiers, we plotted the ROC (Fig. 5) and calculated the area under curve (AUC), for which a greater AUC represents a more accurate test (better classification). For comparison purposes, we focused on the RFCS feature and the HC vs. migraine group. We found an AUC of 0.99 for the Inception-based CNN, a value that indicates an excellent discrimination power.

Discussion

Here, we examined the ability of deep learning-based frameworks to discriminate between MWoA, MWA, and HC using features extracted from rs-fMRI data. Our results indicate that with each of these features—ALFF, ReHo and RFCS—it is possible to achieve relatively high accuracies when classifying the three groups of datasets. Noteworthy, in both CNN models, highest accuracies were achieved when using the RFCS feature. Upon comparing the results obtained using the two different CNNS, we found that the Inception module-based CNN shows a better performance than the

AlexNet-based CNN. Compared with the support vector machine classifier that we previously analyzed (a final classification accuracy of 83.67%) [20], our approach provides preliminary support for deep learning methods combined with the fMRI features as a method for improving the discriminative power for migraine. In fact, we obtained an accuracy as high as 99.25% when using the RFCS feature in deep learning-based frameworks. Resting-state functional connectivity has been defined as the identification of correlation patterns between fluctuations measured in different brain areas [35]. The RFCS measures the average correlation of a given brain region with all other regions. Our results suggest that RFCS holds the potential to increase the translation of fMRI data into clinical diagnosis.

Many studies have reported the existence of strong functional connections in the resting state [36]. These resting-state networks consist of anatomically separated, but functionally linked brain regions that show a high level of ongoing functional connectivity during rest. For example, abnormal functional connectivity has been reported in the prefrontal and temporal regions [11] and in the amygdala and visceroceptive cortex [10] of migraine sufferers (as compared to healthy control individuals). These regions may thereby contribute more in the classification. When measuring the functional connectivity among regions, global alterations emerge that may carry important information for enhancing accuracy rates. This could explain the higher levels of accuracy that we found in the present study. In addition, we found that considering ReHo and ALFF also led to good performance levels. Thus, we conclude that the deep learning-based frameworks can help identify migraine patients when using these fMRI features. We also noticed that classification accuracies were lower when comparing MWoA with MWA, especially in the ALFF group. Perhaps all subjects in the MWoA vs. MWA group experienced headache symptoms and as such, had more similar imaging markers as compared with the HC group.

For all features tested, the Inception module-based CNN resulted in a higher level of accuracy than the AlexNet-based CNN. A considerable amount of research has demonstrated the feasibility and effectiveness of the Inception module by stacking the convolution layers and pooling layers of different scales, which allows for training a high-quality network on training sets of relatively modest sizes. Szegedy et al. [18] exploited Inception models, reaching a multi-crop evaluation top-5 error of 3.5% (a 25% reduction with

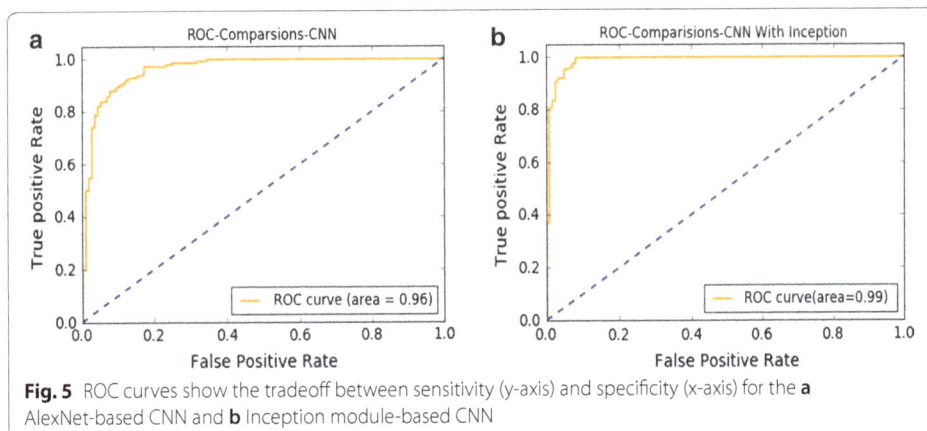

Fig. 5 ROC curves show the tradeoff between sensitivity (y-axis) and specificity (x-axis) for the **a** AlexNet-based CNN and **b** Inception module-based CNN

respect to the best published value). In our models, the Inception module uses parallel 1×1, 3×3, and 5×5 convolutions along with a max-pooling layer in parallel. These characteristics enable our models to capture a variety of features in the parallel and have a filter concatenation layer. This type of structure helps to automatically choose the appropriate kernel size in convolutional layers, as well as to concatenate the outputs of all these parallel layers. This in turn contributes to the combination of different sized features and it is helpful in the classification of migraine. Our approach provides initial evidence for a rapid and accessible method that has the potential to aid in making clinical decisions.

Besides identifying migraine and HC, our method was also able to distinguish between MWoA and MWA data. These two subtypes of migraine likely reflect differences in pathogenesis, and thus, the ability to identify different subtypes of migraine presents an advantage for clinical diagnoses and treatment. Using fMRI, Datta et al. [37] compared the responsiveness of MWA, MWoA and HC to visual stimuli and found that only the occipital and lateral geniculate cortex of MWA patients exhibited high reactivity. This high reactivity in the visual cortex is believed to be directly related to the aura. As the subtypes of migraine differ in their symptoms, choosing the most appropriate treatments for each subtype is a major challenge. Our research makes it possible to identify subtypes of migraine headache based on differences in precipitating factors, an achievement that will prove useful in clinical diagnosis.

Several limitations of this study should be noted. First, we only included 5760 images in the deep learning-based frameworks, the dataset was indeed small. To avoid overfitting, the frameworks we exploited were relatively shallow. In neuroimaging studies using data from migraine patients, it is difficult to obtain a large number of samples. However, the recent trend in the research community toward a greater level of sharing of neuroimaging data should increase the availability of training sets for future studies. The use of larger datasets would be useful to determine whether or not our approach can be applied on a larger scope and be used to make generalizations. Second, although we tried to extract the features, ALFF, ReHo and RFCS features to improve the performance of the classifier, there is no combination of models that relate to specific brain regions. This in fact, should be the aim of further research. Moreover, there is still no definite conclusion as to whether using the original rs-fMRI or the image after feature mapping in the deep learning framework generates better results. Nonetheless, our group has achieved good performances in the discrimination of post-traumatic stress disorder (PTSD) using these three indices [38]. Third, for this method to be applied in clinical practice, it is important to enhance transparency (due to the "black box" models) and generate a higher level of trust. Based on our results of these deep learning-based frameworks, our future goal is to visualize the brain regions that are most affected by migraines. Altogether, this information will ultimately help strengthen our understanding of the pathogenesis of this disorder.

Conclusion

Compared with traditional support vector machine, we were able to improve classification performance using CNN. Moreover, RFCS, ReHo and ALFF, these three functional indices we employed, can be used to represent different degrees of classification features. RFCS, which reflects the global alterations for one subject, carries important information for enhancing accuracy rates and obtaining the optimum classification results. Furthermore, using these three functional measures of rs-fMRI combined with deep learning frameworks is a powerful method that can potentially improve the clinical diagnosis of migraine and other brain disorders.

Abbreviations

CNN: convolutional neural network; MWoA: migraine without aura; MWA: migraine with aura; fMRI: resting-state functional magnetic resonance imaging; ALFF: amplitude of low-frequency fluctuations; ReHo: regional homogeneity; RFCS: regional functional correlation strength; BOLD: blood oxygen level-dependent; ROC: receiver operating characteristic.

Authors' contributions

YH have planned the study with the help of ZJ. YH has executed the experimental work with the help of LQ. All authors have contributed to the writing of the manuscript. All authors read and approved the final manuscript.

Acknowledgements

Not applicable.

Competing interests

The authors declare that they have no competing interests.

Funding

Sichuan Science and Technology Plan Project (JunranZ) Grant No. 2015HH0036; Chengdu Science and Technology Plan Project (JunranZ) Grant No. 2015HM0100561SF; Sichuan University-Luzhou Science and Technology Plan Project (JunranZ) Grant No. 2017CDLZ-G2.

References

1. Lance JW, Goadsby PJ. Mechanism and management of headache. 6th ed. Boston: Butterworth-Heinemann; 1998.
2. Agostoni E, Aliprandi A. The complications of migraine with aura. Neurol Sci. 2006;27(2):91–5. https://doi.org/10.1007/s10072-006-0578-y.
3. Calandre EP, et al. Cognitive disturbances and regional cerebral blood flow abnormalities in migraine patients: their relationship with the clinical manifestations of the illness. Cephalalgia. 2002;22(4):291–302. https://doi.org/10.1046/j.1468-2982.2002.00370.x.
4. Le Pira F, et al. Memory disturbances in migraine with and without aura: a strategy problem? Cephalalgia. 2000;20(5):475–8.
5. Breslau N, et al. Comorbidity of migraine and depression investigating potential etiology and prognosis. Neurology. 2003;60(8):1308–12. https://doi.org/10.1212/01.wnl.0000058907.41080.54.
6. Lipton RB, et al. Migraine diagnosis and treatment: results from the American Migraine Study II. Headache. 2001;41(7):638–45. https://doi.org/10.1046/j.1526-4610.2001.041007638.x.
7. Biswal B, et al. Functional connectivity in the motor cortex of resting human brain using echo-planar MRI. Magn Resonance Med. 1995;34(4):537–41.
8. Blackmon K, et al. Structural evidence for involvement of a left amygdala-orbitofrontal network in subclinical anxiety. Psychiatry Res. 2011;194(3):296–303. https://doi.org/10.1016/j.pscychresns.2011.05.007.
9. Borg C, et al. Impaired emotional processing in a patient with a left posterior insula-SII lesion. Neurocase. 2013;19(6):592–603. https://doi.org/10.1080/13554794.2012.713491.
10. Xue T, et al. Alterations of regional spontaneous neuronal activity and corresponding brain circuit changes during resting state in migraine without aura. NMR Biomed. 2013;26(9):1051–8. https://doi.org/10.1002/nbm.2917.
11. Yu D, et al. Regional homogeneity abnormalities in patients with interictal migraine without aura: a resting-state study. NMR Biomed. 2012;25(5):806–12. https://doi.org/10.1002/nbm.1796.
12. Hinton GE, Osindero S, Teh Y-W. A fast learning algorithm for deep belief nets. Neural Comput. 2006;18(7):1527–54. https://doi.org/10.1162/neco.2006.18.7.1527.
13. LeCun Y, Yoshua B, Geoffrey H. Deep learning. Nature. 2015;521(7553):436. https://doi.org/10.1038/nature14539.

14. Kim J, et al. Deep neural network with weight sparsity control and pre-training extracts hierarchical features and enhances classification performance: evidence from whole-brain resting-state functional connectivity patterns of schizophrenia. Neuroimage. 2016;124:127–46. https://doi.org/10.1016/j.neuroimage.2015.05.018.
15. Suk H-I, et al. State-space model with deep learning for functional dynamics estimation in resting-state fMRI. NeuroImage. 2016;129:292–307. https://doi.org/10.1016/j.neuroimage.2016.01.005.
16. Li R, et al. Deep learning based imaging data completion for improved brain disease diagnosis. In: International conference on medical image computing and computer-assisted intervention. Cham: Springer; 2014.
17. Sarraf S, Ghassem T. Classification of alzheimer's disease using fmri data and deep learning convolutional neural networks. 2016. arXiv preprint arXiv:1603.08631.
18. Szegedy C, et al. Going deeper with convolutions. In: Cvpr; 2015.
19. Payan A, Giovanni M. Predicting Alzheimer's disease: a neuroimaging study with 3D convolutional neural networks. 2015. arXiv preprint arXiv:1502.02506.
20. Zhang Q, et al. Discriminative analysis of migraine without aura: using functional and structural MRI with a multi-feature classification approach. PLoS ONE. 2016;11:9. https://doi.org/10.1371/journal.pone.0163875.
21. Krizhevsky A, Ilya S, and Geoffrey EH. Imagenet classification with deep convolutional neural networks. In: Advances in neural information processing systems. 2012.
22. Olesen Jes. Headache classification committee of the international headache society (IHS) the international classification of headache disorders, asbtracts. Cephalalgia. 2018;38(1):1–211. https://doi.org/10.7892/boris.112425.
23. Lowe MJ, Mock BJ, Sorenson JA. Functional connectivity in single and multislice echoplanar imaging using resting-state fluctuations. Neuroimage. 1998;7(2):119–32.
24. Yu-Feng Z, et al. Altered baseline brain activity in children with ADHD revealed by resting-state functional MRI. Brain Dev. 2007;29(2):83–91. https://doi.org/10.1016/j.braindev.2006.07.002.
25. Dai Z, et al. Discriminative analysis of early Alzheimer's disease using multi-modal imaging and multi-level characterization with multi-classifier (M3). Neuroimage. 2012;59(3):2187–95. https://doi.org/10.1016/j.neuroimage.2011.10.003.
26. Fox MD, et al. The human brain is intrinsically organized into dynamic, anticorrelated functional networks. Proc Natl Acad Sci USA. 2005;102(27):9673–8. https://doi.org/10.1073/pnas.0504136102.
27. Jiang T, et al. Modulation of functional connectivity during the resting state and the motor task. Hum Brain Map. 2004;22(1):63–71. https://doi.org/10.1002/hbm.20012.
28. LeCun Y, et al. Gradient-based learning applied to document recognition. Proc IEEE. 1998;86(11):2278–324. https://doi.org/10.1109/5.726791.
29. Jia Y, et al. Caffe: convolutional architecture for fast feature embedding. In: Proceedings of the ACM international conference on multimedia; 2014.
30. Konda K, et al. Dropout as data augmentation. Stat. 2015;1050:29.
31. Kingma DP, Ba J. Adam: a method for stochastic optimization. In: 3rd international conference for learning representations, San Diego; 2015.
32. Perez L, Jason W. The effectiveness of data augmentation in image classification using deep learning. 2017. arXiv preprint arXiv:1712.04621.
33. Arlot S, Celisse A. A survey of cross-validation procedures for model selection. Stat Surv. 2010;4:40–79. https://doi.org/10.1214/09-SS054.
34. Bradley AP. The use of the area under the ROC curve in the evaluation of machine learning algorithms". Pattern Recognit. 1997;30(7):145–1159. https://doi.org/10.1016/s0031-3203(96)00142-2.
35. Fox MD, Raichle ME. Spontaneous fluctuations in brain activity observed with functional magnetic resonance imaging. Nat Rev Neurosci. 2007;8(9):700. https://doi.org/10.1038/nrn2201.
36. Northoff G, Qin P, Nakao T. Rest-stimulus interaction in the brain: a review. Trends Neurosci. 2010;33(6):277–84. https://doi.org/10.1016/j.tins.2010.02.006.
37. Datta R, et al. Interictal cortical hyperresponsiveness in migraine is directly related to the presence of aura. Cephalalgia. 2013;33(6):365–74. https://doi.org/10.1177/0333102412474503.
38. Zhang Q, et al. Multimodal MRI-based classification of trauma survivors with and without post-traumatic stress disorder. Front Neurosci. 2016;10:292. https://doi.org/10.3389/fnins.2016.00292.

13

Movement artefact removal from NIRS signal using multi-channel IMU data

Masudur R. Siddiquee[*], J. Sebastian Marquez, Roozbeh Atri, Rodrigo Ramon, Robin Perry Mayrand and Ou Bai

*Correspondence:
msidd021@fiu.edu
Florida International
University, Miami, FL 33174,
USA

Abstract

Background: The non-invasive nature of near-infrared spectroscopy (NIRS) makes it a widely accepted method for blood oxygenation measurement in various parts of the human body. One of the main challenges in this method lies in the successful removal of movement artefacts in the detected signal. In this respect, multi-channel inertia measurement unit (IMU) containing accelerometer, gyroscope and magnetometer can be used for better modelling of movement artefact than using accelerometer only, which as a result, movement artefact can be more accurately removed.

Methods: A wearable two-channel continuous wave NIRS system, incorporating an IMU sensor which contain accelerometer, gyroscope and magnetometer in it, was developed to record NIRS signal along with the simultaneous recording of movement artefacts related signal using the IMU. Four healthy subjects volunteered in the recording of the NIRS signals. During the recording from the first two subject, movement artefacts were simulated in one of the NIRS channels by tapping the photodiode sensor nearby. The corresponding IMU data for the simulated movement artefacts were used to estimate the artefacts in the corrupted signal by autoregressive with exogenous input method and subtracted from the corrupted signal to remove the artefacts in the NIRS signal. Signal-to-noise ratio (SNR) improvement was used to evaluate the performance of the movement artefacts removal process. The performance of the movement artefacts estimation and removal were compared using accelerometer only, accelerometer and gyroscope, and accelerometer, gyroscope and magnetometer data from IMU sensor to estimate the artefact in NIRS reading. For the remaining two subjects the NIRS signal was recorded by natural movement artefacts impact and the results of artefacts removal was compared using accelerometer only, accelerometer and gyroscope, and accelerometer, gyroscope and magnetometer data from IMU sensor to estimate the artefact in NIRS reading.

Results: The quantitative and qualitative results revealed that the SNR improvement increases with the number of IMU channels used in the artefacts estimation, and there were approximately 5–11 dB increase in SNR when nine channel IMU data were used rather than using only three channel accelerometer data only. The artefact removal from natural movements also demonstrated that the combination of gyroscope and magnetometer sensors with accelerometer provided better estimation and removal of the movement artefacts, which was revealed by the minimal change of the HbO_2 and Hb level before, during and after movement artefacts occurred in the NIRS signal.

Conclusion: The movement artefacts in NIRS can be more accurately estimated and removed by using accelerometer, gyroscope and magnetograph signals from an integrated IMU sensor than using accelerometer signal only.

Keywords: NIRS, Near infrared spectroscopy, Motion artefacts, Multi-channel IMU, Accelerometer, Gyroscope, Magnetometer, Artefacts removal, SNR improvement, De-noising

Background

The absorption coefficient by human tissue in the near-infrared light region of 700–1000 nm [1] is much lower than other wavelength lights in the spectrum that are harmless to human body. This property leads to the development of near-infrared spectroscopy (NIRS) as a widely used method for detecting oxy- and deoxyhemoglobin level in blood. The NIRS signal can be detected by illuminating the human body with light from the near-infrared region [2, 3]. Moreover, the non-invasiveness, safety and cost-effectiveness [4], make NIRS even more popular than any other methods of detecting blood oxygenation level. The level of blood oxygenation in various parts of the human body convey a great deal of information about various physiological phenomena and processes [5], such as cardiovascular disease [6] and sepsis [7] from muscle oxygenation, cognitive involvement [8, 9] and activation of brain function [10, 11] from cerebral oxygenation.

Movement artefact removal is one of the most challenging parts of any type of bio-signal processing and the NIRS for blood oxygen level detection is no exception. It is not possible to fully restrict a subject from movement, voluntary or involuntary, and thus the acquired bio-signals are contaminated by the movement artefacts in different extents. Sometimes this contamination is so prominent that the subtle changes correspond to physiological changes subdued by the artefacts, thus the usability of the acquired signal mostly depends on the successful removal of the movement artefacts [12]. In the NIRS, the light source and the detectors are directly coupled to the human skin and this coupling is easily altered [13] by movement artefacts, which result in coupling error [14]. This coupling error imposes high uncertainty in the detection of the true changes in the NIRS signal which corresponds to the change in the physiological phenomena [15]. Moreover, from the perspective of the frequency domain, the NIRS signal variation due to the physiological change and the changes due to the movement artefacts are closely overlapped with each other, which makes it harder to separate the movement artefact content from the signal.

Numerous artefacts removal techniques for NIRS have been developed in the last few decades [16–22]. Most of them use the nature of the signal itself and the theoretical assumption of the influence of the movement artefact on the detected signal. A Majority of these methods development direction lie in the fact that detecting any other signal highly correlated to the movement artefacts was not readily available or difficult [23]. But the recent improvements in the Microelectromechanical systems (MEMS) chip components that is capable of registering motion information, i.e. acceleration, yaw, pitch, roll etc., makes it possible to observe motion-related signal concurrently with bio-signals of interest for various biomedical applications [24–29]. With respect to

NIRS signal detection, this advancement in MEMS chip component makes it possible to record movement artefacts related motion data and the NIRS signal simultaneously [30–32]. This additional information related to movement artefact leads to the more effective use of the adaptive noise cancellation technique on NIRS signal. So far, to the best of our knowledge, all the research articles related to adaptive filtering to remove movement artefacts from NIRS signal uses only the three-channel accelerometer data to estimate the movement artefacts [25, 30, 31, 33]. In this study, nine-channel Inertia Measurement Unit (IMU) data, namely accelerometer, gyroscope and magnetometer, are used to estimate the movement artefacts in NIRS signal and subsequently to remove the motion-related movement artefact. The basis of the study is that, the more information correlated to movement artefacts available, the better the estimation of the interfering artefacts contribution on the detected signal, and greater SNR improvement can be achieved. Movement artefacts arise from diverse body movement, which might have very less acceleration but more rotational or directional change, would be better registered by gyroscope and magnetometer, and would result in better estimation of the movement artefacts.

Movement artefacts removal techniques commonly employ the autoregressive modelling (AR) to remove the artefacts from the NIRS signal [22, 34]. In this study, Autoregressive model with exogenous input (ARX) was used as the method to estimate the movement artefacts and the multi-channel IMU data served as the exogenous input to the system.

Adaptive noise cancellation (ANC) was widely used in the various field for noise cancellation [35]. In this approach, one or more additional channels of information that is highly correlated to the interfering noise component in the primary recorded signal is used to remove the noise. The ARX modelling can be assumed as an altered method of ANC, which applies the classical least-squares (LS) algorithm to estimate the noise using exogenous input as the reference source for indirect noise estimation in the observed signal. ARX modelling is extensively used in the problems related to system identifications. In the current study, IMU signals were used as the inputs to the ARX modelling to estimate the movement artefacts in the NIRS signal and then subtracted to estimate the true NIRS signal.

Methods

Data acquisition method

In this study, data acquisition with a single wavelength and a dual wavelength NIR light source were used during the recording from the subjects. The methodology presented in [16] was used to quantitatively assess improvement in the signal quality after the artefacts removal. The methodology requires that two version of the same NIRS channels are positioned as close as possible where one is impacted by movement artefacts and another remains unimpacted. In this respect, the unaffected signal is analogous to the "ground truth" signal presented in [16], which was denote here as "reference ground truth" as the actual "ground truth" cannot be acquired. For the first two subjects the single wavelength LED was used with simulated artefacts, and signal quality improvement in respect to SNR and correlation was used for quantitative assessment. For the remaining two other subjects, a dual wavelength NIR light source was used to record the data with actual movement of the subjects causing the movement artefacts. As the NIRS signal from the last two subjects were recorded using two wavelengths, the NIRS signals can be converted to blood oxygenation concentration changes using the modified Beer–Lambert law [36, 37] and the result of artefact removal can be compared to the expected hemodynamic changes. Considering that the hemodynamic change is minimal during a short duration when movement occurs, the levels of oxygenated (HbO_2) and deoxygenated (Hb) hemoglobin concentration will be stable before and after movement occurs. Based on this, after the movement artefact removal from the contaminated portion of the NIRS signal, a minimal hemodynamic change with stable HbO_2 and Hb levels is expected.

Sensor system

To accomplish the above mentioned NIRS signal acquisition along with the simultaneous recording of multi-channel IMU data, a custom-made wearable NIRS system, based on Texas instrument (TI) CC3200 chip, was developed incorporating other peripheral chips and using the sensor node architecture developed in [38]. The IMU chip used in this architecture is MPU9250 which has 3-axis accelerometer, 3-axis gyroscope and 3-axis magnetometer data acquisition capability with a 16-bit resolution for each channel and the chip was attached to the NIR detector for better registration of the movement artefacts impact at the detector. In this study, each IMU channel was sampled at 62.5 Hz. An 850 nm wavelength LED and another 850–770 nm dual wavelength LED are used as the source of NIR light, and as the detector, a photodiode chip from TI modelled as OPT101 was used, which incorporate the required trans-impedance amplifier in the same chip. To digitize the analog signal from OPT101, a high precision ADC chip from TI, with the part no ADS1292, was used which has 24-bit resolution and high CMMR and support two-channel input. This ADC support multiple sampling rate ranging from 860 to 8000 Hz and we sampled the NIRS signal at 1000 Hz. To minimize the bus contention by the two vital peripheral devices in the system, IMU and ADC, these devices were connected to the main processor unit via the different buses, I2C and SPI respectively. The main processor unit used in this sensor architecture, CC3200, house two separate MCU in the same chip; one is featuring Wi-Fi Internet-On-a-Chip and another one as a typical microcontroller. Thus, the system can simultaneously collect the data from the peripheral devices and transfer those wirelessly through the internet to remote system or any local computer connected to the system using Wi-Fi.

ARX modelling and artefacts removal

The artefacts estimation and removal process is outlined in Fig. 1. Let $s[n]$ denote the true hemodynamic signal which was distorted by the motion artefacts signal $w[n]$. This corrupted hemodynamic signal $x[n]$ detected by the NIRS sensor can be expressed as,

$$x[n] = s[n] + w[n] \tag{1}$$

We used ARX modelling to estimate the motion artefacts in the detected signal. It is a widely used method in system identification task to determine the model structure using the input–output data. In this respect, it uses the least squares method to estimate the

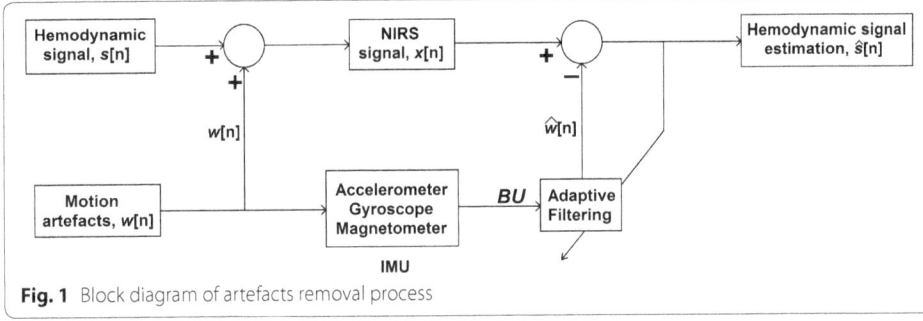

Fig. 1 Block diagram of artefacts removal process

best set of the coefficient of the system model from the input–output data available. In our study, the system resembled a multiple input single output (MISO) model, because the IMU data inputted into the model consisted of multiple channel data. In this case the ARX modelling can be represented using the following equation,

$$
\begin{aligned}
\hat{w}[k] = {}& a_1 x[k-1] + a_2 x[k-2] + \cdots + a_{NA} x[k-NA] \\
& + \boldsymbol{b}_0^T \boldsymbol{u}[k] + \boldsymbol{b}_1^T \boldsymbol{u}[k-1] + \cdots + \boldsymbol{b}_{NB}^T \boldsymbol{u}[k-NB]
\end{aligned}
\tag{2}
$$

where $x[n]$ is the detected NIRS signal and $\boldsymbol{u}[k]$ is the IMU data. Here $\hat{w}[n]$ is the output from the system based on the model coefficients $\boldsymbol{a} = [a_1 a_2 \cdots a_{NA}]$ and $\boldsymbol{B} = [\boldsymbol{b}_0 \boldsymbol{b}_1 \cdots \boldsymbol{b}_{NB}]$ when the input to the model is $\boldsymbol{U} = [\boldsymbol{u}[k] \boldsymbol{u}[k-1] \cdots \boldsymbol{u}[k-NB]]$. The model coefficients $\boldsymbol{a}$ and $\boldsymbol{B}$ selection in ARX modelling can be depicted by the following equation which is also known as least square method,

$$
J(\boldsymbol{a}, \boldsymbol{B}) = \sum_{k=1}^{N} \left(x[k] - \hat{w}[k] \right)^2
\tag{3}
$$

where $\boldsymbol{b}_0^T \boldsymbol{b}_1^T \cdots \boldsymbol{b}_{NB}^T$ are $1 \times L$ coefficient vectors and $\boldsymbol{u}[k]$ is a $L \times 1$ input vector, and the dimension L is the number of IMU data channel used. There were three combination cases for $\boldsymbol{U}$ used in this study, and they were,

Case 1: $\boldsymbol{u}[k] = \left[A_x[k] A_y[k] A_z[k] \right]^T$; $\quad L = 3$

Case 2: $\boldsymbol{u}[k] = \left[A_x[k] A_y[k] A_z[k] G_x[k] G_y[k] G_z[k] \right]^T$; $\quad L = 6$

Case 3: $\boldsymbol{u}[k] = \left[A_x[k] A_y[k] A_z[k] G_x[k] G_y[k] G_z[k] M_x[k] M_y[k] M_z[k] \right]^T$; $\quad L = 9$

For the conventional studies [22, 30, 32, 33] where only three channels Accelerometer data are used for autoregressive modelling, $\boldsymbol{U}$ can be expressed as in case 1 where A_x, A_y and A_z represent the three channels of Accelerometer data. Additional to the conventional study using only three channels Accelerometer data, we extended the study using six channels and nine channels IMU data. In case 2 and 3, G_x, G_y and G_z represent the three channels of Gyroscope data and M_x, M_y and M_z represent the three channels of magnetograph data.

In the artefact estimation process, the portion of the NIRS signal containing movement artefacts was fed to the ARX modelling algorithm as the output and the corresponding multichannel IMU data as the input to determine the model coefficients. For each artefact segment, this operation was iterated using several combinations of model orders from 1 to 10 and the order selected for who's the sum of squared error was the minimum between the estimation and the detected signal. Using the returned coefficient, a simulation model was defined for the current noisy portion of the NIRS signal. In the newly defined simulation model, the IMU signals were used as the input to the model to get the estimation $\hat{w}[n]$. Here, the estimation closely resembled to artefact $w[n]$ as the signal contribution by the true hemodynamic signal $s[n]$ to the detected signal $x[n]$ was very small compared to the artefact's contribution. Thus this estimated signal $\hat{w}[n]$ was then subtracted from the observed signal $x[n]$ to get the estimation of the movement artefacts free hemodynamic signal $\hat{s}[n]$ as per the equation,

$$\hat{s}[n] = x[n] - \hat{w}[n] \tag{4}$$

SNR improvement

As the data acquisition methodology used in this study was very much similar to [16], the SNR calculation was also done using similar formula applied in that study. The difference in the SNR, before and after the artefacts removal, is calculated using the following equation which is described in [39],

$$\Delta SNR = 10 log_{10} \left(\frac{\sigma_x^2}{\sigma_{e_{after}}^2} \right) - 10 log_{10} \left(\frac{\sigma_x^2}{\sigma_{e_{before}}^2} \right) \tag{5}$$

where, σ_x^2 is the variance of the movement artefacts free signal which is the referenced ground truth and $\sigma_{e_{before}}^2$ and $\sigma_{e_{after}}^2$ are the variance of the signal with the movement artefacts before and after the artefacts are removed, respectively.

Subjects and experimental design

Four healthy subjects, ages 22, 25, 27 and 28, with no history of asphyxia or brain disorder, volunteered for this study and a total of twelve sessions of data were collected. The NIRS were recorded from the forehead for simplicity. All the subjects were instructed to sit comfortably during the NIRS recording. As mentioned before, the NIRS signals from the first two subjects were contaminated by simulated movement artefacts. These simulations were done by external tapping on one of the two optical sensors, while the other remained unaffected. The NIRS signal from the other two subjects were collected by dual wavelength NIR light source, the subjects were instructed to move their head to induce natural movement artefact in the NIRS signals.

Results

In this study, the artefacts removal from the NIRS signal was implemented on the raw signal from the optical sensor. Thus, the clean signal (after the movement artefact removal) can be used in any other processing for further study. Figure 2 depicts a representative portion of the raw NIRS and the corresponding IMU data from subject 1. This NIRS signal contains three movement artefact segments which are indicated by vertical lines in the topmost plot in the figure. The simultaneous IMU data of this portion, which are three channels accelerometer, three channels gyroscope and three channels magnetometer data, are plotted in the same figure. It is already mentioned earlier that

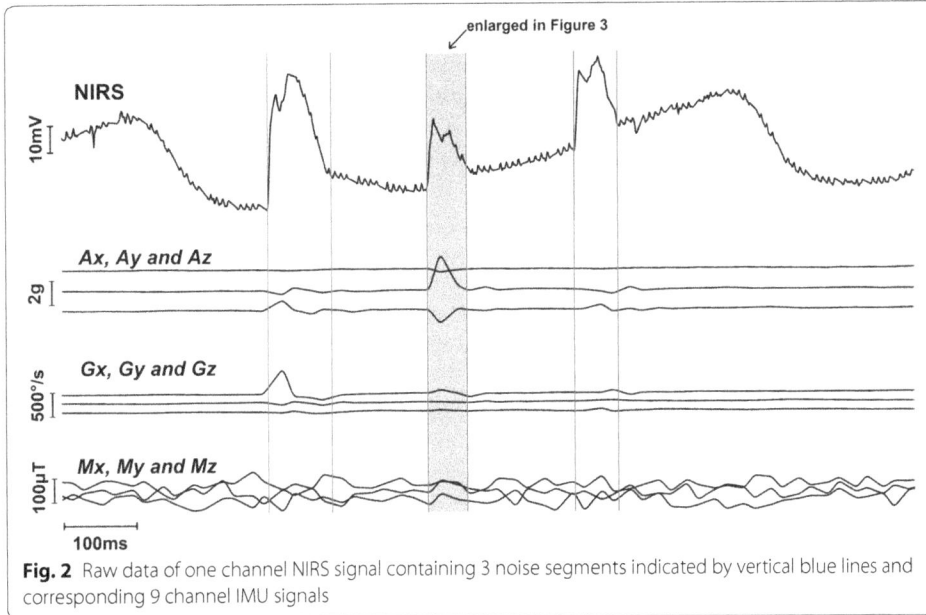

Fig. 2 Raw data of one channel NIRS signal containing 3 noise segments indicated by vertical blue lines and corresponding 9 channel IMU signals

most of the IMU-based movement artefact removal studies used three channels accelerometer data [30, 31, 33], whereas in this study, it was observed from the raw signal that without any processing, the Gyroscope data have the most impact from the motion artefacts and thus highly effective in estimating the artefacts. In the signal portion presented in Fig. 2, the gyroscope data have a more prominent impact on all three movement artefact segments in comparison to the accelerometer and the magnetometer data, which is apparent in Fig. 2.

The movement artefact estimation for the second segment in Fig. 2 is presented in Fig. 3. In Fig. 3, movement artefact containing signal is depicted by the solid black line, and the estimation of the movement artefact is indicated by the dashed blue line. Three estimation results are presented qualitatively in Fig. 3; the plot (a) for the case when only accelerometer data is used to estimate, the plot (b) presents the estimation result when accelerometer and gyroscope data are used in the modelling and lastly bottom-most plot (c) shows the estimation result when all the nine-channel IMU data, namely accelerometer, gyroscope and magnetometer, were used. This qualitative representation indicates that the more IMU channels that were used to model, the better the estimation was, and the best estimation was achieved when all the nine channels of IMU data available were used.

The quantitative metric, to assess the performance of the artefacts removal technique used in this study was the improvement in SNR which was calculated according to the Eq. (5) described in the previous section. This same metric and calculation was also used in several other research related to artefacts removal techniques presented in [16, 22, 32, 39]. In this respect, Table 1 represents the data of SNR improvement's quantitative data for the three movement artefact segments indicated in Fig. 2, which belongs to the data recorded from the subject 1 and for the data from the subject 2. For all the six segments, SNR improvements have been calculated in the cases of

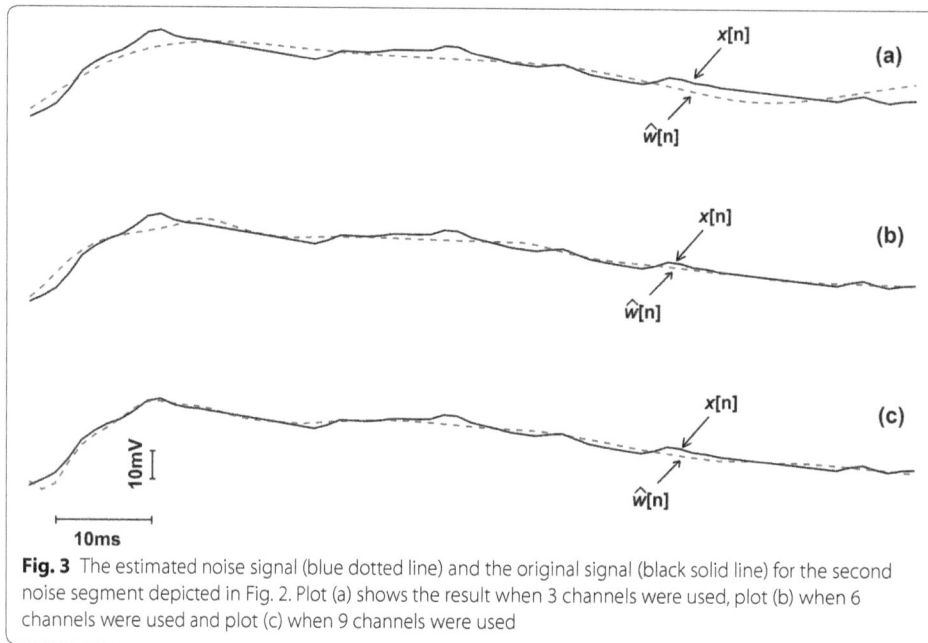

Fig. 3 The estimated noise signal (blue dotted line) and the original signal (black solid line) for the second noise segment depicted in Fig. 2. Plot (a) shows the result when 3 channels were used, plot (b) when 6 channels were used and plot (c) when 9 channels were used

using three channels (only accelerometer data), six channels (accelerometer and gyroscope data) and nine channels (accelerometer, gyroscope and magnetometer). The data presented in Table 1 indicates that for all the movement artefact segments, SNR improvements are higher when gyroscope and magnetometer data are used along with the accelerometer data. Specifically, for the first movement artefact segments from subject 1, we had 11.33 dB SNR when six channels were used, which was 3.41 dB higher than the SNR when only three channels were used, and the SNR was 14.77 dB when nine channels were used which was 6.85 dB higher than the SNR we got for three channels only. Similarly, for the second movement artefact segment, we had 15.11 dB SNR which was 6.18 dB higher than the SNR when we only used three channels IMU data and for the third movement artefact segment, we had 13.92 dB SNR which was 8.61 dB higher than the result for the three channels only case. Additionally, we computed the correlation between the artefacts removed signal and reference

Table 1 SNR improvement of 6 representative segments of noise in the recording from the subject 1 and 2, when various number of IMU channel data were used to remove the artefacts

Subjects	Accel.		Accel. + Gyro.		Accel. + Gyro. + Magn.	
	SNR (dB)	Correlation	SNR (dB)	Correlation	SNR (dB)	Correlation
Sub 1 Seg 1	7.92	0.79	11.33	0.82	14.77	0.83
Sub 1 Seg 2	8.93	0.89	12.66	0.92	15.11	0.93
Sub 1 Seg 3	5.31	0.94	13.92	0.96	13.92	0.96
Sub 2 Seg 1	7.35	0.96	13.44	0.97	19.08	0.97
Sub 2 Seg 2	3.04	0.77	5.29	0.78	11.95	0.98
Sub 2 Seg 3	11.56	0.94	14.31	0.95	17.46	0.95

The correlation coefficient presented in the table is between artefacts removed signal and the ground truth signal

ground truth signal for each movement artefact segments which are also presented in Table 1. For the first movement artefact segment, the correlation coefficient was increased from 0.79 to 0.82 when six channels IMU data were used to remove the movement artefacts than when only three channels data were used, and this value increased to 0.83 when nine channels IMU data were used. In case of the second movement artefact segment, the correlation coefficient was increased from 0.89 to 0.92 and 0.93 when six channels and nine channels IMU data were used respectively to remove the movement artefacts contribution in the signals. For the third movement artefact segment the correlation between the movement artefact removed signal and the reference ground truth signal was 0.94 when only three channel IMU data were used to remove the artefacts and it was 0.96 when six or nine channel IMU data were used for artefacts removal.

In the data recorded from the second subject, another three movement artefact segments were selected and all the similar processing described above were applied on those movement artefact segments. The quantitative SNR improvements for those segments are also presented in the data Table 1 which shows the similar trend in the SNR improvements for subject 1. In case of movement artefact segments from the subject 2, we had 19.08 dB SNR when we used nine channels IMU data which was 11.73 dB higher than the result for three channels usage case for the first movement artefact segment. Likewise, for the second movement artefact segment from subject 2, we had 11.95 dB SNR and for third segment 17.46 dB when nine channels IMU data were used which were 8.91 and 5.90 dB higher than when six and nine channels IMU data were used, respectively. In respect of correction coefficient, the first movement artefact segment for this subject had a coefficient of 0.96 between movement artefact removed signal and referenced ground truth signal when three channels IMU data were used and it 0.97 when six or nine channels IMU data were used. For the second movement artefact segment, the correlation coefficient was increased from 0.77 to 0.78 and 0.99 when we increased the number of IMU channels used in artefact removal to six and nine channels respectively from only three channels data. The last movement artefact segment from this subject had a correlation coefficient of 0.94 between the movement artefact removed signal and the reference ground truth when three channels IMU data were used to remove movement artefacts and 0.95 when six or nine channel IMU data were used.

The qualitative result of removing the artefacts that contaminated the abovementioned three segments present in the recording from subject 1 is presented in Fig. 4 which are plotted with the solid blue lines, whereas the artefacts-free NIRS signals from the other channel are also concurrently plotted in the same figure using black solid lines to indicate the empirical comparisons. Similar to the presentation used in Fig. 3, the result of a various number of IMU channel data usage is presented in Fig. 4, plot (*a*) represent the result when only accelerometer data were used, plot (*b*) when accelerometer along with the gyroscope data were used and finally plot (*c*) depicts the result when accelerometer, gyroscope and magnetometer data were used altogether.

In case of quantitative results, it was easy to determine the best outcome of the artefacts removal as it was obvious from exact values of SNR presented in the data table that the higher the SNR, the better the result is. But in case of empirical result,

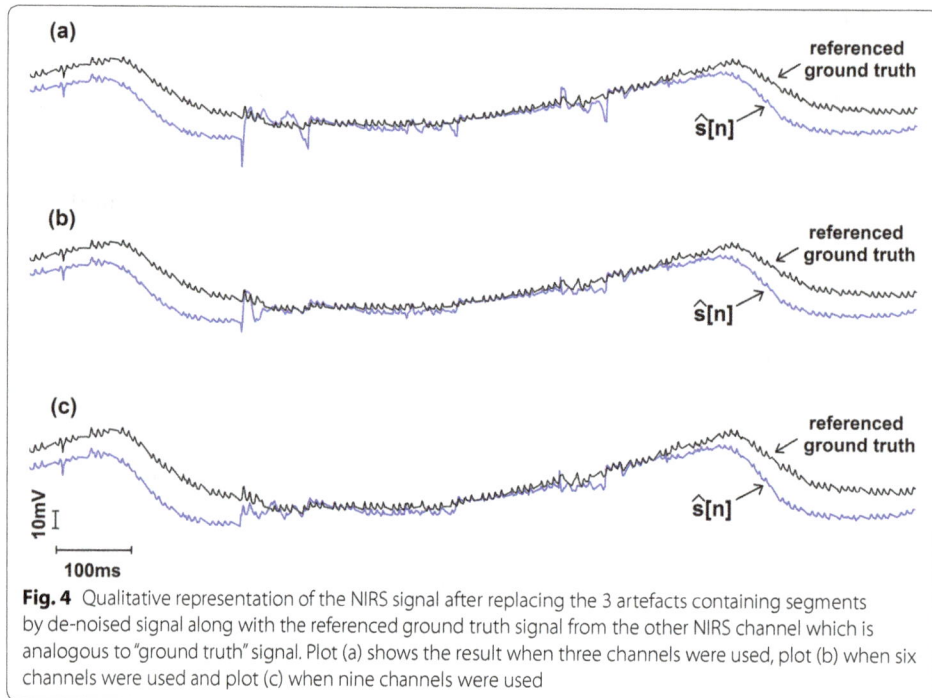

Fig. 4 Qualitative representation of the NIRS signal after replacing the 3 artefacts containing segments by de-noised signal along with the referenced ground truth signal from the other NIRS channel which is analogous to "ground truth" signal. Plot (a) shows the result when three channels were used, plot (b) when six channels were used and plot (c) when nine channels were used

there is no direct way to determine the best. In this respect, in the current study, we assumed the artefact-free NIRS channel to be the reference ground truth signal, to compare empirically. Empirically, we can say that the closer the variance of NIRS signal after the artefacts removal to the variance of the referenced ground truth signal, the better the result is. In that sense, we can say the best result was achieved when nine channels of IMU data were used, which is also consistent with SNR improvement data presented in Table 1 and apparent from the plot (c) in Fig. 4.

The NIRS signal from the subject 3 and 4 were recorded using dual wavelength NIR light source, so that it can be converted to oxygenated (HbO_2) and deoxygenated (Hb) hemoglobin concentration changes with typical processing. The movement artefacts were removed from the raw NIRS signal using IMU signals as described in the previous section and then converted the denoised signals to HbO_2 and Hb changes using typical Beer–Lambert law [36]. Three noise segments data from each of the two subjects are presented in Fig. 5. Each of the six plots in Fig. 5 present the HbO_2 and Hb change when no denoising was used, when only accelerometer signals were used, when accelerometer and gyroscope signal were used and when accelerometer, gyroscope and magnetometer signal were used for artefacts estimation and removal. In all the plots there were substantial changes in HbO_2 and Hb concentration around the movement artefacts containing parts of the signals when no denoising were used. For all the six artefact segments, HbO_2 and Hb changes curve get closer to the minimal change when artefacts estimation and removal was done for the artefacts containing part of the NIRS signal, and the case of estimating the artefacts by accelerometer, gyroscope and magnetometer result better than the case of using only accelerometer for the estimation and removal of the movement artefacts.

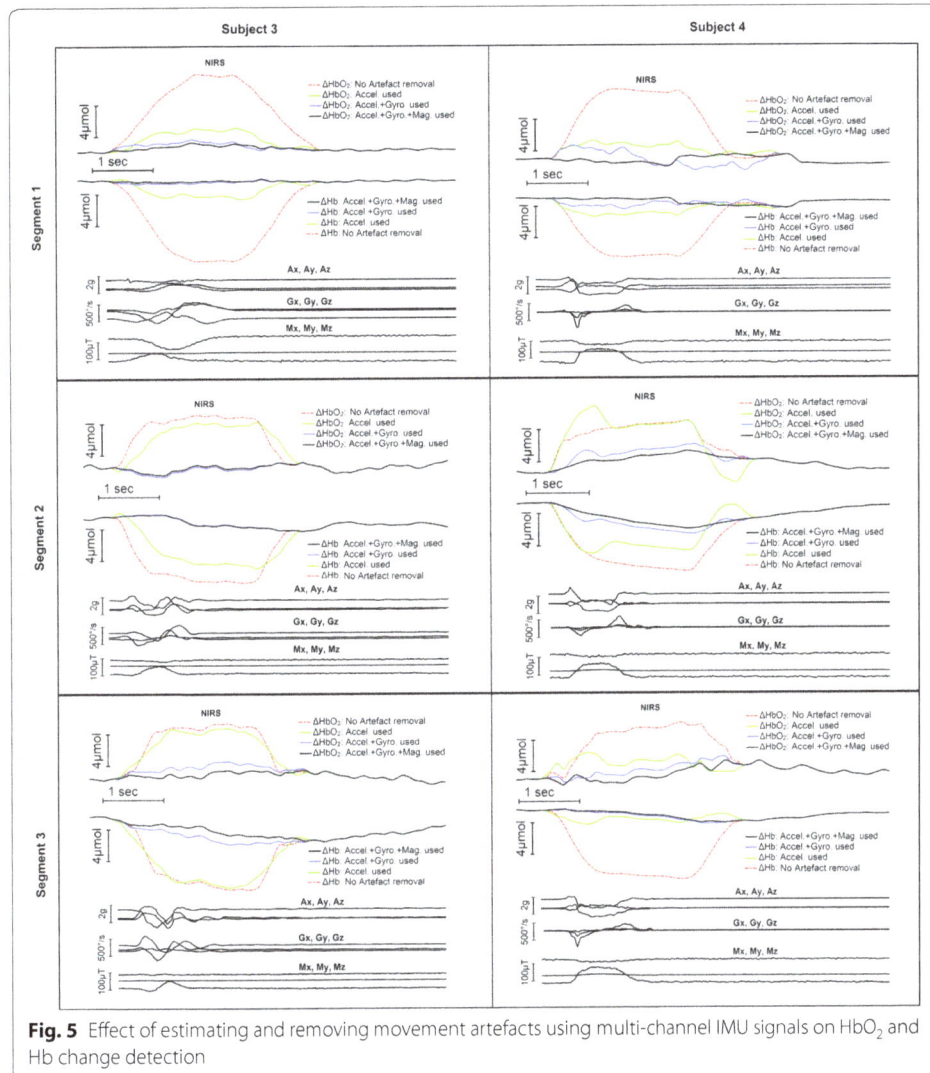

Fig. 5 Effect of estimating and removing movement artefacts using multi-channel IMU signals on HbO$_2$ and Hb change detection

Discussion

In this paper, a method of using multi-channel IMU data to successfully remove movement artefacts from NIRS signal has been presented. The qualitative and quantitative results show that implementation of multi-channel IMU data resulted in more accurate modelling of motion artefacts in NIRS, and thus we obtained more accurate motion artefacts free NIRS signal which can be used to detect physiological changes accurately.

After the first massive production of micro-electromechanical-system (MEMS) chip based accelerometer in about 1993 [40] it was extensively used in other fields such as automobile, aerodynamics and so on; but there was a sloth progress in effectively using such devices in the field of bio-signal acquisition due to the reduction of comfort of the subject [32] which results from the extra wiring and placement of this additional sensors. Moreover, placing the accelerometer sensor close to the NIR sensor was also another challenge as the NIRS system used in most of the research, utilized the optical fiber based light transport system [41] to and from the subject body. This close placement of the accelerometer to the light coupling to the subject body is necessary to record motion artefacts caused by the subject movement as well as due to the NIR sensor shifting [32]. These challenges were mitigated in this study by careful selection of the components used in the design of the system and making the system as wearable as possible, for instance, the IMU chip, MPU9250, used in the system is only 3 mm by 3 mm in dimension and this IMU was attached to the body of the NIR detector sensor OPT101 chip to record the true motion of this detector as well as the movement of the subject body. The entire wireless NIRS system used in this study has a dimension of only 30 mm by 48 mm and power by a small lithium-ion battery, which makes it a true wearable NIRS data acquisition system. Considering the subtle details like those presented above could be helpful in designing wearable NIRS system incorporating sophisticated chip like IMU which will improve the overall system performance.

The main challenge related to the hardware of the system in this study came from the strategy used in the acquisition of the multi-channel NIRS signals. As already mentioned earlier, to use the effective method of quantitative evaluation of the result of movement artefacts removal success, the two NIRS detectors were placed as close as possible with the target of recording a very similar version of NIRS signal where one of them are intentional movements' artefacts induced. Due to this closeness, inducing artefacts in one NIRS channel while leaving another channel undisturbed was tough and a lot of attempts took place to achieve this data acquisition strategic goal.

The effectiveness of the regressive modelling, like the ARX used in this study to estimate motion artefacts, highly depends on how much motion related information present in the exogenous input. The raw data presented in Fig. 2 has an important finding in this context, the first and the last artefact segments have a high impact on the gyroscope data whereas, for the middle artefact segment, the accelerometer data have a prominent impact. This finding implies that, using more channels of IMU data increases the probability of capturing the motion-related data in at least some of the channels which in turn increase successful estimation of the movement artefacts. In this respect, the SNR improvement results presented in Table 1 depicts that SNR improvement increased if the number of IMU data channel used in the modelling were increased except for the movement artefact segment 3 from the subject 1. For the movement artefact segment 3 from subject 1, the SNR improvement value for the six channel and nine channel IMU data, are same, which is due to the fact that the last three channel of the IMU data (the magnetometer data) had insignificant contribution on the artefact estimation. This insignificant contribution might have two reasons, either the artefact was indifferent to the variable the sensor was sensing or the sensor itself was not sensitive enough to detect the subtle change in that variable. On the other hand, for the movement artefact segment 2 from subject 2 in Table 1, the magnetometer data has a high contribution to the SNR improvement result. Besides the SNR improvement results, correlation coefficients between the movement artefact removed signal and the referenced ground truth signal are presented in Table 1 for both subjects 1 and 2 for each of the movement artefact segments as another quantitative improvement indicators. This indicator signifies how much alike the signals are in respect of covariance by a single unitless quantity ranging from -1 to $+1$

where values closer to $+1$ indicates stronger correlation between the signals [42]. For all the movement artefact segments the correlation coefficients increase towards $+1$ when six channels of IMU data were used than only when three channels IMU data were used to estimate and remove the movement artefacts. When nine channel IMU data were used to estimate and remove the movement artefacts, the correlation coefficients remained same for some of the segments which is analogues to the case of SNR improvements after artefact removal from the third movement artefact segment from the first subject.

The data from the subject three and four are presented in Fig. 5. In the experiments with these two subjects, the movement artefacts in the NIRS were induced by natural head movements rather than the simulated movement artefacts as presented for the first two subjects. Furthermore, the NIRS signals were converted into oxygenated (HbO_2) and deoxygenated (Hb) hemoglobin change using Beer-Lambert law [36]. In contrast to the experiments with simulated artefacts where validation of the artefacts removal was assessed by the SNR and correlation improvement, the same validation method cannot be used in case of natural movement artefacts removal due to the lack of any reference ground truth signal. Considering that there was a minimal hemodynamic change during a short duration of the natural movement occurrence, a minimal change in concentration of the HbO_2 and Hb was used to determine the performance of the artefacts removal associated with natural movements. The result presented in Fig. 5 demonstrates that with additional gyroscope sensors and magnetometer, the artefact in NIRS signals can be better removed as revealed by the minimal change of HbO_2 and Hb signals before, during, and after the movement artefacts occurred. This suggests that, in addition to the accelerometer in the IMU sensor, gyroscopes and magnetometer in the IMU are complementary to the accelerometer for a better modelling of the movement artefacts, which as a result, leads to better removal of the movement artefact in NIRS signals. A previous study [21] showed that the level of negative correlation between Hb and HbO_2 get reduced when movement occur. In case of this study, the duration of the natural head movement resulting artefact in NIRS signals was relatively short and hypothetically, there would not be any significant hemodynamic change during this short period. Based on this assumption, the distorted waveform of Hb and HbO_2 signals as depicted in Fig. 5, was due to the artefacts contaminated in the NIRS signal. As the distorted Hb and HbO_2 waveform was due to the artefacts, the distortion should be independent to the real hemodynamic changes of Hb and HbO_2, which might be positively or negatively correlated. In the experiments performed in this study, the data showed a major negative correlation between the oxyhaemoglobin and deoxyhemoglobin signal; however, after removing the distortion induced by the artefact, the level of Hb and HbO_2 kept the same as the baseline, which was in agreement with the hypothesis.

It is to be noted here, the selection of the movement artefact segments were done manually by keeping track of the time of movement artefact occurrence and later visual inspections on the raw IMU data and the raw NIRS signal. This manual detection of the segment will be automated in future based on IMU and NIRS signal feature changes. Though this study was not purposed to evaluate the quality of NIR signal from the custom-made system, we did perform preliminary evaluation—the raw NIRS signal was visually inspected for the presence of heart beating signal, showing whether the signal was correctly acquired. However, the accuracy and quality of the NIR signals using the custom-made system needs further well-controlled test, particularly, the correlation with brain functions. This will be the goal for the future study.

Although the current results presented in this paper showed a significant improvement in artefacts removal, there are still a lot of scopes to improve the developed technique. From the system identification point of view, any movement artefact impact on the NIRS signal is unpredictable in nature as they might differ in amplitudes, directivities, latencies, frequency contents and so on [41]; moreover, it has been observed in this research that various IMU channel might have different level of artefacts impact in different cases, which is another variability probably arise from sensor or from the nature of the artefact itself.

Conclusion

In the previous studies, accelerometer was used in adaptive filtering for movement artefacts registering and estimating its impact in NIRS signal. The theoretical application to accelerometer-based motion artefact removal is effective in mechanical systems, but the organic movements of a human subject are not only subjected to linear movements, but simultaneous rotation and multi-directional displacements. These motions are roughly captured by the accelerometer but are effortlessly quantified with the use of the additional magnetometer and gyroscope. Thus, movement artefacts related signals detected by other sensors from IMU, along with the accelerometer signal, result in better estimation of the movement artefacts in the detected NIRS signal. In this study, the result showed that using the accelerometer, gyroscope and magnetometer signals from IMU sensor provide more accurate modelling of motion artefacts and thus improves the SNR improvement yields.

Authors' contributions

MS was responsible for writing the manuscript. MS, SB and RA were responsible for data acquisition, data analysis and algorithmic development. MS, RR and RM contributed in data acquisition system hardware and software development. OB was responsible for overall planning of the study. All authors read and approved the final manuscript.

Acknowledgements

Not applicable.

Competing interests

The authors declare that they have no competing interests.

Funding

This research is partly supported by National Science Foundation (CNS-1552163).

References

1. Villringer A, Chance B. Non-invasive optical spectroscopy and imaging of human brain function. Trends Neurosci. 1997;20:435–42.
2. Ferrari M, Quaresima V. A brief review on the history of human functional near-infrared spectroscopy (fNIRS) development and fields of application. Neuroimage. 2012;63:921–35. https://doi.org/10.1016/j.neuroimage.2012.03.049.

3. Scholkmann F, Kleiser S, Metz AJ, Zimmermann R, Mata Pavia J, Wolf U, et al. A review on continuous wave functional near-infrared spectroscopy and imaging instrumentation and methodology. Neuroimage. 2014;85:6–27. https://doi.org/10.1016/j.neuroimage.2013.05.004.

4. Scheeren TWL, Schober P, Schwarte LA. Monitoring tissue oxygenation by near infrared spectroscopy (NIRS): background and current applications. J Clin Monit Comput. 2012;26:279–87. https://doi.org/10.1007/s10877-012-9348-y.

5. Jobsis F. Noninvasive, infrared monitoring of cerebral and myocardial oxygen sufficiency and circulatory parameters. Science (80-). 1977;198:1264–7. https://doi.org/10.1126/science.929199.

6. Oemrawsingh RM, Cheng JM, García-García HM, van Geuns R-J, de Boer SPM, Simsek C, et al. Near-infrared spectroscopy predicts cardiovascular outcome in patients with coronary artery disease. J Am Coll Cardiol. 2014;64:2510–8.

7. Skarda DE, Mulier KE, Myers DE, Taylor JH, Beilman GJ. Dynamic near-infrared spectroscopy measurements in patients with severe sepsis. Shock. 2007;27:348–53.

8. León-Carrion J, Damas-López J, Martín-Rodríguez JF, Domínguez-Roldán JM, Murillo-Cabezas F, Barroso y Martin JM, et al. The hemodynamics of cognitive control: the level of concentration of oxygenated hemoglobin in the superior prefrontal cortex varies as a function of performance in a modified Stroop task. Behav Brain Res. 2008;193:248–56.

9. Cui X, Bray S, Bryant DM, Glover GH, Reiss AL. A quantitative comparison of NIRS and fMRI across multiple cognitive tasks. Neuroimage. 2011;54:2808–21. https://doi.org/10.1016/j.neuroimage.2010.10.069.

10. Villringer A, Planck J, Hock C, Schleinkofer L, Dirnagl U. Near infrared spectroscopy (NIRS): A new tool to study hemodynamic changes during activation of brain function in human adults. Neurosci Lett. 1993;154:101–4.

11. Holper L, Muehlemann T, Scholkmann F, Eng K, Kiper D, Wolf M. Testing the potential of a virtual reality neurorehabilitation system during performance of observation, imagery and imitation of motor actions recorded by wireless functional near-infrared spectroscopy (fNIRS). J Neuroeng Rehabil. 2010;7:1–13.

12. Salehizadeh SMA. Motion and noise artifact detection and vital signal reconstruction in ECG/PPG based wearable devices. Dr Diss. 2015. http://digitalcommons.uconn.edu/dissertations/980. Accessed 26 Feb 2018.

13. Noponen TEJ, Kotilahti K, Nissilä I, Kajava T, Meriläinen PT. Effects of improper source coupling in frequency-domain near-infrared spectroscopy. Phys Med Biol. 2010;55:2941–60.

14. Schweiger M, Nissilä I, Boas DA, Arridge SR. Image reconstruction in optical tomography in the presence of coupling errors. Appl Opt. 2007;46(14):2743–56.

15. Scholkmann F, Metz AJ, Wolf M. Measuring tissue hemodynamics and oxygenation by continuous-wave functional near-infrared spectroscopy—how robust are the different calculation methods against movement artifacts? Physiol Meas. 2014;35:717–34.

16. Sweeney KT, Ayaz H, Ward TE, Izzetoglu M, McLoone SF, Onaral B. A methodology for validating artifact removal techniques for physiological signals. IEEE Trans Inf Technol Biomed. 2012;16:918–26.

17. Scholkmann F, Spichtig S, Muehlemann T, Wolf M. How to detect and reduce movement artifacts in near-infrared imaging using moving standard deviation and spline interpolation. Physiol Meas. 2010;31:649–62.

18. Molavi B, Dumont GA. Wavelet-based motion artifact removal for functional near-infrared spectroscopy. Physiol Meas. 2012;33:259–70.

19. Robertson FC, Douglas TS, Meintjes EM. Motion artifact removal for functional near infrared spectroscopy: a comparison of methods. IEEE Trans Biomed Eng. 2010;57:1377–87.

20. Barker JW, Aarabi A, Huppert TJ. Autoregressive model based algorithm for correcting motion and serially correlated errors in fNIRS. Biomed Opt Express. 2013;4:1366. https://www.osapublishing.org/boe/abstract.cfm?uri=boe-4-8-1366. Accessed 26 Feb 2018.

21. Cui X, Bray S, Reiss AL. Functional near infrared spectroscopy (NIRS) signal improvement based on negative correlation between oxygenated and deoxygenated hemoglobin dynamics. Neuroimage. 2010;49:3039–46. https://doi.org/10.1016/j.neuroimage.2009.11.050.

22. Izzetoglu M, Chitrapu P, Bunce S, Onaral B. Motion artifact cancellation in NIR spectroscopy using discrete Kalman filtering. Biomed Eng Online. 2010;9:16.

23. Izzetoglu M, Devaraj A, Bunce S, Onaral B. Motion artifact cancellation in NIR spectroscopy using wiener filtering. IEEE Trans Biomed Eng. 2005;52:2–5.

24. Hossen A, Muthuraman M, Raethjen J, Deuschl G, Heute U. Discrimination of parkinsonian tremor from essential tremor by implementation of a wavelet-based soft-decision technique on emg and accelerometer signals. Biomed Signal Process. 2010;5:181–8. https://doi.org/10.1016/j.bspc.2010.02.005.

25. Lawoyin S, Fei D-Y, Bai O. Accelerometer-based steering-wheel movement monitoring for drowsy-driving detection. Proc Inst Mech Eng Part D J Automob Eng. 2015;229:163–73. https://doi.org/10.1177/0954407014536148.

26. Lawoyin S, Fei D-Y, Bai O, Liu X. Evaluating the efficacy of an accelerometer–based method for drowsy driving detection. Int J Veh Saf. 2015;8:165–79. http://dx.doi.org/10.1504/IJVS.2015.068691, http://inderscience.metapress.com/link.asp?target=contribution&id=T3M776626325760K, https://trid.trb.org/view/1350249.

27. Bai O, Atri R, Marquez JS, Fei D-Y. Characterization of lower limb activity during gait using wearable, multi-channel surface EMG and IMU sensors. Int Electr Eng Congr. 2017;1–4. http://ieeexplore.ieee.org/document/8075872/. Accessed 26 Feb 2018.

28. Lawoyin S, Liu X, Fei DY, Bai O. Detection methods for a low-cost accelerometer-based approach for driver drowsiness detection. Conf Proc IEEE Int Conf Syst Man Cybern. 2014;2014:1636–41.

29. Lawoyin SA, Fei D-Y, Bai O. A novel application of inertial measurement units (IMUs) as vehicular technologies for Drowsy driving detection via steering wheel movement. Open J Saf Sci Technol. 2014;04:166–77. https://doi.org/10.4236/ojsst.2014.44018.

30. Virtanen J, Noponen T, Kotilahti K, Virtanen J, Ilmoniemi RJ. Accelerometer-based method for correcting signal baseline changes caused by motion artifacts in medical near-infrared spectroscopy. J Biomed Opt. 2011;16:087005. https://doi.org/10.1117/1.3606576.

31. Kim SH, Ryoo DW, Bae C. Adaptive noise cancellation using accelerometers for the PPG signal from forehead. Conf Proc IEEE Eng Med Biol Soc. 2007;2007:2564–7.

32. Izzetoglu M, Devaraj A, Bunce S, Onaral B. Motion artifact cancellation in NIR spectroscopy using Wiener filtering. IEEE Trans Biomed Eng. 2005;52:934–8.

33. Kim CK, Lee S, Koh D, Kim BM. Development of wireless NIRS system with dynamic removal of motion artifacts. Biomed Eng Lett. 2011;1:254–9.
34. Cooper RJ, Selb J, Gagnon L, Phillip D, Schytz HW, Iversen HK, et al. A systematic comparison of motion artifact correction techniques for functional near-infrared spectroscopy. Front Neurosci. 2012;6:1–10.
35. Widrow B, Williams CS, Glover JR, McCool JM, Hearn RH, Zeidler JR, et al. Adaptive noise cancelling: principles and applications. Proc IEEE. 1975;63:1692–716.
36. Cope M. The development of a near infrared spectroscopy system and its application for non invasive monitory of cerebral blood and tissue oxygenation in the newborn infants. Doctoral thesis, University of London. 1991. p. 1–342. http://discovery.ucl.ac.uk/1317956/. Accessed 26 Feb 2018.
37. Wray S, Cope M, Delpy DT, Wyatt JS, Reynolds EOR. Characterization of the near infrared absorption spectra of cytochrome aa3 and haemoglobin for the non-invasive monitoring of cerebral oxygenation. Biochim Biophys Acta Bioenergy. 1988;933:184–92.
38. Wang Y, Zheng Y, Bai O, Wang Q, Liu D, Liu X, et al. A multifunctional wireless body area sensors network with real time embedded data analysis. Circuits Syst Conf BioCAS. 2016;2016(2017):508–11.
39. Sweeney KT, McLoone SF, Ward TE. The use of ensemble empirical mode decomposition with canonical correlation analysis as a novel artifact removal technique. IEEE Trans Biomed Eng. 2013;60:97–105.
40. Boser BE, Howe RT. Surface micromachined accelerometers. IEEE J Solid-State Circ. 1996;31:366–75.
41. Brigadoi S, Ceccherini L, Cutini S, Scarpa F, Scatturin P, Selb J, et al. Motion artifacts in functional near-infrared spectroscopy: a comparison of motion correction techniques applied to real cognitive data. Neuroimage. 2014;85:181–91. https://doi.org/10.1016/j.neuroimage.2013.04.082.
42. Walplte RE, Myers RH, Myers Key SL. Probability and statistics for Engineers and Scientists. 9th ed. New Jersey: Prentice Hall; 2011.

Permissions

List of Contributors

Yang Zhixiang, Xiang Jibing, Ge Bisheng, Xu Ming and Liu Deyu
Department of Pediatrics, Lixian People's Hospital in Hunan, Lixian 415500, China

Wang Cheng
Department of Pediatric Cardiovasology, Children's Medical Center, The Second Xiangya Hospital of Central South University, Institute of Pediatrics of Central South University, Changsha 410011, China

Ga Young Kim and Yoo Na Hwang
Department of Medical Biotechnology, Dongguk University-Bio Medi Campus, 32, Dongguk-ro, Ilsandong-gu, Goyang, Gyeonggi-do 10326, Republic of Korea

Sung Min Kim
Department of Medical Biotechnology, Dongguk University-Bio Medi Campus, 32, Dongguk-ro, Ilsandong-gu, Goyang, Gyeonggi-do 10326, Republic of Korea
Department of Medical Devices Industry, 26, Pil-dong 3-ga, Jung-gu,Seoul 04620, Republic of Korea

Ju Hwan Lee
Department of Medical Devices Industry, 26, Pil-dong 3-ga, Jung-gu,Seoul 04620, Republic of Korea

Min-Ho Jun, Young Ju Jeon, Jung-Hee Cho and Young-Min Kim
Future Medicine Division, Korea Institute of Oriental Medicine (KIOM), 1672 Yuseongdaero, Yuseong-gu, Deajeon 34054, Republic of Korea

Haiyan Li, Tangyu Wang, Jun Wu, Pengfei Yu, Lei Guo, Jianhua Chen and Yufeng Zhang
School of Information Science and Engineering, Electronic Engineering, Yunnan University, Chenggong District, Kunming 650000, China

Yiying Tang
Breast Surgery Department, The Third Hospital Affiliated to the Medical University of Kunming, Kunming 605118, Yunnan Province, China

He Wang, Jingna Jin, Xin Wang, Ying Li and Zhipeng Liu
Institute of Biomedical Engineering, Chinese Academy of Medical Science & Peking Union Medical College, Tianjin 300192, China

Tao Yin
Institute of Biomedical Engineering, Chinese Academy of Medical Science & Peking Union Medical College, Tianjin 300192, China
Neuroscience Center, Chinese Academy of Medical Science & Peking Union Medical College, Beijing 100730, China

Oleksandr Makeyev
Department of Mathematics, Diné College, 1 Circle Dr, Tsaile, AZ 86556, USA

Wenbao Wu
Department of Acupuncture, Longyan First Hospital, Longyan 364000, China

Wei Zeng
School of Physics and Mechanical & Electrical Engineering, Longyan University, Longyan 364012, China

Limin Ma
Department of Orthopaedic Surgery, Guangzhou General Hospital of Guangzhou Military Command, Guangzhou 510010, China

Chengzhi Yuan
Department of Mechanical, Industrial and Systems Engineering, University of Rhode Island, Kingston, RI 02881, USA

Yu Zhang
Department of Orthopedics, Guangdong General Hospital, Guangdong Academy of Medical Sciences, Guangzhou 510080, China

Kenneth J. Hunt
Division of Mechanical Engineering, Department of Engineering and Information Technology, Institute for Rehabilitation and Performance Technology, Bern University of Applied Sciences, 3400 Burgdorf, Switzerland

Jittima Saengsuwan
Department of Rehabilitation Medicine, Srinagarind Hospital and Faculty of Medicine, Khon Kaen University, Khon Kaen, Thailand
Exercise and Sport Sciences Development and Research Group, Khon Kaen University, Khon Kaen, Thailand

Xiao Wang, Liuye Yao, Yuemei Zhao, Lidong Xing, Zhiyu Qian, Weitao Li and Yamin Yang
College of Automation Engineering, Nanjing University of Aeronautics and Astronautics, No. 29 Jiangjun Avenue, Jiangning District, Nanjing 211106, China

Sara Marceglia
Dipartimento di Ingegneria e Architettura, Università degli Studi di Trieste, Via A. Valerio 10, 34127 Trieste, Italy
Fondazione IRCCS Ca' Granda Ospedale Maggiore Policlinico, Milan, Italy

Michael Rigby
Health Information Strategy, Keele University, Keele, UK

Albert Alonso
Hospital Clinic Barcelona, Barcelona, Spain

Debbie Keeling
University of Sussex, Brighton, UK

Lutz Kubitschke
Empirica Communications and Technology Research, Bonn, Germany

Giuseppe Pozzi
Dipartimento di Elettronica, Informazione e Bioingegneria, Politecnico di Milano, Milan, Italy

Hongze Pan, Zhi Xu, Hong Yan, Yue Gao, Zhanghuang Chen, Jinzhong Song and Yu Zhang
China Astronaut Researching and Training Center, Beijing, China

Hao Yang, Junran Zhang, Qihong Liu and Yi Wang
Department of Medical Information Engineering, School of Electrical Engineering and Information, Sichuan University, Chengdu, Sichuan, China

Masudur R. Siddiquee, J. Sebastian Marquez, Roozbeh Atri, Rodrigo Ramon, Robin Perry Mayrand and Ou Bai
Florida International University, Miami, FL 33174, USA

Index